Texts in Computer Science

Series editors

David Gries
Fred B. Schneider

More information about this series at http://www.springer.com/series/3191

Gerard O'Regan

Guide to Discrete Mathematics

An Accessible Introduction to the History, Theory, Logic and Applications

 Springer

Gerard O'Regan
SQC Consulting
Mallow, Cork
Ireland

ISSN 1868-0941   ISSN 1868-095X (electronic)
Texts in Computer Science
ISBN 978-3-319-83080-3   ISBN 978-3-319-44561-8 (eBook)
DOI 10.1007/978-3-319-44561-8

Printed on acid-free paper

This Springer imprint is published by Springer Nature
The registered company is Springer International Publishing AG
The registered company address is: Gewerbestrasse 11, 6330 Cham, Switzerland

To Lizbeth Román Padilla (Liz)
For sincere friendship

Preface

Overview

The objective of this book is to give the reader a flavor of discrete mathematics and its applications to the computing field. The goal is provide a broad and accessible guide to the fundamentals of discrete mathematics, and to show how it may be applied to various areas in computing such as cryptography, coding theory, formal methods, language theory, computability, artificial intelligence, theory of databases, and software reliability. The emphasis is on both theory and applications, rather than on the study of mathematics for its own sake.

There are many existing books on discrete mathematics, and while many of these provide more in-depth coverage on selected topics, this book is different in that it aims to provide a broad and accessible guide to the reader, and to show the rich applications of discrete mathematics in a wide number of areas in the computing field.

Each chapter of this book could potentially be a book in its own right, and so there are limits to the depth of coverage for each chapter. However, the author hopes that this book will motivate and stimulate the reader, and encourage further study of the more advanced texts.

Organization and Features

The first chapter discusses the contributions made by early civilizations to computing. This includes works done by the Babylonians, Egyptians, and Greeks. The Egyptians applied mathematics to solving practical problems such as the construction of pyramids. The Greeks made major contributions to mathematics and geometry.

Chapter 2 provides an introduction to fundamental building blocks in discrete mathematics including sets, relations and functions. A set is a collection of well-defined objects and it may be finite or infinite. A relation between two sets A and B indicates a relationship between members of the two sets, and is a subset of the Cartesian product of the two sets. A function is a special type of relation such that for each element in A there is at most one element in the co-domain B. Functions may be partial or total and injective, surjective, or bijective.

Chapter 3 presents the fundamentals of number theory, and discusses prime number theory and the greatest common divisor and the least common multiple of two numbers. We also discuss the representation of numbers on a computer.

Chapter 4 discusses mathematical induction and recursion. Induction is a common proof technique in mathematics, and there are two parts to a proof by induction (the base case and the inductive step). We discuss strong and weak induction, and we discuss how recursion is used to define sets, sequences, and functions. This leads us to structural induction, which is used to prove properties of recursively defined structures.

Chapter 5 discusses sequences and series, and permutations and combinations. Arithmetic and geometric sequences and series and applications of geometric sequences and series to the calculation of compound interest and annuities are discussed.

Chapter 6 discusses algebra and simple and simultaneous equations, including the method of elimination and the method of substitution to solve simultaneous equations. We show how quadratic equations may be solved by factorization, completing the square or using the quadratic formula. We present the laws of logarithms and indices. We discuss various structures in abstract algebra, including monoids, groups, rings, integral domains, fields, and vector spaces.

Chapter 7 discusses automata theory, including finite-state machines, pushdown automata, and Turing machines. Finite-state machines are abstract machines that are in only one state at a time, and the input symbol causes a transition from the current state to the next state. Pushdown automata have greater computational power than finite-state machines, and they contain extra memory in the form of a stack from which symbols may be pushed or popped. The Turing machine is the most powerful model for computation, and this theoretical machine is equivalent to an actual computer in the sense that it can compute exactly the same set of functions.

Chapter 8 discusses matrices including 2×2 and general $m \times n$ matrices. Various operations such as the addition and multiplication of matrices are considered, and the determinant and the inverse of a matrix are discussed. The application of matrices to solving a set of linear equations using Gaussian elimination is considered.

Chapter 9 discusses graph theory where a graph $G = (V, E)$ consists of vertices and edges. It is a practical branch of mathematics that deals with the arrangements of vertices and edges between them, and it has been applied to practical problems such as the modeling of computer networks, determining the shortest driving route between two cities, and the traveling salesman problem.

Chapter 10 discusses cryptography, which is an important application of number theory. The code breaking work done at Bletchley Park in England during the Second World War is discussed, and the fundamentals of cryptography, including private and public key cryptosystems, are discussed.

Chapter 11 presents coding theory and concerns error detection and error correction codes. The underlying mathematics of coding theory is abstract algebra, and this includes group theory, ring theory, fields, and vector spaces.

Chapter 12 discusses language theory and grammars, parse trees, and derivations from a grammar. The important area of programming language semantics is discussed, including axiomatic, denotational, and operational semantics.

Chapter 13 discusses computability and decidability. The Church–Turing thesis states that anything that is computable is computable by a Turing machine. Church and Turing showed that mathematics is not decidable, in that there is no mechanical procedure (i.e., algorithm) to determine whether an arbitrary mathematical proposition is true or false, and so the only way is to determine the truth or falsity of a statement is by trying to solve the problem.

Chapter 14 presents a short history of logic and Greek contributions to syllogistic logic, stoic logic, fallacies, and paradoxes. Boole's symbolic logic and its application to digital computing, and Frege's work on predicate logic are discussed.

Chapter 15 provides an introduction to propositional and predicate logic. Propositional logic may be used to encode simple arguments that are expressed in natural language, and to determine their validity. The nature of mathematical proof along with proof by truth tables, semantic tableaux, and natural deduction is discussed. Predicate logic allows complex facts about the world to be represented, and new facts may be determined via deductive reasoning. Predicate calculus includes predicates, variables, and quantifiers, and a predicate is a characteristic or property that the subject of a statement can have.

Chapter 16 presents some advanced topics in logic including fuzzy logic, temporal logic, intuitionistic logic, undefined values, theorem provers, and the applications of logic to AI. Fuzzy logic is an extension of classical logic that acts as a mathematical model for vagueness. Temporal logic is concerned with the expression of properties that have time dependencies, and it allows temporal properties about the past, present, and future to be expressed. Intuitionism was a controversial theory on the foundations of mathematics based on a rejection of the law of the excluded middle, and an insistence on constructive existence. We discuss three approaches to deal with undefined values, including the logic of partial functions; Dijkstra's approach with his cand and cor operators; and Parnas' approach which preserves a classical two-valued logic.

Chapter 17 provides an introduction to the important field of software engineering. The birth of the discipline was at the Garmisch conference in Germany in the late 1960s. The extent to which mathematics should be employed in software engineering is discussed, and this remains a topic of active debate.

Chapter 18 discusses formal methods, which consist of a set of mathematic techniques that provide an extra level of confidence in the correctness of the software. They may be employed to formally state the requirements of the proposed system, and to derive a program from its mathematical specification. They may be employed to provide a rigorous proof that the implemented program satisfies its specification. They have been mainly applied to the safety critical field.

Chapter 19 presents the Z specification language, which is one of the most widely used formal methods. It was developed at Oxford University in the U.K.

Chapter 20 discusses probability and statistics and includes a discussion on discrete random variables; probability distributions; sample spaces; sampling; the abuse of statistics; variance and standard deviation; and hypothesis testing. The applications of probability to the software reliability field and queuing theory are briefly discussed.

Audience

The audience of this book includes computer science students who wish to gain a broad and accessible overview of discrete mathematics and its applications to the computing field. The book will also be of interest to students of mathematics who are curious as to how discrete mathematics is applied to the computing field. The book will also be of interest to the motivated general reader.

Acknowledgments

I am deeply indebted to family and friends who supported my efforts in this endeavor. I would like to thank Lizbeth Román Padilla (Liz) for sincere friendship over the years, and I wish her continued success with her Bayesian statistics. I would like to thank the team at Springer, and especially Wayne Wheeler and Simon Rees.

Cork, Ireland   Gerard O'Regan

Contents

List of Figures

Mathematics in Civilization

1

Key Topics

Babylonian Mathematics
Egyptian Civilisation
Greek and Roman Civilisation
Islamic Civilisation
Counting and Numbers
Solving Practical Problems
Syllogistic Logic
Algorithm
Early Ciphers

1.1 Introduction

It is difficult to think of western society today without modern technology. The last decades of the twentieth century have witnessed a proliferation of high-tech computers, mobile phones, text messaging, the Internet and the World Wide Web. Software is now pervasive, and it is an integral part of automobiles, airplanes, televisions and mobile communication. The pace of change as a result of all this new technology has been extraordinary. Today consumers may book flights over the World Wide Web as well as keep in contact with the family members in any part of the world via e-mail or mobile phone. In previous generations, communication often involved writing letters that took months to reach the recipient.

© Springer International Publishing Switzerland 2016
G. O'Regan, *Guide to Discrete Mathematics*, Texts in Computer Science,
DOI 10.1007/978-3-319-44561-8_1

Communication improved with the telegrams and the telephone in the late nineteenth century. Communication today is instantaneous with text messaging, mobile phones and e-mail, and the new generation probably views the world of their parents and grandparents as being old fashioned.

The new technologies have led to major benefits[1] to society and to improvements in the standard of living for many citizens in the western world. It has also reduced the necessity for humans to perform some of the more tedious or dangerous manual tasks, as computers may now automate many of these. The increase in productivity due to the more advanced computerized technologies has allowed humans, at least in theory, the freedom to engage in more creative and rewarding tasks.

Early societies had a limited vocabulary for counting: e.g. 'one, two, three, many' is associated with some primitive societies, and indicates primitive computation and scientific ability. It suggests that there was no need for more sophisticated arithmetic in the primitive culture as the problems dealt with were elementary. These early societies would typically have employed their fingers for counting, and as humans have five fingers on each hand and five toes on each foot then the obvious bases would have been 5, 10 and 20. Traces of the earlier use of the base 20 system are still apparent in modern languages such as English and French. This includes phrases such as 'three score' in English and '*quatre vingt*' in French.

The decimal system (base 10) is used today in western society, but the base 60 was common in computation *circa* 1500 B.C. One example of the use of base 60 today is the subdivision of hours into 60 min, and the subdivision of minutes into 60 s. The base 60 system (i.e. the sexagesimal system) is inherited from the Babylonians [1]. The Babylonians were able to represent arbitrarily large numbers or fractions with just two symbols. The binary (base 2) and hexadecimal (base 16) systems play a key role in computing (as the machine instructions that computers understand are in binary code).

The achievements of some of these ancient societies were spectacular. The archaeological remains of ancient Egypt such as the pyramids at Giza and the temples of Karnak and Abu Simbel are impressive. These monuments provide an indication of the engineering sophistication of the ancient Egyptian civilization. The objects found in the tomb of Tutankhamun[2] are now displayed in the Egyptian museum in Cairo, and demonstrate the artistic skill of the Egyptians.

---

[1]Of course, it is essential that the population of the world moves towards more sustainable development to ensure the long-term survival of the planet for future generations. This involves finding technological and other solutions to reduce greenhouse gas emissions as well as moving to a carbon neutral way of life. The solution to the environmental issues will be a major challenge for the twenty first century.

[2]Tutankhamun was a minor Egyptian pharaoh who reigned after the controversial rule of Akenaten. Tutankamun's tomb was discovered by Howard Carter in the Valley of the Kings, and the tomb was intact. The quality of the workmanship of the artefacts found in the tomb is extraordinary and a visit to the Egyptian museum in Cairo is memorable.

The Greeks made major contributions to western civilization including contributions to Mathematics, Philosophy, Logic, Drama, Architecture, Biology and Democracy.[3] The Greek philosophers considered fundamental questions such as ethics, the nature of being, how to live a good life, and the nature of justice and politics. The Greek philosophers include Parmenides, Heraclitus, Socrates, Plato and Aristotle. The Greeks invented democracy and their democracy was radically different from today's representative democracy.[4] The sophistication of Greek architecture and sculpture is evident from the Parthenon on the Acropolis, and the Elgin marbles[5] that are housed today in the British Museum, London.

The Hellenistic[6] period commenced with Alexander the Great and led to the spread of Greek culture throughout most of the known world. The city of Alexandria became a centre of learning and knowledge during the Hellenistic period. Its scholars included Euclid who provided a systematic foundation for geometry. His work is known as 'The Elements', and consists of 13 books. The early books are concerned with the construction of geometric figures, number theory and solid geometry.

There are many words of Greek origin that are part of the English language. These include words such as psychology that is derived from two Greek words: psyche (ψυχε) and logos (λογος). The Greek word 'psyche' means mind or soul, and the word 'logos' means an account or discourse. Other examples are anthropology derived from 'anthropos (αντροπος) and 'logos' (λογος).

The Romans were influenced by Greeks culture. The Romans built aqueducts, viaducts, and amphitheatres. They also developed the Julian calendar, formulated laws (lex); and maintained peace throughout the Roman Empire (pax Romano). The ruins of Pompeii and Herculaneum demonstrate their engineering capability. Their

---

[3]The origin of the word "democracy' is from demos (δημος) meaning people and kratos (κρατος) meaning rule. That is, it means rule by the people. It was introduced into Athens following the reforms introduced by Cleisthenes. He divided the Athenian city state into thirty areas. Twenty of these areas were inland or along the coast and ten were in Attica itself. Fishermen lived mainly in the ten coastal areas; farmers in the ten inland areas; and various tradesmen in Attica. Cleisthenes introduced ten new clans where the members of each clan came from one coastal area, one inland area on one area in Attica. He then introduced a Boule (or assembly) which consisted of 500 members (50 from each clan). Each clan ruled for 1/10 th of the year.

[4]The Athenian democracy involved the full participations of the citizens (i.e., the male adult members of the city state who were not slaves) whereas in representative democracy the citizens elect representatives to rule and represent their interests. The Athenian democracy was chaotic and could also be easily influenced by individuals who were skilled in rhetoric. There were teachers (known as the Sophists) who taught wealthy citizens rhetoric in return for a fee. The origin of the word 'sophist' is the Greek word σοφος meaning wisdom. One of the most well known of the sophists was Protagorus. The problems with the Athenian democracy led philosophers such as Plato to consider alternate solutions such as rule by philosopher kings. This totalitarian utopian state is described in Plato's Republic.

[5]The Elgin marbles are named after Lord Elgin who moved them from the Parthenon in Athens to London in 1806. The marbles show the Pan-Athenaic festival that was held in Athens in honour of the goddess Athena after whom Athens is named.

[6]The origin of the word Hellenistic is from Hellene ('Ελλην) meaning Greek.

numbering system is still employed in clocks and for page numbering in documents. However, it is cumbersome for serious computation. The collapse of the Roman Empire in Western Europe led to a decline in knowledge and learning in Europe. However, the eastern part of the Roman Empire continued at Constantinople until it was sacked by the Ottomans in 1453.

1.2 The Babylonians

The Babylonian[7] civilization flourished in Mesopotamia (in modern Iraq) from about 2000 B.C., until about 300 B.C. Various clay cuneiform tablets containing mathematical texts were discovered and later deciphered in the nineteenth century [2]. These included tables for multiplication, division, squares, cubes and square roots and the measurement of area and length. Their calculations allowed the solution of a linear equation and one root of a quadratic equation to be determined. The late Babylonian period (c. 300 B.C.) includes work on astronomy.

They recorded their mathematics on soft clay using a wedge shaped instrument to form impressions of the *cuneiform* numbers. The clay tablets were then baked in an oven or by the heat of the sun. They employed just two symbols (1 and 10) to represent numbers, and these symbols were then combined to form all other numbers. They employed a positional number system[8] and used the base 60 system. The symbol representing 1 could also (depending on the context) represent 60, 60^2, 60^3, etc. It could also mean 1/60, 1/3600, and so on. There was no zero employed in the system and there was no decimal point (no 'sexagesimal point'), and therefore the context was essential.

$$\text{\Large\Yup \quad \triangleleft \quad \Yup}$$

The example above illustrates the cuneiform notation and represents the number $60 + 10 + 1 = 71$. The Babylonians used the base 60 system, and this base is still in use today in the division of hours into minutes and the division of minutes into seconds. One possible explanation for the use of the base 60 notation is the ease of dividing 60 into parts. It is divisible by 2,3,4,5,6,10,12,15,20 and 30. They were able to represent large and small numbers and had no difficulty in working with fractions (in base 60) and in multiplying fractions. The Babylonians maintained tables of reciprocals (i.e. $1/n$, $n = 1$, ... 59) apart from numbers like 7, 11, etc., which cannot be written as a finite sexagesimal expansion (i.e. 7, 11, etc., are not of the form $2^\alpha 3^\beta 5^\gamma$).

---

[7]The hanging gardens of Babylon were one of the seven wonders of the ancient world.

[8]A positional numbering system is a number system where each position is related to the next by a constant multiplier. The decimal system is an example: e.g., $546 = 5 * 10^2 + 4 * 10^1 + 6$.

Fig. 1.1 The Plimpton 322 Tablet

The modern sexagesimal notation [1] 1;24,51,10 represents the number 1 + 24/60 + 51/3600 + 10/216,000 = 1 + 0.4 + 0.0141666 + 0.0000462 = 1.4142129. This is the Babylonian representation of the square root of 2. They performed multiplication as follows: e.g. consider 20 * sqrt(2) = (20) * (1;24,51,10)

$$20 * 1 = 20$$
$$20 * ;24 = 20 * 24/60 = 8$$
$$20 * 51/3600 = 51/180 = 17/60 = ;17$$
$$20 * 10/216,000 = 3/3600 + 20/216,000 = ;0,3,20$$

Hence, the product 20 * sqrt (2) = 20; + 8; +;17 +;0,3,20 = 28;17,3,20

The Babylonians appear to have been aware of Pythagoras's Theorem about 1000 years before the time of Pythagoras. The Plimpton 322 tablet (Fig. 1.1) records various Pythagorean triples, i.e. triples of numbers (a, b, c) where $a^2 + b^2 = c^2$. It dates from approximately 1700 B.C.

They developed an algebra to assist with problem solving, and their algebra allowed problems involving length, breadth and area to be discussed and solved. They did not employ notation for representation of unknown values (e.g. let x be the length and y be the breadth), and instead they used words like 'length' and 'breadth'. They were familiar with and used square roots in their calculations, and they were familiar with techniques that allowed one root of a quadratic equation to be solved.

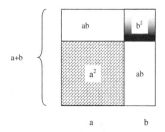

Fig. 1.2 Geometric representation of $(a + b)^2 = (a^2 + 2ab + b^2)$

They were familiar with various mathematical identities such as $(a + b)^2 = (a^2 + 2ab + b^2)$ as illustrated geometrically in Fig. 1.2. They also worked on astronomical problems, and they had mathematical theories of the cosmos to make predictions of when eclipses and other astronomical events would occur. They were also interested in astrology, and they associated various deities with the heavenly bodies such as the planets, as well as the sun and moon. They associated various cluster of stars with familiar creatures such as lions, goats and so on.

The Babylonians used counting boards to assist with counting and simple calculations. A counting board is an early version of the abacus, and it was usually made of wood or stone. The counting board contained grooves that allowed beads, or stones could be moved along the groove. The abacus differs from counting boards in that the beads in abaci contain holes that enable them to be placed in a particular rod of the abacus.

1.3 The Egyptians

The Egyptian Civilization developed along the Nile from about 4000 B.C. and the pyramids were built around 3000 B.C. They used mathematics to solve practical problems such as measuring time, measuring the annual Nile flooding, calculating the area of land, book keeping and accounting and calculating taxes. They developed a calendar circa 4000 B.C., which consisted of 12 months with each month having 30 days. There were then five extra feast days to give 365 days in a year. Egyptian writing commenced around 3000 B.C., and is recorded on the walls of temples and tombs.[9] A reed like parchment termed 'papyrus' was used for writing, and three Egyptian writing scripts were employed. These were hieroglyphics, the hieratic script, and the demotic script.

Hieroglyphs are little pictures and are used to represent words, alphabetic characters as well as syllables or sounds. Champollion deciphered hieroglyphics with his work on the Rosetta stone. This object was discovered during the

---

[9]The decorations of the tombs in the Valley of the Kings record the life of the pharaoh including his exploits and successes in battle.

ᘖ	𓂁	𓏏	𓏤	∩		
100,000	10,000	1000	100	10	1

Fig. 1.3 Egyptian numerals

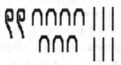

Fig. 1.4 Egyptian representation of a number

Napoleonic campaign in Egypt, and it is now in the British Museum in London. It contains three scripts: Hieroglyphics, Demotic script and Greek. The key to its decipherment was that the Rosetta stone contained just one name 'Ptolemy' in the Greek text, and this was identified with the hieroglyphic characters in the cartouche[10] of the hieroglyphics. There was just one cartouche on the Rosetta stone, and Champollion inferred that the cartouche represented the name 'Ptolemy'. He was familiar with another multilingual object that contained two names in the cartouche. One he recognized as Ptolemy and the other he deduced from the Greek text as 'Cleopatra'. This led to the breakthrough in the translation of the hieroglyphics [1].

The Rhind Papyrus is a famous Egyptian papyrus on mathematics. The Scottish Egyptologist, Henry Rhind, purchased it in 1858, and it is a copy created by an Egyptian scribe called Ahmose[11] around 1832 B.C. It contains examples of many kinds of arithmetic and geometric problems, and students may have used it as a textbook to develop their mathematical knowledge. This would allow them to participate in the pharaoh's building programme.

The Egyptians were familiar with geometry, arithmetic and elementary algebra. They had techniques to find solutions to problems with one or two unknowns. A base 10 number system was employed with separate symbols for one, ten, a hundred, a thousand, a ten thousand, a hundred thousand, and so on. These hieroglyphic symbols are represented in Fig. 1.3.

For example, the representation of the number 276 in Egyptian Hieroglyphics is described in Fig. 1.4.

---

[10]The cartouche surrounded a group of hieroglyphic symbols enclosed by an oval shape. Champollion's insight was that the group of hieroglyphic symbols represented the name of the Ptolemaic pharaoh 'Ptolemy'.

[11]The Rhind papyrus is sometimes referred to as the Ahmes papyrus in honour of the scribe who wrote it in 1832 B.C.

Fig. 1.5 Egyptian representation of a fraction

The addition of two numerals is straightforward and involves adding the individual symbols, and where there are ten copies of a symbol it is then replaced by a single symbol of the next higher value. The Egyptian employed unit fractions (e.g. $1/n$ where n is an integer). These were represented in hieroglyphs by placing the symbol representing a 'mouth' above the number. The symbol 'mouth' represents part of the number. For example, the representation of the number $1/276$ is described in Fig. 1.5.

The problems on the papyrus included the determination of the angle of the slope of the pyramid's face. They were familiar with trigonometry including sine, cosine, tangent and cotangent, and they knew how to build right angles into their structures by using the ratio 3:4:5. The Rhind papyrus also considered problems such as the calculation of the number of bricks required for part of a building project. Multiplication and division was cumbersome in Egyptian mathematics as they could only multiply and divide by two.

Suppose they wished to multiply a number n by 7. Then $n * 7$ is determined by $n * 2 + n * 2 + n * 2 + n$. Similarly, if they wished to divide 27 by 7 they would note that $7 * 2 + 7 = 21$ and that $27 - 21 = 6$ and that therefore the answer was $3(6/7)$. Egyptian mathematics was cumbersome and the writing of their mathematics was long and repetitive. For example, they wrote a number such as 22 by $10 + 10 + 1 + 1$.

The Egyptians calculated the approximate area of a circle by calculating the area of a square 8/9 of the diameter of a circle. That is, instead of calculating the area in terms of our familiar πr^2 their approximate calculation yielded $(8/9 * 2r)^2 = (256/81)$ r^2 or $3.16 \ r^2$. Their approximation of π was $256/81$ or 3.16. They were able to calculate the area of a triangle and volumes. The Moscow papyrus includes a problem to calculate the volume of the frustum. The formula for the volume of a frustum of a square pyramid[12] was given by $V = (1/3) \ h(b_1^2 + b_1 b_2 + b_2^2)$ and when b_2 is 0 then the well-known formula for the volume of a pyramid is given: i.e. $1/3 \ hb_1^2$.

1.4 The Greeks

The Greeks made major contributions to western civilization including mathematics, logic, astronomy, philosophy, politics, drama and architecture. The Greek world of 500 B.C. consisted of several independent city-states such as Athens and Sparta, and various city-states in Asia Minor. The Greek polis (πολισ) or city-state

---

[12]The length of a side of the bottom base of the pyramid is b_1 and the length of a side of the top base is b_2.

tended to be quite small, and consisted of the Greek city and a certain amount of territory outside the city-state. Each city-state had political structures for its citizens, and some were oligarchs where political power was maintained in the hands of a few individuals or aristocratic families. Others were ruled by tyrants (or sole rulers), who sometimes took power by force, but who often had a lot of support from the public. The tyrants included people such as Solon, Peisistratus and Cleisthenes in Athens.

The reforms by Cleisthenes led to the introduction of the Athenian democracy. Power was placed in the hands of the citizens who were male (women or slaves did not participate in the Athenian democracy). It was an extremely liberal democracy where citizens voted on all important issues. Often, this led to disastrous results as speakers who were skilled in rhetoric could exert significant influence. This later led to Plato to advocate rule by philosopher kings rather than by democracy.[13]

Early Greek mathematics commenced approximately 500–600 B.C., with work done by Pythagoras and Thales. Pythagoras was a philosopher and mathematician who had spent time in Egypt becoming familiar with Egyptian mathematics. He lived on the island of Samos, and formed a secret society known as the Pythagoreans. They included men and women and believed in the transmigration of souls, and that number was the essence of all things. They discovered the mathematics for harmony in music with the relationship between musical notes being expressed in numerical ratios of small whole numbers. Pythagoras is credited with the discovery of Pythagoras's Theorem, although the Babylonians probably knew this theorem about 1000 years earlier. The Pythagorean society was dealt a major blow[14] by the discovery of the incommensurability of the square root of 2: i.e. there are no numbers p, q such that $\sqrt{2} = p/q$.

Thales was a sixth century (B.C.) philosopher from Miletus in Asia Minor who made contributions to philosophy, geometry and astronomy. His contributions to philosophy are mainly in the area of metaphysics, and he was concerned with questions on the nature of the world. His objective was to give a natural or scientific explanation of the cosmos, rather than relying on the traditional supernatural explanation of creation in Greek mythology. He believed that there was single substance that was the underlying constituent of the world, and he believed that this substance was water.

He also contributed to mathematics [3], and a well-known theorem in Euclidean geometry is named after him. It states that if A, B and C are points on a circle, and where the line AC is a diameter of the circle, then the angle $<ABC$ is a right angle.

The rise of Macedonia led to the Greek city-states being conquered by Philip of Macedonia in the fourth century B.C. His son, Alexander the Great, defeated the Persian Empire and extended his empire to include most of the known world. This

---

[13]Plato's Republic describes his utopian state, and seems to be based on the austere Spartan model.

[14]The Pythagoreans took a vow of silence with respect to the discovery of incommensurable numbers. However, one member of the society is said to have shared the secret result with others outside the sect, and an apocryphal account is that he was thrown into a lake for his betrayal and drowned. The Pythagoreans obviously took Mathematics seriously back then.

led to the Hellenistic Age with Greek language and culture spread throughout the known world. Alexander founded the city of Alexandra, and it became a major centre of learning. However, Alexander's reign was very short as he died at the young age of 33 in 323 B.C.

Euclid lived in Alexandria during the early Hellenistic period and he is considered as the father of geometry and the deductive method in mathematics. His systematic treatment of geometry and number theory is published in the 13 books of the Elements [4]. It starts from five axioms, five postulates and twenty-three definitions to logically derive a comprehensive set of theorems. His method of proof was often *constructive* in that, as well as demonstrating the truth of a theorem the proof would often include the construction of the required entity. He also used *indirect proof* as to show that there are an infinite number of primes

1. Suppose there are a finite number of primes (say n primes).
2. Multiply all n primes together and add 1 to form N.

$$(N = p_1 * p_2 * \ldots * p_n + 1)$$

3. N is not divisible by $p_1, p_2, \ldots, p_n$ as dividing by any of these gives a remainder of one.
4. Therefore, N must either be prime or divisible by some other prime that was not included in the list.
5. Therefore, there must be at least $n + 1$ primes.
6. This is a contradiction as it was assumed that there was a finite number of primes n.
7. Therefore, the assumption that there are a finite number of primes is false.
8. Therefore, there are an infinite number of primes.

Euclidean geometry included the parallel postulate (or Euclid's fifth postulate). This postulate generated interest, as many mathematicians believed that it was unnecessary and could be proved as a theorem. It states that:

Definition 1.1 (*Parallel Postulate*) If a line segment intersects two straight lines forming two interior angles on the same side that sum to less than two right angles, then the two lines, if extended indefinitely, meet on that side on which the angles sum to less than two right angles.

This postulate was later proved to be independent of the other postulates, with the development of non-Euclidean geometries in the nineteenth century. These include the *hyperbolic geometry* discovered independently by Bolyai and Lobachevsky, and *elliptic geometry* developed by Riemann. The standard model of Riemannian geometry is the sphere where lines are great circles.

Euclid's Elements is a systematic development of geometry starting from the small set of axioms, postulates and definitions, leading to theorems logically derived from the axioms and postulates. Euclid's deductive method influenced later mathematicians and scientists. There are some jumps in reasoning and the German mathematician, David Hilbert, later added extra axioms to address this.

The Elements contain many well-known mathematical results such as Pythagoras's Theorem, Thales Theorem, Sum of Angles in a Triangle, Prime Numbers, Greatest Common Divisor and Least Common Multiple, Euclidean Algorithm, Areas and Volumes, Tangents to a point and Algebra.

The Euclidean algorithm is one of the oldest known algorithms and is employed to produce the greatest common divisor of two numbers. It is presented in the Elements but was known well before Euclid. The algorithm to determine the gcd of two natural numbers, a and b, is given by

1. Check if b is zero. If so, then a is the gcd.
2. Otherwise, the gcd (a, b) is given by gcd $(b, a \bmod b)$.

It is also possible to determine integers p and q such that $ap + bq = \gcd(a, b)$.

The proof of the Euclidean algorithm is as follows. Suppose a and b are two positive numbers whose gcd has to be determined, and let r be the remainder when a is divided by b.

1. Clearly $a = qb + r$ where q is the quotient of the division.
2. Any common divisor of a and b is also a divisor or r (since $r = a - qb$).
3. Similarly, any common divisor of b and r will also divide a.
4. Therefore, the greatest common divisor of a and b is the same as the greatest common divisor of b and r.
5. The number r is smaller than b and we will reach $r = 0$ in finitely many steps.
6. The process continues until $r = 0$.

Comment 1.1
Algorithms are fundamental in computing as they define the procedure by which a problem is solved. A computer program implements the algorithm in some programming language.

Eratosthenes was a Hellenistic mathematician and scientist who worked at the library in Alexandria, which was the largest library in the ancient world. It was built during the Hellenistic period in the third century B.C. and destroyed by fire in 391 A.D.

Eratosthenes devised a system of latitude and longitude, and became the first person to estimate of the size of the circumference of the Earth (Fig. 1.6). His calculation proceeded as follows:

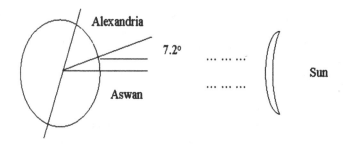

Fig. 1.6 Eratosthenes measurement of the circumference of the earth

1. On the summer solstice at noon in the town of Aswan[15] on the Tropic of Cancer in Egypt the Sun appears directly overhead.
2. Eratosthenes believed that the Earth was a sphere.
3. He assumed that rays of light came from the Sun in parallel beams and reached the Earth at the same time.
4. At the same time in Alexandria he had measured that the sun would be 7.2° south of the zenith.
5. He assumed that Alexandria was directly North of Aswan.
6. He concluded that the distance from Alexandria to Aswan was 7.2/360 of the circumference of the Earth.
7. Distance between Alexandria and Aswan was 5000 stadia (approximately 800 km).
8. He established a value of 252,000 stadia or approximately 40,320 km.

Eratosthenes's calculation was an impressive result for 200 B.C. The errors in his calculation were due to

1. Aswan is not exactly on the Tropic of Cancer but it is actually 55 km North of it.
2. Alexandria is not exactly North of Aswan and there is a difference of 3° longitude.
3. The distance between Aswan and Alexandria is 729 km not 800 km.
4. Angles in antiquity could not be measured with a high degree of precision.
5. The angular distance is actually 7.08° and not 7.2°.

Eratosthenes also calculated the approximate distance to the Moon and Sun and he also produced maps of the known world. He developed a very useful algorithm for determining all of the prime numbers up to a specified integer. The method is known as the Sieve of Eratosthenes and the steps are as follows:

[15]The town of Aswan is famous today for the Aswan high dam, which was built in the 1960s. There was an older Aswan dam built by the British in the late nineteenth century. The new dam led to a rise in the water level of Lake Nasser and flooding of archaeological sites along the Nile. Several archaeological sites such as Abu Simbel and the temple of Philae were relocated to higher ground.

1. Write a list of the numbers from 2 to the largest number that you wish to test for primality. This first list is called A.
2. A second list, called as B, is created to list the primes. It is initially empty.
3. The number 2 is the first prime number and is added to the list of primes in B.
4. Strike off (or remove) 2 and all multiples of 2 from List A.
5. The first remaining number in List A is a prime number and this prime number is added to List B.
6. Strike off (or remove) this number and all multiples of this number from List A.
7. Repeat steps 5 through 7 until no more numbers are left in List A.

Comment 1.2
The Sieve of Eratosthenes method is a well-known algorithm for determining prime numbers.

Archimedes was a Hellenistic mathematician, astronomer and philosopher who lived in Syracuse in the third century B.C. He discovered the law of buoyancy known as Archimedes's principle:

The buoyancy force is equal to the weight of the displaced fluid.

He is believed to have discovered the principle while sitting in his bath He was so overwhelmed with his discovery that he rushed out onto the streets of Syracuse shouting '*Eureka*', but forgot to put on his clothes to announce the discovery.

The weight of the displaced liquid will be proportional to the volume of the displaced liquid. Therefore, if two objects have the same mass, the one with greater volume (or smaller density) has greater buoyancy. An object will float if its buoyancy force (i.e. the weight of liquid displaced) exceeds the downward force of gravity (i.e. its weight). If the object has exactly the same density as the liquid, then it will stay still, neither sinking nor floating upwards.

For example, a rock is generally a very dense material and will generally not displace its own weight. Therefore, a rock will sink to the bottom as the downward weight exceeds the buoyancy weight. However, if the weight of the object is less than the liquid it would displace then it floats at a level where it displaces the same weight of liquid as the weight of the object.

Archimedes (Fig. 1.7) was born in Syracuse[16] in the third century B.C. He was a leading scientist in the Greco-Roman world, and he is credited with designing several innovative machines.

His inventions include the 'Archimedes Screw' which was a screw pump that is still used today in pumping liquids and solids. Another of his inventions was the 'Archimedes Claw', which was a weapon used to defend the city of Syracuse. It was also known as the 'ship shaker' and it consisted of a crane arm from which a large metal hook was suspended. The claw would swing up and drop down on the attacking ship. It would then lift it out of the water and possibly sink it. Another of

---

[16]Sysacuse is located on the island of Sicily in Southern Italy.

Fig. 1.7 Archimedes in
thought by Fetti

his inventions was said to be the 'Archimedes Heat Ray'. This device is said to have consisted of a number of mirrors that allowed sunlight to be focused on an enemy ship thereby causing it to go on fire.

He made good contributions to mathematics including developing a good approximation to π, as well as contributions to the positional numbering system, geometric series, and to maths physics. He also solved several interesting problems: e.g. the calculation of the composition of cattle in the herd of the Sun god by solving a number of simultaneous Diophantine equations. The herd consisted of bulls and cows with one part of the herd consisting of white, second part black, third spotted and the fourth brown. Various constraints were then expressed in Diophantine equations and the problem was to determine the precise composition of the herd. Diophantine equations are named after Diophantus who worked on number theory in the third century.

There is a well-known anecdote concerning Archimedes and the crown of King Hiero II. The king wished to determine whether his new crown was made entirely of solid gold, and that the goldsmith had not added substitute silver. Archimedes was required to solve the problem without damaging the crown, and as he was taking a bath he realized that if the crown was placed in water that the water displaced would give him the volume of the crown. From this he could then determine the density of the crown and therefore whether it consisted entirely of gold.

Archimedes also calculated an upper bound of the number of grains of sands in the known universe. The largest number in common use at the time was a myriad myriad (100 million), where a myriad is 10,000. Archimedes' numbering system goes up to $8 * 10^{16}$ and he also developed the laws of exponents: i.e. $10^a\, 10^b = 10^{a+b}$. His calculation of the upper bound includes not only the grains of sand on each beach but on the earth filled with sand and the known universe filled with sand. His final estimate of the upper bound for the number of grains of sand in a filled universe was 10^{64}.

Fig. 1.8 Plato and Aristotle

It is possible that he may have developed the odometer,[17] and this instrument could calculate the total distance travelled on a journey. An odometer is described by the Roman engineer Vitruvius around 25 B.C. It employed a wheel with a diameter of 4 feet, and the wheel turned 400 times in every mile.[18] The device included gears and pebbles and a 400-tooth cogwheel that turned once every mile and caused one pebble to drop into a box. The total distance travelled was determined by counting the pebbles in the box.

Aristotle was born in Macedonia and became a student of Plato in Athens (Fig. 1.8). Plato had founded a school (known as Plato's academy) in Athens in the fourth century B.C., and this school remained open until 529 A.D. Aristotle founded his own school (known as the Lyceum) in Athens. He was also the tutor of Alexander the Great. He made contributions to physics, biology, logic, politics, ethics and metaphysics.

Aristotle's starting point to the acquisition of knowledge was the senses, as he believed that these were essential to acquire knowledge. This position is the opposite from Plato who argued that the senses deceive and should not be relied upon. Plato's writings are mainly in dialogues involving his former mentor Socrates.[19]

---

[17]The origin of the word 'odometer' is from the Greek words 'οδοζ (meaning journey) and μετρον meaning (measure).

[18]The figures given here are for the distance of one Roman mile. This is given by $\pi 4 * 400 = 12.56 * 400 = 5024$ (which is less than 5280 feet for a standard mile in the Imperial system).

[19]Socrates was a moral philosopher who deeply influenced Plato. His method of enquiry into philosophical problems and ethics was by questioning. Socrates himself maintained that he knew nothing (Socratic ignorance). However, from his questioning it became apparent that those who thought they were clever were not really that clever after all. His approach obviously would not have made him very popular with the citizens of Athens. Socrates had consulted the oracle at Delphi to find out who was the wisest of all men, and he was informed that there was no one wiser

Aristotle made important contributions to formal reasoning with his development of syllogistic logic. His collected works on logic is called the Organon and it was used in his school in Athens. Syllogistic logic (also known as term logic) consists of reasoning with two premises and one conclusion. Each premise consists of two terms and there is a common middle term. The conclusion links the two unrelated terms from the premises. For example

Premise 1   All Greeks are Mortal
Premise 2   Socrates is a Greek
_____

Conclusion   Socrates is Mortal

The common middle term is 'Greek', which appears in the two premises. The two unrelated terms from the premises are 'Socrates' and 'Mortal'. The relationship between the terms in the first premise is that of the universal: i.e. anything or any person that is a Greek is mortal. The relationship between the terms in the second premise is that of the particular: i.e. Socrates is a person that is a Greek. The conclusion from the two premises is that Socrates is mortal: i.e. a particular relationship between the two unrelated terms 'Socrates' and 'Mortal'.

The syllogism above is a valid syllogistic argument. Aristotle studied the various possible syllogistic arguments and determined those that were valid and invalid. Syllogistic logic is described in more detail in Chap. 14. Aristotle's work was highly regarded in classical and medieval times, and Kant believed that there was nothing else to invent in Logic. There was another competing system of logic proposed by the Stoics in Hellenistic times: i.e. an early form of propositional logic that was developed by Chrysippus[20] in the third century B.C. Aristotelian logic is mainly of historical interest today.

Aquinas,[21] a thirteenth century Christian theologian and philosopher, was deeply influenced by Aristotle, and referred to him as the philosopher. Aquinas was an empiricist (i.e. he believed that all knowledge was gained by sense experience), and he used some of Aristotle's arguments to offer five proofs of the existence of God. These arguments included the Cosmological argument and the Design argument. The Cosmological argument used Aristotle's ideas on the scientific method and causation. Aquinas argued that there was a first cause and he deduced that this first cause is God.

1. Every effect has a cause
2. Nothing can cause itself

_____

(Footnote 19 continued)
than him. Socrates was sentenced to death for allegedly corrupting the youth of Athens, and the sentence was carried out by Socrates being forced to take hemlock (a type of poison). The juice of the hemlock plant was prepared for Socrates to drink.

[20]Chrysippus was the head of the Stoics in the third century B.C.

[21]Aquinus's (or St. Thomas's) most famous work is Sumna Theologicae.

Fig. 1.9 Julius Caesar

3. A causal chain cannot be of infinite length
4. Therefore, there must be a first cause.

The Antikythera [5] was an ancient mechanical device that is believed to have been designed to calculate astronomical positions. It was discovered in 1902 in a wreck off the Greek island of Antikythera, and dates from about 80 B.C. It is one of the oldest known geared devices, and it is believed that it was used for calculating the position of the Sun, Moon, Stars and Planets for a particular date entered.

The Romans appear to have been aware of a device similar to the Antikythera that was capable of calculating the position of the planets. The island of Antikythera was well known in the Greek and Roman period for its displays of mechanical engineering.

1.5 The Romans

Rome is said to have been founded[22] by Romulus and Remus about 750 B.C. Early Rome covered a small part of Italy but it gradually expanded in size and importance. It destroyed Carthage[23] in 146 B.C. to become the major power in the Mediterranean. The Romans colonized the Hellenistic world, and they were influenced by Greek culture and mathematics. Julius Caesar conquered the Gauls in 58 B.C. (Fig. 1.9).

---

[22]The Aenid by Virgil suggests that the Romans were descended from survivors of the Trojan war, and that Aeneas brought surviving Trojans to Rome after the fall of Troy.

[23]Carthage was located in Tunisia, and the wars between Rome and Carthage are known as the Punic wars. Hannibal was one of the great Carthaginan military commanders, and during the second Punic war, he brought his army to Spain, marched through Spain and crossed the Pyrnees. He then marched along southern France and crossed the Alps into Northern Italy. His army also consisted of war elephants. Rome finally defeated Carthage and destroyed the city.

Fig. 1.10 Roman numbers

I = 1
V = 5
X = 10
L = 50
C = 100
D = 500
M = 1000

The Gauls consisted of several disunited Celtic[24] tribes. Vercingetorix succeeded in uniting them, but he was defeated by at the siege of Alesia in 52 B.C.

The Roman number system uses letters to represented numbers and a number consists of a sequence of letters. The evaluation rules specify that if a number follows a smaller number then the smaller number is subtracted from the larger number: e.g. IX represents 9 and XL represents 40. Similarly, if a smaller number followed a larger number they were generally added: e.g. MCC represents 1200. They had no zero in their number system (Fig. 1.10).

The use of Roman numerals was cumbersome in calculation, and an abacus was often employed. An abacus is a device that is usually of wood and has a frame that holds rods with freely sliding beads mounted on them. It is used as a tool to assist calculation, and it is useful for keeping track of the sums and the carries of calculations.

It consists of several columns in which beads or pebbles are placed. Each column represented powers of 10: i.e. 10^0, 10^1, 10^2, 10^3, etc. The column to the far right represents one; the column to the left 10; next column to the left 100; and so on. Pebbles[25] (calculi) were placed in the columns to represent different numbers: e.g. the number represented by an abacus with four pebbles on the far right; two pebbles in the column to the left; and three pebbles in the next column to the left is 324. The calculations were performed by moving pebbles from column to column.

Merchants introduced a set of weights and measures (including the *libra* for weights and the *pes* for lengths). They developed an early banking system to provide loans for business, and commenced minting money about 290 B.C. The Romans also made contributions to calendars, and Julius Caesar introduced the Julian calendar in 45 B.C. It has a regular year of 365 days divided into 12 months and a leap day is added to February every four years. It remained in use up to the

---

[24]The Celtic period commenced around 1000 B.C. in Hallstaat (near Salzburg in Austria). The Celts were skilled in working with Iron and Bronze, and they gradually expanded into Europe. They eventually reached Britain and Ireland around 600 B.C. The early Celtic period was known as the 'Hallstaat period' and the later Celtic period is known as 'La Téne'. The later La Téne period is characterized by the quality of ornamentation produced. The Celtic museum in Hallein in Austria provides valuable information and artefacts on the Celtic period. The Celtic language would have similarities to the Irish language. However, the Celts did not employ writing, and the Ogham writing used in Ireland was developed in the early Christian period.

[25]The origin of the word 'Calculus' is from Latin and means a small stone or pebble used for counting.

Alphabet Symbol	abcde fghij klmno pqrst uvwxyz
Cipher Symbol	dfegh ijklm nopqr stuvw xyzabc

Fig. 1.11 Caesar cipher

twentieth century, but has since been replaced by the Gregorian calendar. The problem with the Julian calendar is that too many leap years are added over time. The Gregorian calendar was first introduced in 1582.

The Romans employed the mathematics that had been developed by the Greeks. Caesar employed a substitution cipher on his military campaigns to enable important messages to be communicated safely. It involves the substitution of each letter in the plaintext (i.e. the original message) by a letter a fixed number of positions down in the alphabet. For example, a shift of three positions causes the letter B to be replaced by E, the letter C by F, and so on. It is easily broken, as the frequency distribution of letters may be employed to determine the mapping. The cipher is defined in Fig. 1.11.

The process of enciphering a message (i.e. plaintext) involves looking up each letter in the plaintext and writing down the corresponding cipher letter. The decryption involves the reverse operation: i.e. for each cipher letter the corresponding plaintext letter is identified from the table.

The encryption may also be represented using modular arithmetic,[26] with the numbers 0–25 representing the alphabet letters, and addition (modulo 26) is used to perform the encryption.

The emperor Augustus[27] employed a similar substitution cipher (with a shift key of 1). The Caesar cipher remained in use up to the early twentieth century. However, by then frequency analysis techniques were available to break the cipher.

1.6 Islamic Influence

Islamic mathematics refers to mathematics developed in the Islamic world from the birth of Islam in the early seventh century up until the seventeenth century. The Islamic world commenced with the prophet Mohammed in Mecca, and spread throughout the Middle East, North Africa and Spain. The Golden Age of Islamic civilization was from 750 A.D. to 1250 A.D., and during this period enlightened

---

[26]Modular arithmetic is discussed in chapter seven.

[27]Augustus was the first Roman emperor and his reign ushered in a period of peace and stability following the bitter civil wars. He was the adopted son of Julius Caesar and was called Octavion before he became emperor. The earlier civil wars were between Caesar and Pompey, and following Caesar's assassination civil war broke out between Mark Anthony and Octavion. Octavion defeated Anthony and Cleopatra at the battle of Actium, and became the first Roman emperor, Augusus.

Fig. 1.12 Mohammed
Al-Khwarizmi

caliphs recognized the value of knowledge, and sponsored scholars to come to Baghdad to gather and translate the existing world knowledge into Arabic.

This led to the preservation of the Greek texts during the Dark ages in Europe. Further, the Islamic cities of Baghdad, Cordoba and Cairo became key intellectual centres, and scholars added to existing knowledge (e.g. in mathematics, astronomy, medicine and philosophy), as well as translating the known knowledge into Arabic.

The Islamic mathematicians and scholars were based in several countries in the Middle East, North Africa and Spain. Early work commenced in Baghdad, and the mathematicians were also influenced by the work of Hindu mathematicians who had introduced the decimal system and decimal numerals. Among the well-known Islamic scholars are Ibn Al Haytham, a tenth century Iraqi scientist; Mohammed Al-Khwarizmi (Fig. 1.12), a ninth Persian mathematician; Abd Al Rahman al Sufi, a Persian astronomer who discovered the Andromeda galaxy; Ibn Al Nafis, a Syrian who did work on circulation in medicine; Averroes, who was an Aristotelian philosopher from Cordoba in Spain; Avicenna who was a Persian philosopher; and Omar Khayyman who was a Persian Mathematician and poet.

Many caliphs (Muslim rulers) were enlightened and encouraged scholarship in mathematics and science. They has setup a centre for translation and research in Baghdad, and existing Greek texts such as the works of Euclid, Archimedes, Apollonius and Diophantus were translated into Arabic. Al-Khwarizmi made contributions to early classical algebra, and the word algebra comes from the Arabic word '*al jabr*' that appears in a textbook by Al-Khwarizmi. The origin of the word *algorithm* is from the name of the Islamic scholar 'Al-Khwarizmi'.

Education was important during the Golden Age, and the Al Azhar University in Cairo (Fig. 1.13) was established in 970 A.D., and the Al-Qarawiyyin University in Fez, Morocco was established in 859 A.D. The Islamic World has created beautiful architecture and art including the ninth century Great Mosque of Samarra in Iraq; the tenth century Great Mosque of Cordoba; and the eleventh century Alhambra in Grenada.

Fig. 1.13 Al Azhar University, Cairo

The Moors[28] invaded Spain in the eighth century A.D., and they ruled large parts of the Peninsula for several centuries. Moorish Spain became a centre of learning, and this led to Islamic and other scholars coming to study at the universities in Spain. Many texts on Islamic mathematics were translated from Arabic into Latin, and these were invaluable in the renaissance in European learning and mathematics from the thirteenth century. The Moorish influence[29] in Spain continued until the time of the Catholic Monarchs[30] in the fifth century. Ferdinand and Isabella united Spain, defeated the Moors in Andalusia, and expelled them from Spain.

The Islamic contribution to algebra was an advance on the achievements of the Greeks. They developed a broader theory that treated rational and irrational numbers as algebraic objects, and moved away from the Greek concept of mathematics as being essentially Geometry. Later Islamic scholars applied algebra to arithmetic and geometry, and studied curves using equations. This included contributions to

---

[28]The origin of the word 'Moor' is from the Greek work μυορος̌ meaning very dark. It referred to the fact that many of the original Moors who came to Spain were from Egypt, Tunisia and other parts of North Africa.

[29]The Moorish influence includes the construction of various castles (*alcazar*), fortresses (*alcalzaba*) and mosques. One of the most striking Islamic sites in Spain is the palace of Alhambra in Granada, and it represents the zenith of Islamic art.

[30]The Catholic Monarchs refer to Ferdinand of Aragon and Isabella of Castille who married in 1469. They captured Granada (the last remaining part of Spain controlled by the Moors) in 1492.

reduce geometric problems such as duplicating the cube to algebraic problems. Eventually this led to the use of symbols in the fifteenth century such as

$$x^n \cdot x^m = x^{m+n}.$$

The poet Omar Khayman was also a mathematician who did work on the classification of cubic equations with geometric solutions. Other scholars made contributions to the theory of numbers: e.g. a theorem that allows pairs of amicable numbers to be found. Amicable numbers are two numbers such that each is the sum of the proper divisors of the other. They were aware of Wilson's theory in number theory: i.e. for p prime then p divides $(p - 1)! + 1$.

The Islamic world was tolerant of other religious belief systems during the Golden Age, and there was freedom of expression provided that it did not infringe on the rights of others. It began to come to an end following the Mongol invasion and sack of Baghdad in the late 1250s and the Crusades. It continued to some extent until the conquest by Ferdinand and Isabella of Andalusia in the late fifteenth century.

1.7   Chinese and Indian Mathematics

The development of mathematics commenced in China about 1000 B.C., and was independent of developments in other countries. The emphasis was on problem solving rather than on conducting formal proofs. It was concerned with finding the solution to practical problems such as the calendar, the prediction of the positions of the heavenly bodies, land measurement, conducting trade and the calculation of taxes.

The Chinese employed counting boards as mechanical aids for calculation from the fourth century B.C. These are similar to abaci and are usually made of wood or metal, and contained carved grooves between which beads, pebbles or metal discs were moved.

Early Chinese mathematics was written on bamboo strips and included work on arithmetic and astronomy. The Chinese method of learning and calculation in mathematics was learning by analogy. This involves a person acquiring knowledge from observation of how a problem is solved, and then applying this knowledge for problem solving to similar kinds of problems.

They had their version of Pythagoras's Theorem and applied it to practical problems. They were familiar with the Chinese remainder theorem, the formula for finding the area of a triangle, as well as showing how polynomial equations (up to degree ten) could be solved. They showed how geometric problems could be solved by algebra, how roots of polynomials could be solved, how quadratic and simultaneous equations could be solved, and how the area of various geometric shapes such as rectangles, trapezia and circles could be computed. Chinese mathematicians were familiar with the formula to calculate the volume of a sphere. The best approximation that the Chinese had to π was 3.14159, and this was obtained by approximations from inscribing regular polygons with 3×2^n sides in a circle.

The Chinese made contributions to number theory including the summation of arithmetic series and solving simultaneous congruences. The Chinese remainder theorem deals with finding the solutions to a set of simultaneous congruences in modular arithmetic. Chinese astronomers made accurate observations, which were used to produce a new calendar in the sixth century. This was known as the Taming Calendar and it was based on a cycle of 391 years.

Indian mathematicians have made important contributions such as the development of the decimal notation for numbers that is now used throughout the world. This was developed in India sometime between 400 B.C. and 400 A.D. Indian mathematicians also invented zero and negative numbers, and also did early work on the trigonometric functions of sine and cosine. The knowledge of the decimal numerals reached Europe through Arabic mathematicians, and the resulting system is known as the Hindu–Arabic numeral system.

The Sulva Sutras is a Hindu text that documents Indian mathematics and it dates from about 400 B.C. They were familiar with the statement and proof of Pythagoras's theorem, Rational numbers, quadratic equations, as well as the calculation of the square root of 2 to five decimal places.

1.8 Review Questions

1. Discuss the strengths and weaknesses of the various numbering system.
2. Describe the ciphers used during the Roman civilization and write a program to implement one of these.
3. Discuss the nature of an algorithm and its importance in computing.
4. Discuss the working of an abacus and its application to calculation.
5. What are the differences between syllogistic logic and stoic logic?
6. Describe the main achievements of the Islamic world in mathematics.

1.9 Summary

Software is pervasive in the modern world, and it has transformed the world in which we live in. New technology has led to improvements in all aspects of our lives including medicine, transport, education, and so on. The pace of change of new technology is relentless, with new versions of technology products becoming available several times a year.

This chapter considered some of the contributions of early civilizations to computing. We commenced our journey with an examination of some of the contributions of the Babylonians. We then moved forward to consider some of the achievements of the Egyptians, the Greek and Romans; Islamic scholars; and the Indians and Chinese.

The Babylonians recorded their mathematical knowledge on clay cuneiform tablets. These tablets included tables for multiplication, division, squares and square roots and the calculation of area. They were familiar with techniques that allowed the solution of a linear equation and one root of a quadratic equation to be determined.

The Egyptian civilization developed along the River Nile, and they applied their knowledge of mathematics to solve practical problem such as measuring the annual Nile flooding, and constructing temples and pyramids.

The Greeks and the later Hellenistic period made important contributions to western civilization. Their contributions to mathematics included the Euclidean algorithm, which is used to determine the greatest common divisor of two numbers. Eratosthenes developed an algorithm to determine the prime numbers up to a given number. Archimedes invented the 'Archimedes Screw', the 'Archimedes Claw', and a type of heat ray.

The Islamic civilization helped to preserve western knowledge that was lost during the dark ages in Europe, and they also continued to develop mathematics and algebra. Hindu mathematicians introduced the decimal notation that is familiar today. Islamic mathematicians adopted it and the resulting system is known as the Hindu–Arabic system.

References

1. Mathematics in Civilisation. H.L. Resnikoff and R.O.Wells. Dover Publications. 1984.
2. History of mathematics. D.E. Smith. Volume 1. Dover Publications, New York. 1923.
3. The Heritage of Thales. W.S. Anglin and J. Lambek. Springer Verlag. New York. 1995.
4. Euclid. The Thirteen Books of the Elements. Vol. 1. Translated by Sir Thomas Heath. Dover Publications, 1956. (First published in 1925).
5. An Ancient Greek Computer. Derek J. de Solla Price. *Scientific American*, pp. 60–67. June 1959.

Sets, Relations and Functions

2

Key Topics

Sets
Set Operations
Russell's Paradox
Computer Representation of sets
Relations
Composition of Relations
Reflexive, Symmetric and Transitive Relations
Relational Database Management System
Functions
Partial and Total Functions
Injective, Surjective and Bijective Functions
Functional Programming

2.1 Introduction

This chapter provides an introduction to fundamental building blocks in mathematics such as sets, relations and functions. Sets are collections of well-defined objects; relations indicate relationships between members of two sets A and B; and

© Springer International Publishing Switzerland 2016
G. O'Regan, *Guide to Discrete Mathematics*, Texts in Computer Science,
DOI 10.1007/978-3-319-44561-8_2

functions are a special type of relation where there is exactly (or at most)[1] one relationship for each element $a \in A$ with an element in B.

A set is a collection of well-defined objects that contain no duplicates. The term 'well defined' means that for a given value it is possible to determine whether or not it is a member of the set. There are many examples of sets such as the set of natural numbers $\mathbb{N}$, the set of integer numbers $\mathbb{Z}$, and the set of rational numbers $\mathbb{Q}$. The natural numbers $\mathbb{N}$ is an infinite set consisting of the numbers $\{1, 2, \dots\}$. Venn diagrams may be used to represent sets pictorially.

A binary relation $R\ (A, B)$ where A and B are sets is a subset of the Cartesian product $(A \times B)$ of A and B. The domain of the relation is A and the codomain of the relation is B. The notation aRb signifies that there is a relation between a and b and that $(a, b) \in R$. An n-ary relation $R\ (A_1, A_2, \dots A_n)$ is a subset of $(A_1 \times A_2 \times \dots \times A_n)$. However, an n-ary relation may also be regarded as a binary relation $R(A, B)$ with $A = A_1 \times A_2 \times \dots \times A_{n-1}$ and $B = A_n$.

Functions may be total or partial. A total function $f\colon A \to B$ is a special relation such that for each element $a \in A$ there is exactly one element $b \in B$. This is written as $f(a) = b$. A partial function differs from a total function in that the function may be undefined for one or more values of A. The domain of a function (denoted by **dom** f) is the set of values in A for which the partial function is defined. The domain of the function is A provided that f is a total function. The codomain of the function is B.

2.2 Set Theory

A set is a fundamental building block in mathematics, and it is defined as a collection of well-defined objects. The elements in a set are of the same kind, and they are distinct with no repetition of the same element in the set.[2] Most sets encountered in computer science are finite, as computers can only deal with finite entities. Venn diagrams[3] are often employed to give a pictorial representation of a set, and they may be used to illustrate various set operations such as set union, intersection and set difference.

There are many well-known examples of sets including the set of natural numbers denoted by $\mathbb{N}$; the set of integers denoted by $\mathbb{Z}$; the set of rational numbers is denoted by $\mathbb{Q}$; the set of real numbers denoted by $\mathbb{R}$; and the set of complex numbers denoted by $\mathbb{C}$.

---

[1]We distinguish between total and partial functions. A total function $f\colon A \to B$ is defined for every element in A whereas a partial function may be undefined for one or more values in A.

[2]There are mathematical objects known as multi-sets or bags that allow duplication of elements. For example, a bag of marbles may contain three green marbles, two blue and one red marble.

[3]The British logician, John Venn, invented the Venn diagram. It provides a visual representation of a set and the various set theoretical operations. Their use is limited to the representation of two or three sets as they become cumbersome with a larger number of sets.

Example 2.1 The following are examples of sets.

- The books on the shelves in a library
- The books that are currently overdue from the library
- The customers of a bank
- The bank accounts in a bank
- The set of Natural Numbers $\mathbb{N} = \{1, 2, 3, \ldots\}$
- The Integer Numbers $\mathbb{Z} = \{\ldots, -3, -2, -1, 0, 1, 2, 3, \ldots\}$
- The non-negative integers $\mathbb{Z}^{+} = \{0, 1, 2, 3, \ldots\}$
- The set of Prime Numbers $= \{2, 3, 5, 7, 11, 13, 17, \ldots\}$
- The Rational Numbers is the set of quotients of integers

$$\mathbb{Q} = \{p/q : p, q \in \mathbb{Z} \text{ and } q \neq 0\}$$

A finite set may be defined by listing all of its elements. For example, the set $A = \{2, 4, 6, 8, 10\}$ is the set of all even natural numbers less than or equal to 10. The order in which the elements are listed is not relevant: i.e. the set $\{2, 4, 6, 8, 10\}$ is the same as the set $\{8, 4, 2, 10, 6\}$.

Sets may be defined by using a predicate to constrain set membership. For example, the set $S = \{n: \mathbb{N}: n \leq 10 \wedge n \bmod 2 = 0\}$ also represents the set $\{2, 4, 6, 8, 10\}$. That is, the use of a predicate allows a new set to be created from an existing set by using the predicate to restrict membership of the set. The set of even natural numbers may be defined by a predicate over the set of natural numbers that restricts membership to the even numbers. It is defined by

$$\text{Evens} = \{x | x \in \mathbb{N} \wedge even(x)\}.$$

In this example, *even(x)* is a predicate that is true if x is even and false otherwise. In general, $A = \{x \in E \mid P(x)\}$ denotes a set A formed from a set E using the predicate P to restrict membership of A to those elements of E for which the predicate is true.

The elements of a finite set S are denoted by $\{x_1, x_2, \ldots x_n\}$. The expression $x \in S$ denotes that the element x is a member of the set S, whereas the expression $x \notin S$ indicates that x is not a member of the set S.

A set S is a subset of a set T (denoted $S \subseteq T$) if whenever $s \in S$ then $s \in T$, and in this case the set T is said to be a superset of S (denoted $T \supseteq S$). Two sets S and T are said to be equal if they contain identical elements: i.e. $S = T$ if and only if $S \subseteq T$ and $T \subseteq S$. A set S is a proper subset of a set T (denoted $S \subset T$) if $S \subseteq T$ and $S \neq T$. That is, every element of S is an element of T and there is at least one element in T that is not an element of S. In this case, T is a proper superset of S (denoted $T \supset S$).

The empty set (denoted by $\varnothing$ or $\{\}$) represents the set that has no elements. Clearly $\varnothing$ is a subset of every set. The singleton set containing just one element x is denoted by $\{x\}$, and clearly $x \in \{x\}$ and $x \neq \{x\}$. Clearly, $y \in \{x\}$ if and only if $x = y$.

Example 2.2

(i) $\{1, 2\} \subseteq \{1, 2, 3\}$
(ii) $\varnothing \subset \mathbb{N} \subset \mathbb{Z} \subset \mathbb{Q} \subset \mathbb{R} \subset \mathbb{C}$

The cardinality (or size) of a finite set S defines the number of elements present in the set. It is denoted by $|S|$. The cardinality of an infinite[4] set S is written as $|S| = \infty$.

Example 2.3

(i)   Given A = $\{2, 4, 5, 8, 10\}$ then $|A| = 5$.
(ii) Given A = $\{x \in \mathbb{Z}: x^2 = 9\}$ then $|A| = 2$
(iii) Given A = $\{x \in \mathbb{Z}: x^2 = -9\}$ then $|A| = 0$.

2.2.1 Set Theoretical Operations

Several set theoretical operations are considered in this section. These include the Cartesian product operation; the power set of a set; the set union operation; the set intersection operation; the set difference operation; and the symmetric difference operation.

Cartesian Product

The Cartesian product allows a new set to be created from existing sets. The Cartesian[5] product of two sets S and T (denoted $S \times T$) is the set of ordered pairs $\{(s, t) \mid s \in S, t \in T\}$. Clearly, $S \times T \neq T \times S$ and so the Cartesian product of two sets is not commutative. Two ordered pairs (s_1, t_1) and (s_2, t_2) are considered equal if and only if $s_1 = s_2$ and $t_1 = t_2$.

The Cartesian product may be extended to that of n sets $S_1, S_2, ..., S_n$. The Cartesian product $S_1 \times S_2 \times ... \times S_n$ is the set of ordered tuples $\{(s_1, s_2, ..., s_n) \mid s_1$

---

[4]The natural numbers, integers and rational numbers are countable sets whereas the real and complex numbers are uncountable sets.

[5]Cartesian product is named after René Descartes who was a famous 17th French mathematician and philosopher. He invented the Cartesian coordinates system that links geometry and algebra, and allows geometric shapes to be defined by algebraic equations.

$\in S_1, s_2 \in S_2, \ldots, s_n \in S_n\}$. Two ordered n-tuples $(s_1, s_2, \ldots, s_n)$ and $(s_1', s_2', \ldots, s_n')$ are considered equal if and only if $s_1 = s_1', s_2, = s_2', \ldots, s_n = s_n'$.

The Cartesian product may also be applied to a single set S to create ordered n-tuples of S: i.e. $S^n = S \times S \times \ldots \times S$ (n-times).

Power Set

The power set of a set A (denoted $\mathbb{P}A$) denotes the set of subsets of A. For example, the power set of the set $A = \{1, 2, 3\}$ has 8 elements and is given by

$$\mathbb{P}A = \{\varnothing, \{1\}, \{2\}, \{3\}, \{1,2\}, \{1,3\}, \{2,3\}, \{1,2,3\}\}.$$

There are $2^3 = 8$ elements in the power set of $A = \{1, 2, 3\}$ and the cardinality of A is 3. In general, there are $2^{|A|}$ elements in the power set of A.

Theorem 2.1 (Cardinality of Power Set of A) *There are $2^{|A|}$ elements in the power set of A*

Proof Let $|A| = n$ then the cardinality of the subsets of A are subsets of size 0, 1, $\ldots, n$. There are $\binom{n}{k}$ subsets of A of size k.[6] Therefore, the total number of subsets of A is the total number of subsets of size 0, 1, 2, ... up to n. That is

$$|\mathbb{P}A| = \sum_{k=0}^{n} \binom{n}{k}$$

The Binomial Theorem (we prove it in Example 4.2 in Chap. 4) states that

$$(1+x)^n = \sum_{k=0}^{n} \binom{n}{k} x^k$$

Therefore, putting $x = 1$ we get that

$$2^n = (1+1)^n = \sum_{k=0}^{n} \binom{n}{k} 1^k = |\mathbb{P}A|$$

Union and Intersection Operations

The union of two sets A and B is denoted by $A \cup B$. It results in a set that contains all of the members of A and of B and is defined by

$$A \cup B = \{r | r \in A \text{ or } r \in B\}.$$

For example, suppose $A = \{1, 2, 3\}$ and $B = \{2, 3, 4\}$ then $A \cup B = \{1, 2, 3, 4\}$. Set union is a commutative operation: i.e. $A \cup B = B \cup A$. Venn Diagrams are used to illustrate these operations pictorially.

---

[6]We discuss permutations and combinations in Chap. 5.

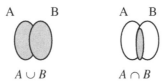

$$A \cup B \qquad A \cap B$$

The intersection of two sets A and B is denoted by $A \cap B$. It results in a set containing the elements that A and B have in common and is defined by

$$A \cap B = \{r | r \in A \text{ and } r \in B\}.$$

Suppose $A = \{1, 2, 3\}$ and $B = \{2, 3, 4\}$ then $A \cap B = \{2, 3\}$. Set intersection is a commutative operation: i.e. $A \cap B = B \cap A$.

Union and intersection are binary operations but may be extended to more generalized union and intersection operations. For example

$\cup_{i=1}^{n} A_i$ denotes the union of n sets.

$\cap_{i=1}^{n} A_i$ denotes the intersection of n sets

Set Difference Operations

The set difference operation $A \backslash B$ yields the elements in A that are not in B. It is defined by

$$A \backslash B = \{a | a \in A \text{ and } a \notin B\}.$$

For A and B defined as $A = \{1, 2\}$ and $B = \{2, 3\}$ we have $A \backslash B = \{1\}$ and $B \backslash A = \{3\}$. Clearly, set difference is not commutative: i.e. $A \backslash B \neq B \backslash A$. Clearly, $A \backslash A = \emptyset$ and $A \backslash \emptyset = A$.

The symmetric difference of two sets A and B is denoted by $A \Delta B$ and is given by

$$A \Delta B = A \backslash B \cup B \backslash A$$

The symmetric difference operation is commutative: i.e. $A \Delta B = B \Delta A$. Venn diagrams are used to illustrate these operations pictorially.

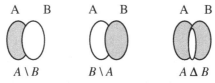

$$A \backslash B \qquad B \backslash A \qquad A \Delta B$$

The complement of a set A (with respect to the universal set U) is the elements in the universal set that are not in A. It is denoted by A^c (or A') and is defined as

$$A^c = \{u | u \in U \text{ and } u \notin A\} = U \backslash A$$

The complement of the set A is illustrated by the shaded area below

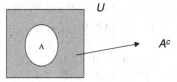

2.2.2 Properties of Set Theoretical Operations

The set union and set intersection properties are commutative and associative. Their properties are listed in Table 2.1.

These properties may be seen to be true with Venn diagrams, and we give a proof of the distributive property (this proof uses logic which is discussed in Chaps. 14–16).

Proof of Properties (Distributive Property)
To show $A \cap (B \cup C) = (A \cap B) \cup (A \cap C)$
  Suppose $x \in A \cap (B \cup C)$ then

$$x \in A \wedge x \in (B \cup C)$$

$$\Rightarrow x \in A \wedge (x \in B \vee x \in C)$$

Table 2.1 Properties of set operations

Property	Description
Commutative	Union and intersection operations are commutative: i.e. $S \cup T = T \cup S$ $S \cap T = T \cap S$
Associative	Union and intersection operations are associative: i.e. $R \cup (S \cup T) = (R \cup S) \cup T$ $R \cap (S \cap T) = (R \cap S) \cap T$
Identity	The identity under set union is the empty set $\varnothing$, and the identity under intersection is the universal set U. $S \cup \varnothing = \varnothing \cup S = S$ $S \cap U = U \cap S = S$
Distributive	The union operator distributes over the intersection operator and vice versa. $R \cap (S \cup T) = (R \cap S) \cup (R \cap T)$ $R \cup (S \cap T) = (R \cup S) \cap (R \cup T)$.
DeMorgan's[a] Law	The complement of $S \cup T$ is given by $(S \cup T)^c = S^c \cap T^c$ The complement of $S \cap T$ is given by $(S \cap T)^c = S^c \cup T^c$

[a]De Morgan's law is named after Augustus De Morgan, a nineteenth century English mathematician who was a contemporary of George Boole

$$\Rightarrow (x \in A \land x \in B) \lor (x \in A \land x \in C)$$
$$\Rightarrow x \in (A \cap B) \lor x \in (A \cap C)$$
$$\Rightarrow x \in (A \cap B) \cup (A \cap C)$$

Therefore, $A \cap (B \cup C) \subseteq (A \cap B) \cup (A \cap C)$
Similarly $(A \cap B) \cup (A \cap C) \subseteq A \cap (B \cup C)$
Therefore, $A \cap (B \cup C) = (A \cap B) \cup (A \cap C)$

2.2.3 Russell's Paradox

Bertrand Russell (Fig. 2.1) was a famous British logician, mathematician and philosopher. He was the co-author with Alfred Whitehead of *Principia Mathematica*, which aimed to derive all of the truths of mathematics from logic. Russell's Paradox was discovered by Bertrand Russell in 1901, and showed that the system of logicism being proposed by Frege (discussed in Chap. 14) contained a contradiction.

Question (Posed by Russell to Frege)
Is the set of all sets that do not contain themselves as members a set?

Russell's Paradox
Let $A = \{S$ a set and $S \notin S\}$. Is $A \in A$? Then $A \in A \Rightarrow A \notin A$ and vice versa. Therefore, a contradiction arises in either case and there is no such set A.

Two ways of avoiding the paradox were developed in 1908, and these were Russell's theory of types and Zermelo set theory. Russell's theory of types was a response to the paradox by arguing that the set of all sets is ill formed. Russell developed a hierarchy with individual elements the lowest level; sets of elements at the next level; sets of sets of elements at the next level; and so on. It is then prohibited for a set to contain members of different types.

A set of elements has a different type from its elements, and one cannot speak of the set of all sets that do not contain themselves as members as these are of different

Fig. 2.1 Bertrand russell

types. The other way of avoiding the paradox was Zermelo's axiomatization of set theory.

Remark Russell's paradox may also be illustrated by the story of a town that has exactly one barber who is male. *The barber shaves all and only those men in town who do not shave themselves.* The question is who shaves the barber.

If the barber does not shave himself then according to the rule he is shaved by the barber (i.e. himself). If he shaves himself then according to the rule he is not shaved by the barber (i.e. himself).

The paradox occurs due to self-reference in the statement and a logical examination shows that the statement is a contradiction.

2.2.4 Computer Representation of Sets

Sets are fundamental building blocks in mathematics, and so the question arises as to how a set is stored and manipulated in a computer. The representation of a set M on a computer requires a change from the normal view that the order of the elements of the set is irrelevant, and we will need to assume a definite order in the underlying universal set $\mathscr{M}$ from which the set M is defined.

That is, a set is always defined in a computer program with respect to an underlying universal set, and the elements in the universal set are listed in a definite order. Any set M arising in the program that is defined with respect to this universal set $\mathscr{M}$ is a subset of $\mathscr{M}$. Next, we show how the set M is stored internally on the computer.

The set M is represented in a computer as a string of binary digits $b_1 b_2 \ldots b_n$ where n is the cardinality of the universal set $\mathscr{M}$. The bits b_i (where i ranges over the values 1, 2, $\ldots n$) are determined according to the rule

$b_i = 1$ if ith element of $\mathscr{M}$ is in M
$b_i = 0$ if ith element of $\mathscr{M}$ is not in M

For example, if $\mathscr{M} = \{1, 2, \ldots 10\}$ then the representation of $M = \{1, 2, 5, 8\}$ is given by the bit string 1100100100 where this is given by looking at each element of $\mathscr{M}$ in turn and writing down 1 if it is in M and 0 otherwise.

Similarly, the bit string 0100101100 represents the set $M = \{2, 5, 7, 8\}$, and this is determined by writing down the corresponding element in $\mathscr{M}$ that corresponds to a 1 in the bit string.

Clearly, there is a one-to-one correspondence between the subsets of $\mathscr{M}$ and all possible n-bit strings. Further, the set theoretical operations of set union, intersection and complement can be carried out directly with the bit strings (provided that the sets involved are defined with respect to the same universal set). This involves a bitwise 'or' operation for set union; a bitwise 'and' operation for set intersection; and a bitwise 'not' operation for the set complement operation.

2.3 Relations

A binary relation $R(A, B)$ where A and B are sets is a subset of $A \times B$: i.e. $R \subseteq A \times B$. The domain of the relation is A and the codomain of the relation is B. The notation aRb signifies that $(a, b) \in R$.

A binary relation $R(A, A)$ is a relation between A and A. This type of relation may always be composed with itself, and its inverse is also a binary relation on A. The identity relation on A is defined by $a\ i_A a$ for all $a \in A$.

Example 2.4 There are many examples of relations

(i)   The relation on a set of students in a class where $(a, b) \in R$ if the height of a is greater than the height of b.
(ii) The relation between A and B where $A = \{0, 1, 2\}$ and $B = \{3, 4, 5\}$ with R given by

$$R = \{(0, 3), (0, 4), (1, 4)\}$$

(iii) The relation less than $(<)$ between and $\mathbb{R}$ and $\mathbb{R}$ is given by

$$\{(x, y) \in \mathbb{R}^2 : x < y\}$$

(iv) A bank may represent the relationship between the set of accounts and the set of customers by a relation. The implementation of a bank account will often be a positive integer with at most eight decimal digits.
      The relationship between accounts and customers may be done with a relation $R \subseteq A \times B$, with the set A chosen to be the set of natural numbers, and the set B chosen to be the set of all human beings alive or dead. The set A could also be chosen to be $A = \{n \in \mathbb{N}: n < 10^8\}$

A relation $R(A, B)$ may be represented pictorially. This is referred to as the graph of the relation, and it is illustrated in the diagram below. An arrow from x to y is drawn if (x, y) is in the relation. Thus for the height relation R given by $\{(a, p), (a, r), (b, q)\}$ an arrow is drawn from a to p, from a to r and from b to q to indicate that (a, p), (a, r) and (b, q) are in the relation R.

The pictorial representation of the relation makes it easy to see that the height of a is greater than the height of p and r; and that the height of b is greater than the height of q.

An *n*-ary relation $R(A_1, A_2, \ldots A_n)$ is a subset of $(A_1 \times A_2 \times \ldots \times A_n)$. However, an *n*-ary relation may also be regarded as a binary relation $R(A, B)$ with $A = A_1 \times A_2 \times \ldots \times A_{n-1}$ and $B = A_n$.

2.3.1 Reflexive, Symmetric and Transitive Relations

There are various types of relations including reflexive, symmetric and transitive relations.

(i)   A relation on a set A is *reflexive* if $(a, a) \in R$ for all $a \in A$.
(ii) A relation R is *symmetric* if whenever $(a, b) \in R$ then $(b, a) \in R$.
(iii) A relation is *transitive* if whenever $(a, b) \in R$ and $(b, c) \in R$ then $(a, c) \in R$.

A relation that is reflexive, symmetric and transitive is termed an *equivalence relation*.

Example 2.5 (**Reflexive Relation**) A relation is reflexive if each element possesses an edge looping around on itself. The relation in Fig. 2.2 is reflexive.

Example 2.6 (**Symmetric Relation**) The graph of a symmetric relation will show for every arrow from a to b an opposite arrow from b to a. The relation in Fig. 2.3 is symmetric: i.e. whenever $(a, b) \in R$ then $(b, a) \in R$.

Example 2.7 (**Transitive relation**) The graph of a transitive relation will show that whenever there is an arrow from a to b and an arrow from b to c that there is an arrow from a to c. The relation in Fig. 2.4 is transitive: i.e. whenever $(a, b) \in R$ and $(b, c) \in R$ then $(a, c) \in R$.

Example 2.8 (**Equivalence relation**) The relation on the set of integers $\mathbb{Z}$ defined by $(a, b) \in R$ if $a - b = 2k$ for some $k \in \mathbb{Z}$ is an equivalence relation, and it partitions the set of integers into two equivalence classes: i.e. the even and odd integers.

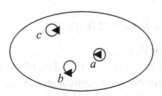

Fig. 2.2 Reflexive relation

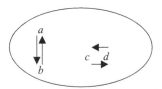

Fig. 2.3 Symmetric relation

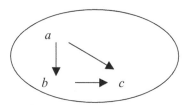

Fig. 2.4 Transitive relation

Domain and Range of Relation

The domain of a relation $R\ (A, B)$ is given by $\{a \in A \mid \exists b \in B \text{ and } (a, b) \in R\}$. It is denoted by **dom** R. The domain of the relation $R = \{(a, p), (a, r), (b, q)\}$ is $\{a, b\}$.

The range of a relation $R\ (A, B)$ is given by $\{b \in B \mid \exists a \in A \text{ and } (a, b) \in R\}$. It is denoted by **rng** R. The range of the relation $R = \{(a, p), (a, r), (b, q)\}$ is $\{p, q, r\}$.

Inverse of a Relation

Suppose $R \subseteq A \times B$ is a relation between A and B then the inverse relation $R^{-1} \subseteq B \times A$ is defined as the relation between B and A and is given by

$$b\ R^{-1}\ a \text{ if and only if } a\ R\ b$$

That is

$$R^{-1} = \{(b, a) \in B \times A : (a, b) \in R\}$$

Example 2.9 Let R be the relation between $\mathbb{Z}$ and $\mathbb{Z}^+$ defined by mRn if and only if $m^2 = n$. Then $R = \{(m, n) \in \mathbb{Z} \times \mathbb{Z}^+ : m^2 = n\}$ and $R^{-1} = \{(n, m) \in \mathbb{Z}^+ \times \mathbb{Z} : m^2 = n\}$.

For example, $-3\ R\ 9$, $-4\ R\ 16$, $0\ R\ 0$, $16\ R^{-1} - 4$, $9\ R^{-1} - 3$, etc.

Partitions and Equivalence Relations

An equivalence relation on A leads to a partition of A, and vice versa for every partition of A there is a corresponding equivalence relation.

Let A be a finite set and let $A_1, A_2, ..., A_n$ be subsets of A such $A_i \neq \emptyset$ for all i, $A_i \cap A_j = \emptyset$ if $i \neq j$ and $A = \cup_i^n A_i = A_1 \cup A_2 \cup ... \cup A_n$. The sets A_i partition the set A, and these sets are called the classes of the partition (Fig. 2.5).

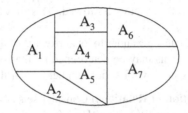

Fig. 2.5 Partitions of A

Theorem 2.2 (Equivalence Relation and Partitions) *An equivalence relation on A gives rise to a partition of A where the equivalence classes are given by Class (a) = {x | x ∈ A and (a, x) ∈ R}. Similarly, a partition gives rise to an equivalence relation R, where (a, b) ∈ R if and only if a and b are in the same partition.*

Proof Clearly, $a \in$ Class(a) since R is reflexive and clearly the union of the equivalence classes is A. Next, we show that two equivalence classes are either equal or disjoint.

Suppose Class(a) ∩ Class(b) ≠ Ø. Let $x \in$ Class(a) ∩ Class(b) and so (a, x) and (b, x) ∈ R. By the symmetric property (x, b) ∈ R and since R is transitive from (a, x) and (x, b) in R we deduce that (a, b) ∈ R. Therefore $b \in$ Class(a). Suppose y is an arbitrary member of Class (b) then (b, y) ∈ R therefore from (a, b) and (b, y) in R we deduce that (a, y) is in R. Therefore since y was an arbitrary member of Class(a) we deduce that Class(b) ⊆ Class(a). Similarly, Class(a) ⊆ Class(b) and so Class(a) = Class(b).

This proves the first part of the theorem and for the second part we define a relation R such that (a, b) ∈ R if a and b are in the same partition. It is clear that this is an equivalence relation.

2.3.2 Composition of Relations

The composition of two relations $R_1(A, B)$ and $R_2(B, C)$ is given by $R_2 \circ R_1$ where (a, c) ∈ $R_2 \circ R_1$ if and only there exists $b \in B$ such that (a, b) ∈ R_1 and (b, c) ∈ R_2. The composition of relations is associative: i.e.

$$(R_3 \circ R_2) \circ R_1 = R_3 \circ (R_2 \circ R_1)$$

Example 2.10 (**Composition of Relations**) Consider a library that maintains two files. The first file maintains the serial number s of each book as well as the details of the author a of the book. This may be represented by the relation $R_1 = sR_1a$. The second file maintains the library card number c of its borrowers and the serial number s of any books that they have borrowed. This may be represented by the relation $R_2 = c\ R_2s$.

The library wishes to issue a reminder to its borrowers of the authors of all books currently on loan to them. This may be determined by the composition of R_1 o R_2: i.e. $c\ R_1$ o $R_2\ a$ if there is book with serial number s such that $c\ R_2\ s$ and $s\ R_1\ a$.

Example 2.11 (**Composition of Relations**) Consider sets $A = \{a, b, c\}$, $B = \{d, e, f\}$, $C = \{g, h, i\}$ and relations $R(A, B) = \{(a, d), (a, f), (b, d), (c, e)\}$ and $S(B, C) = \{(d, h), (d, i), (e, g), (e, h)\}$. Then we graph these relations and show how to determine the composition pictorially.

S o R is determined by choosing $x \in A$ and $y \in C$ and checking if there is a route from x to y in the graph (Fig. 2.6). If so, we join x to y in S o R. For example, if we consider a and h we see that there is a path from a to d and from d to h and therefore (a, h) is in the composition of S and R.

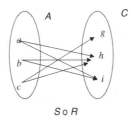

S o R

The union of two relations $R_1(A, B)$ and $R_2(A, B)$ is meaningful (as these are both subsets of $A \times B$). The union $R_1 \cup R_2$ is defined as $(a, b) \in R_1 \cup R_2$ if and only if $(a, b) \in R_1$ or $(a, b) \in R_2$.

Similarly, the intersection of R_1 and R_2 ($R_1 \cap R_2$) is meaningful and is defined as $(a, b) \in R_1 \cap R_2$ if and only if $(a, b) \in R_1$ and $(a, b) \in R_2$. The relation R_1 is a subset of R_2 ($R_1 \subseteq R_2$) if whenever $(a, b) \in R_1$ then $(a, b) \in R_2$.

The inverse of the relation R was discussed earlier and is given by the relation R^{-1} where $R^{-1} = \{(b, a) \mid (a, b) \in R\}$.

The composition of R and R^{-1} yields: R^{-1} o $R = \{(a, a) \mid a \in$ dom R$\} = i_A$ and R o $R^{-1} = \{(b, b) \mid b \in$ **dom** $R^{-1}\} = i_B$.

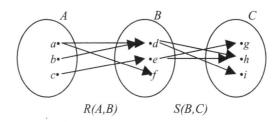

$R(A,B)$ $S(B,C)$

Fig. 2.6 Composition of relations

2.3.3 Binary Relations

A binary relation R on A is a relation between A and A, and a binary relation can always be composed with itself. Its inverse is a binary relation on the same set. The following are all relations on A:

$$R^2 = R \circ R$$
$$R^3 = (R \circ R) \circ R$$
$$R^0 = i_A (\text{identity relation})$$
$$R^{-2} = R^{-1} \circ R^{-1}$$

Example 2.12 Let R be the binary relation on the set of all people P such that $(a, b) \in R$ if a is a parent of b. Then the relation R^n is interpreted as

R is the parent relationship
R^2 is the grandparent relationship
R^3 is the great grandparent relationship.
R^{-1} is the child relationship.
R^{-2} is the grandchild relationship.
R^{-3} is the great grandchild relationship

This can be generalized to a relation R^n on A where $R^n = R \circ R \circ \ldots \circ R$ (n-times). The transitive closure of the relation R on A is given by

$$R^* = \cup_{i=0}^{\infty} R^i = R^0 \cup R^1 \cup R^2 \cup \ldots R^n \cup \ldots$$

where R^0 is the reflexive relation containing only each element in the domain of R: i.e. $R^0 = i_A = \{(a, a) \mid a \in \textbf{dom } R\}$.

The positive transitive closure is similar to the transitive closure except that it does not contain R^0. It is given by

$$R^+ = \cup_{i=1}^{\infty} R^i = R^1 \cup R^2 \cup \ldots \cup R^n \cup \ldots$$

$a R^+ b$ if and only if $a R^n b$ for some $n > 0$: i.e. there exists $c_1, c_2 \ldots c_n \in A$ such that

$$a R c_1, c_1 R c_2, \ldots, c_n R b$$

Parnas[7] introduced the concept of the limited domain relation (LD-relation), and a LD relation L consists of an ordered pair (R_L, C_L) where R_L is a relation and C_L is a subset of Dom R_L. The relation R_L is on a set U and C_L is termed the competence

---

[7]Parnas made important contributions to software engineering in the 1970s. He invented information hiding which is used in object-oriented design.

set of the LD relation *L*. A description of LD relations and a discussion of their properties are in Chap. 2 of [1].

The importance of LD relations is that they may be used to describe program execution. The relation component of the LD relation *L* describes a set of states such that if execution starts in state *x* it may terminate in state *y*. The set U is the set of states. The competence set of *L* is such that if execution starts in a state that is in the competence set then it is guaranteed to terminate.

2.3.4 Applications of Relations

A relational database management system (RDBMS) is a system that manages data using the relational model, and examples of such systems include RDMS developed at MIT in the 1970s; Ingres developed at the University of California, Berkeley in the mid-1970s; Oracle developed in the late 1970s; DB2; Informix; and Microsoft SQL Server.

A relation is defined as a set of tuples and is usually represented by a table. A table is data organized in rows and columns, with the data in each column of the table of the same data type. Constraints may be employed to provide restrictions on the kinds of data that may be stored in the relations. Constraints are Boolean expressions which indicate whether the constraint holds or not, and are a way of implementing business rules in the database.

Relations have one or more keys associated with them, and the *key uniquely identifies the row of the table*. An index is a way of providing fast access to the data in a relational database, as it allows the tuple in a relation to be looked up directly (using the index) rather than checking all of the tuples in the relation.

The Structured Query Language (SQL) is a computer language that tells the relational database what to retrieve and how to display it. A stored procedure is executable code that is associated with the database, and it is used to perform common operations on the database.

The concept of a relational database was first described in a paper '*A Relational Model of Data for Large Shared Data Banks*' by Codd [2]. A relational database is a database that conforms to the relational model, and it may be defined as a set of relations (or tables).

Codd (Fig. 2.7) developed the *relational data base model* in the late 1960s, and today, this is the standard way that information is organized and retrieved from computers. Relational databases are at the heart of systems from hospitals' patient records to airline flight and schedule information.

A binary relation R(A, B) where A and B are sets is a subset of the Cartesian product (A × B) of A and B. The domain of the relation is A, and the codomain of the relation is B. The notation aRb signifies that there is a relation between a and b and that $(a, b) \in$ R. An *n*-ary relation R $(A_1, A_2, \ldots A_n)$ is a subset of the Cartesian product of the *n* sets: i.e. a subset of $(A_1 \times A_2 \times \ldots \times A_n)$. However, an

Fig. 2.7 Edgar Codd

n-ary relation may also be regarded as a binary relation R(A, B) with A = A$_1$ × A$_2$ × ... × A$_{n-1}$ and B = A$_n$.

The data in the relational model are represented as a mathematical *n*-ary relation. In other words, a relation is defined as a set of *n*-tuples, and is usually represented by a table. A table is a visual representation of the relation, and the data are organized in rows and columns. The data stored in each column of the table are of the same data type.

The basic relational building block is the domain or data type (often called just type). Each row of the table represents one *n*-tuple (one tuple) of the relation, and the number of tuples in the relation is the cardinality of the relation. Consider the PART relation taken from [3], where this relation consists of a heading and the body. There are five data types representing part numbers, part names, part colours, part weights, and locations in which the parts are stored. The body consists of a set of *n*-tuples, and the PART relation given in Fig. 2.8 is of cardinality six.

For more information on the relational model and databases see [4]

2.4 Functions

A function *f*: *A* → *B* is a special relation such that for each element *a* ∈ A there is exactly (or at most)[8] one element *b* ∈ B. This is written as *f*(*a*) = *b*.

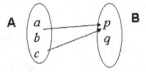

---

[8]We distinguish between total and partial functions. A total function is defined for all elements in the domain whereas a partial function may be undefined for one or more elements in the domain.

P#	PName	Colour	Weight	City
P1	Nut	Red	12	London
P2	Bolt	Green	17	Paris
P3	Screw	Blue	17	Rome
P4	Screw	Red	14	London
P5	Cam	Blue	12	Paris
P6	Cog	Red	19	London

Fig. 2.8 PART relation

A function is a relation but not every relation is a function. For example, the relation in the diagram below is not a function since there are two arrows from the element $a \in A$.

The domain of the function (denoted by **dom** f) is the set of values in A for which the function is defined. The domain of the function is A provided that f is a total function. The codomain of the function is B. The range of the function (denoted **rng** f) is a subset of the codomain and consists of

$$\mathbf{rng}\,f = \{r \,|\, r \in B \text{ such that } f(a) = r \text{ for some } a \in A\}.$$

Functions may be partial or total. A *partial function* (or partial mapping) may be undefined for some values of A, and partial functions arise regularly in the computing field (Fig. 2.9). *Total functions* are defined for every value in A and many functions encountered in mathematics are total.

Example 2.13 (**Functions**) Functions are an essential part of mathematics and computer science, and there are many well-known functions such as the trigonometric functions $\sin(x)$, $\cos(x)$, and $\tan(x)$; the logarithmic function $\ln(x)$; the exponential functions e^x; and polynomial functions.

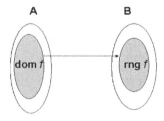

Fig. 2.9 Domain and range of a partial function

(i) Consider the partial function $f: \mathbb{R} \to \mathbb{R}$ where

$$f(x) = 1/x \quad (\text{where } x \neq 0).$$

This partial function is defined everywhere except for $x = 0$

(ii) Consider the function $f: \mathbb{R} \to \mathbb{R}$ where

$$f(x) = x^2$$

Then this function is defined for all $x \in \mathbb{R}$

Partial functions often arise in computing as a program may be undefined or fail to terminate for several values of its arguments (e.g. infinite loops). Care is required to ensure that the partial function is defined for the argument to which it is to be applied.

Consider a program P that has one natural number as its input and which for some input values will never terminate. Suppose that if it terminates it prints a single real result and halts. Then P can be regarded as a partial mapping from $\mathbb{N}$ to $\mathbb{R}$.

$$P : \mathbb{N} \to \mathbb{R}$$

Example 2.14 How many total functions $f: A \to B$ are there from A to B (where A and B are finite sets)?

Each element of A maps to any element of B, i.e. there are $|B|$ choices for each element $a \in A$. Since there are $|A|$ elements in A the number of total functions is given by

$$|B| \, |B| \ldots |B| \quad (|A| \text{ times})$$
$$= |B|^{|A|} \qquad \text{total functions between A and B.}$$

Example 2.15 How many partial functions $f: A \to B$ are there from A to B (where A and B are finite sets) ?

Each element of A may map to any element of B or to no element of B (as it may be undefined for that element of A). In other words, there are $|B| + 1$ choices for each element of A. As there are $|A|$ elements in A, the number of distinct partial functions between A and B is given by

$$(|B|+1)(|B|+1)\ldots(|B|+1) \quad (|A| \text{ times})$$
$$= (|B|+1)^{|A|}$$

Two partial functions f and g are equal if

1. dom f = dom g
2. $f(a) = g(a)$ for all $a \in$ dom f.

A function f is less defined than a function g ($f \subseteq g$) if the domain of f is a subset of the domain of g, and the functions agree for every value on the domain of f

1. dom $f \subseteq$ dom g
2. $f(a) = g(a)$ for all $a \in$ dom f.

The composition of functions is similar to the composition of relations. Suppose $f: A \rightarrow B$ and $g: B \rightarrow C$ then $g \circ f: A \rightarrow C$ is a function, and this is written as $g \circ f$ (x) or $g(f(x))$ for $x \in A$.

The composition of functions is not commutative and this can be seen by an example. Consider the function $f: \mathbb{R} \rightarrow \mathbb{R}$ such that $f(x) = x^2$ and the function g: $\mathbb{R} \rightarrow \mathbb{R}$ such that $g(x) = x + 2$. Then

$$g \circ f(x) = g(x^2) = x^2 + 2.$$
$$f \circ g(x) = f(x+2) = (x+2)^2 = x^2 + 4x + 4.$$

Clearly, $g \circ f(x) \neq f \circ g(x)$ and so composition of functions is not commutative. The composition of functions is associative, as the composition of relations is associative and every function is a relation. For $f: A \rightarrow B$, $g: B \rightarrow C$, and $h: C \rightarrow D$ we have

$$h \circ (g \circ f) = (h \circ g) \circ f$$

A function $f: A \rightarrow B$ is *injective* (*one to one*) if

$$f(a_1) = f(a_2) \Rightarrow a_1 = a_2.$$

For example, consider the function $f: \mathbb{R} \rightarrow \mathbb{R}$ with $f(x) = x^2$. Then $f(3) = f(-3) = 9$ and so this function is not one to one.

A function $f: A \rightarrow CB$ is *surjective* (*onto*) if given any $b \in B$ there exists an $a \in A$ such that $f(a) = b$ (Fig. 2.10). Consider the function $f: \mathbb{R} \rightarrow \mathbb{R}$ with $f(x) = x + 1$. Clearly, given any $r \in \mathbb{R}$ then $f(r - 1) = r$ and so f is onto.

A function is *bijective* if it is one to one and onto (Fig. 2.11). That is, there is a one-to-one correspondence between the elements in A and B for each $b \in B$ there is a unique $a \in A$ such that $f(a) = b$.

The inverse of a relation was discussed earlier and the relational inverse of a function $f: A \rightarrow B$ clearly exists. The relational inverse of the function may or may not be a function.

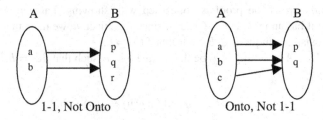

Fig. 2.10 Injective and surjective functions

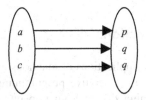

Fig. 2.11 Bijective function (One to one and Onto)

However, if the relational inverse is a function it is denoted by f^{-1}: B → A. A total function has an inverse if and only if it is bijective whereas a partial function has an inverse if and only if it is injective.

The identity function 1_A: A → A is a function such that $1_A(a) = a$ for all $a \in A$. Clearly, when the inverse of the function exists then we have that $f^{-1} \circ f = 1_A$ and $f \circ f^{-1} = 1_B$.

Theorem 2.3 (Inverse of Function) *A total function has an inverse if and only if it is bijective.*

Proof Suppose f: A → B has an inverse f^{-1}. Then we show that f is bijective.

We first show that f is one to one.
Suppose $f(x_1) = f(x_2)$ then

$$f^{-1}(f(x_1)) = f^{-1}(f(x_2))$$
$$\Rightarrow f^{-1} \circ f(x_1) = f^{-1} \circ f(x_2)$$
$$\Rightarrow 1_A(x_1) = 1_A(x_2)$$
$$\Rightarrow x_1 = x_2$$

Next we first show that f is onto. Let $b \in B$ and let $a = f^{-1}(b)$ then

$$f(a) = f(f^{-1}(b)) = b \text{ and so } f \text{ is surjective}$$

The second part of the proof is concerned with showing that if $f: A \rightarrow B$ is bijective then it has an inverse f^{-1}. Clearly, since f is bijective we have that for each $a \in A$ there exists a unique $b \in B$ such that $f(a) = b$.

Define $g: B \rightarrow A$ by letting $g(b)$ be the unique a in A such that $f(a) = b$. Then we have

$$g \, of(a) = g(b) = a \text{ and } f \, o \, g(b) = f(a) = b.$$

Therefore, g is the inverse of f.

2.5 Application of Functions

In this section, we discuss the applications of functions to functional programming, which is quite distinct from the imperative programming languages used in computing. Functional programming differs from imperative programming in that it involves the evaluation of mathematical functions, whereas imperative programming involves the execution of sequential (or iterative) commands that change the state. For example, the assignment statement alters the value of a variable, and the value of a given variable x may change during program execution.

There are no changes of state for functional programs, and the fact that the value of x will always be the same makes it easier to reason about functional programs than imperative programs. Functional programming languages provide *referential transparency*: i.e. equals may be substituted for equals, and if two expressions have equal values, then one can be substituted for the other in any larger expression without affecting the result of the computation.

Functional programming languages use higher order functions,[9] recursion, lazy and eager evaluation, monads,[10] and Hindley–Milner type inference systems.[11] These languages are mainly been used in academia, but there has been some industrial use, including the use of Erlang for concurrent applications in industry. Alonzo Church developed Lambda calculus in the 1930s, and it provides an abstract framework for describing mathematical functions and their evaluation. It provides the foundation for functional programming languages. Church employed lambda calculus to prove that there is no solution to the decision problem for first-order arithmetic in 1936 (discussed in Chap. 13).

---

[9]Higher order functions are functions take functions as arguments or return a function as a result. They are known as operators (or functionals) in mathematics, and one example is the derivative function dy/dx that takes a function as an argument and returns a function as a result.

[10]Monads are used in functional programming to express input and output operations without introducing side effects. The Haskell functional programming language makes use of uses this feature.

[11]This is the most common algorithm used to perform type inference. Type inference is concerned with determining the type of the value derived from the eventual evaluation of an expression.

Lambda calculus uses transformation rules, and one of these rules is variable substitution. The original calculus developed by Church was untyped, but typed lambda calculi have since been developed. Any computable function can be expressed and evaluated using lambda calculus, but there is no general algorithm to determine whether two arbitrary lambda calculus expressions are equivalent. Lambda calculus influenced functional programming languages such as Lisp, ML and Haskell.

Functional programming uses the notion of higher order functions. Higher order takes other functions as arguments, and may return functions as results. The derivative function $d/dx\, f(x) = f'(x)$ is a higher order function. It takes a function as an argument and returns a function as a result. For example, the derivative of the function $Sin(x)$ is given by $Cos(x)$. Higher order functions allow currying which is a technique developed by Schönfinkel. It allows a function with several arguments to be applied to each of its arguments one at a time, with each application returning a new (higher order) function that accepts the next argument. This allows a function of n-arguments to be treated as n applications of a function with 1-argument.

John McCarthy developed LISP at MIT in the late 1950s, and this language includes many of the features found in modern functional programming languages.[12] Scheme built upon the ideas in LISP. Kenneth Iverson developed APL[13] in the early 1960s, and this language influenced Backus's FP programming language. Robin Milner designed the ML programming language in the early 1970s. David Turner developed Miranda in the mid-1980s. The Haskell programming language was released in the late 1980s.

Miranda Functional Programming Language
Miranda was developed by David Turner at the University of Kent in the mid-1980s [5]. It is a non-strict functional programming language: i.e. the arguments to a function are not evaluated until they are actually required within the function being called. This is also known as lazy evaluation, and one of its main advantages is that it allows an infinite data structures to be passed as an argument to a function. Miranda is a pure functional language in that there are no side effect features in the language. The language has been used for

- Rapid prototyping
- Specification language
- Teaching Language

A Miranda program is a collection of equations that define various functions and data structures. It is a strongly typed language with a terse notation.

---

[12]Lisp is a multi-paradigm language rather than a functional programming language.

[13]Iverson received the Turing Award in 1979 for his contributions to programming language and mathematical notation. The title of his Turing award paper was 'Notation as a tool of thought'.

$$z = \mathrm{sqr}\, p / \mathrm{sqr}\, q$$
$$\mathrm{sqr}\, k = k * k$$
$$p = a + b$$
$$q = a - b$$
$$a = 10$$
$$b = 5$$

The scope of a formal parameter (e.g. the parameter k above in the function sqr) is limited to the definition of the function in which it occurs.

One of the most common data structures used in Miranda is the list. The empty list is denoted by [ ], and an example of a list of integers is given by [1, 3, 4, 8]. Lists may be appended to by using the '++' operator. For example

$$[1, 3, 5] ++ [2, 4] = [1, 3, 5, 2, 4].$$

The length of a list is given by the '#' operator

$$\#[1, 3] = 2$$

The infix operator ':' is employed to prefix an element to the front of a list. For example

$$5 : [2, 4, 6]\ \text{is equal to}\ [5, 2, 4, 6]$$

The subscript operator '!' is employed for subscripting: For example

$$\mathrm{Nums} = [5, 2, 4, 6] \qquad \text{then} \quad \mathrm{Nums}!0\ \text{is}\ 5.$$

The elements of a list are required to be of the same type. A sequence of elements that contains mixed types is called a tuple. A tuple is written as follows:

$$\mathrm{Employee} = (\text{``Holmes''}, \text{``222 Baker St. London''}, 50, \text{``Detective''})$$

A tuple is similar to a record in Pascal whereas lists are similar to arrays. Tuples cannot be subscripted but their elements may be extracted by pattern matching. Pattern matching is illustrated by the well-known example of the factorial function

$$\mathrm{fac}\, 0 = 1$$
$$\mathrm{fac}(n + 1) = (n + 1) * \mathrm{fac}\, n$$

The definition of the factorial function uses two equations, distinguished by the use of different patterns in the formal parameters. Another example of pattern matching is the reverse function on lists

$$\text{reverse} \; [] = []$$
$$\text{reverse} \; (a : x) = \text{reverse} \; x ++ [a]$$

Miranda is a higher order language, and it allows functions to be passed as parameters and returned as results. Currying is allowed and this allows a function of n-arguments to be treated as n applications of a function with 1-argument. Function application is left associative: i.e. f x y means (f x) y. That is, the result of applying the function f to x is a function, and this function is then applied to y. Every function with two or more arguments in Miranda is a higher order function.

2.6 Review Questions

1. What is a set? A relation? A function?
2. Explain the difference between a partial and a total function.
3. Explain the difference between a relation and a function.
4. Determine $A \times B$ where $A = \{a, b, c, d\}$ and $B = \{1, 2, 3\}$
5. Determine the symmetric difference $A \triangle B$ where $A = \{a, b, c, d\}$ and $B = \{c, d, e\}$
6. What is the graph of the relation $\leq$ on the set $A = \{2, 3, 4\}$.
7. What is the composition of S and R (i.e. S o R), where R is a relation between A and B, and S is a relation between B and C. The sets A, B, C are defined as $A = \{a, b, c, d\}$, $B = \{e, f, g\}$, $C = \{h, i, j, k\}$ and $R = \{(a, e), (b, e), (b, g), (c, e), (d, f)\}$ with $S = \{(e, h), (e, k), (f, j), (f, k), (g, h)\}$
8. What is the domain and range of the relation R where $R = \{(a, p), (a, r), (b, q)\}$.
9. Determine the inverse relation R^{-1} where $R = \{(a, 2), (a, 5), (b, 3), (b, 4), (c, 1)\}$.
10. Determine the inverse of the function $f: \mathbb{R} \times \mathbb{R} \to \mathbb{R}$ defined by

$$f(x) = \frac{x - 2}{x - 3} (x \neq 3) \quad \text{and} f(3) = 1$$

11. Give examples of injective, surjective and bijective functions.

12. Let $n \geq 2$ be a fixed integer. Consider the relation $\equiv$ defined by $\{(p, q): p, q \in \mathbb{Z}, n \mid (q - p)\}$

    a. Show $\equiv$ is an equivalence relation.
    b. What are the equivalence classes of this relation?

13. Describe the differences between imperative programming languages and functional programming languages.

2.7 Summary

This chapter provided an introduction to set theory, relations and functions. Sets are collections of well-defined objects; a relation between A and B indicates relationships between members of the sets A and B; and functions are a special type of relation where there is at most one relationship for each element $a \in A$ with an element in B.

A set is a collection of well-defined objects that contain no duplicates. There are many examples of sets such as the set of natural numbers $\mathbb{N}$, the integer numbers $\mathbb{Z}$ and so on.

The Cartesian product allows a new set to be created from existing sets. The Cartesian product of two sets S and T (denoted $S \times T$) is the set of ordered pairs $\{(s, t) \mid s \in S, t \in T\}$.

A binary relation $R (A, B)$ is a subset of the Cartesian product $(A \times B)$ of A and B where A and B are sets. The domain of the relation is A and the codomain of the relation is B. The notation aRb signifies that there is a relation between a and b and that $(a, b) \in R$. An n-ary relation $R (A_1, A_2, \ldots A_n)$ is a subset of $(A_1 \times A_2 \times \ldots \times A_n)$.

A total function $f: A \rightarrow B$ is a special relation such that for each element $a \in A$ there is exactly one element $b \in B$. This is written as $f(a) = b$. A function is a relation but not every relation is a function.

The domain of the function (denoted by **dom** f) is the set of values in A for which the function is defined. The domain of the function is A provided that f is a total function. The codomain of the function is B.

Functional programming is quite distinct from imperative programming in that there is no change of state, and the value of the variable x remains the same during program execution. This makes functional programs easier to reason about than imperative programs.

References

1. Software Fundamentals. Collected Papers by David *L.* Parnas. Edited by Daniel Hoffman and David Weiss. Addison Wesley. 2001.
2. A Relational Model of Data for Large Shared Data Banks. E.F. Codd. Communications of the ACM 13 (6): 377–387. 1970.
3. An Introduction to Database Systems. 3rd Edition. C.J. Date. The Systems Programming Series. 1981.
4. Introduction to the History of Computing. Gerard O'Regan. Springer Verlag. 2016.
5. Miranda. David Turner. Proceedings IFIP Conference, Nancy France, Springer LNCS (201). September 1985.

Number Theory

3

Key Topics

Square, rectangular and triangular Numbers
Prime numbers
Pythagorean triples
Mersenne primes
Division algorithm
Perfect and amicable numbers
Greatest common divisor
Least common multiples
Euclid's algorithm
Modular arithmetic
Binary numbers
Computer representation of numbers

3.1 Introduction

Number theory is the branch of mathematics that is concerned with the mathematical properties of the natural numbers and integers. These include properties such as the parity of a number; divisibility; additive, and multiplicative properties; whether a number is prime or composite; the prime factors of a number; the greatest common divisor and least common multiple of two numbers; and so on.

© Springer International Publishing Switzerland 2016
G. O'Regan, *Guide to Discrete Mathematics*, Texts in Computer Science,
DOI 10.1007/978-3-319-44561-8_3

Fig. 3.1 Pierre de Fermat

Number theory has many applications in computing including cryptography and coding theory. For example, the RSA public key cryptographic system relies on its security due to the infeasibility of the integer factorization problem for large numbers.

There are several unsolved problems in number theory and especially in prime number theory. For example, Goldbach's[1] Conjecture states that every even integer greater than two is the sum of two primes, and this result has not been proved to date. Fermat's[2] last Theorem (Fig. 3.1) states that there is no integer solution to $x^n + y^n = z^n$ for $n > 2$, and this result remained unproved for over three hundred years until Andrew Wiles finally proved it in the mid-1990s.

The natural numbers $\mathbb{N}$ consist of the numbers $\{1, 2, 3, \ldots\}$. The integer numbers $\mathbb{Z}$ consist of $\{\ldots -2, -1, 0, 1, 2, \ldots\}$. The rational numbers $\mathbb{Q}$ consist of all numbers of the form $\{p/q$ where p and q are integers and $q \neq 0\}$. The real numbers $\mathbb{R}$ is defined to be the set of converging sequences of rational numbers and they are a superset of the rational numbers. They contain the rational and irrational numbers. The complex numbers $\mathbb{C}$ consist of all numbers of the form $\{a + bi$ where $a, b \in \mathbb{R}$ and $i = \sqrt{-1}\}$.

Pythagorean triples (Fig. 3.2) are combinations of three whole numbers that satisfy Pythagoras's equation $x^2 + y^2 = z^2$. There are an infinite number of such triples, and an example of such a triple is 3, 4, 5 since $3^2 + 4^2 = 5^2$.

---

[1]Goldbach was an eighteenth century German mathematician and Goldbach's conjecture has been verified to be true for all integers $n < 12 \times 10^{17}$.

[2]Pierre de Fermat was a 17th French civil servant and amateur mathematician. He occasionally wrote to contemporary mathematicians announcing his latest theorem without providing the accompanying proof and inviting them to find the proof. The fact that he never revealed his proofs caused a lot of frustration among his contemporaries, and in his announcement of his famous last theorem he stated that he had a wonderful proof that was too large to include in the margin. He corresponded with Pascal and they did some early work on the mathematical rules of games of chance and early probability theory. He also did some early work on the Calculus.

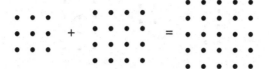

Fig. 3.2 Pythagorean triples

The Pythagoreans discovered the mathematical relationship between the harmony of music and numbers, and their philosophy was that numbers are hidden in everything from music to science and nature. This led to their philosophy that 'everything is number'.

3.2 Elementary Number Theory

A square number (Fig. 3.3) is an integer that is the square of another integer. For example, the number 4 is a square number since $4 = 2^2$. Similarly, the number 9 and the number 16 are square numbers. A number n is a square number if and only if one can arrange the n points in a square. For example, the square numbers 4, 9, 16 are represented in squares as follows:

The square of an odd number is odd, whereas the square of an even number is even. This is clear since an even number is of the form $n = 2k$ for some k, and so $n^2 = 4k^2$ which is even. Similarly, an odd number is of the form $n = 2k + 1$ and so $n^2 = 4k^2 + 4k + 1$ which is odd.

A rectangular number (Fig. 3.4) n may be represented by a vertical and horizontal rectangle of n points. For example, the number 6 may be represented by a rectangle with length 3 and breadth 2, or a rectangle with length 2 and breadth 3. Similarly, the number 12 can be represented by a 4×3 or a 3×4 rectangle.

A triangular number (Fig. 3.5) n may be represented by an equilateral triangle of n points. It is the sum of k natural numbers from 1 to k. = That is,

$$n = 1 + 2 + \cdots + k$$

Fig. 3.3 Square numbers

Fig. 3.4 Rectangular numbers

Fig. 3.5 Triangular numbers

Parity of Integers

The parity of an integer refers to whether the integer is odd or even. An integer n is odd if there is a remainder of one when it is divided by two, and it is of the form $n = 2k + 1$. Otherwise, the number is even and of the form $n = 2k$.

The sum of two numbers is even if both are even or both are odd. The product of two numbers is even if at least one of the numbers is even. These properties are expressed as

$$\text{even} \pm \text{even} = \text{even}$$
$$\text{even} \pm \text{odd} = \text{odd}$$
$$\text{odd} \pm \text{odd} = \text{even}$$
$$\text{even} \times \text{even} = \text{even}$$
$$\text{even} \times \text{odd} = \text{even}$$
$$\text{odd} \times \text{odd} = \text{odd}$$

Divisors

Let a and b be integers with $a \neq 0$ then a is said to be a divisor of b (denoted by $a|b$) if there exists an integer k such that $b = ka$.

A divisor of n is called a *trivial divisor* if it is either 1 or n itself; otherwise it is called a *nontrivial divisor*. A *proper divisor* of n is a divisor of n other than n itself.

Definition (Prime Number)

A *prime number* is a number whose only divisors are trivial. There are an infinite number of prime numbers.

The *fundamental theorem of arithmetic* states that every integer number can be factored as the product of prime numbers.

Mesenne Primes

Mersenne primes are prime numbers of the form $2^p - 1$, where p is a prime. They are named after Marin Mersenne (Fig. 3.6) who was a 17th French monk, philosopher and mathematician. Mersenne did some early work in identifying primes of this format, and there are 47 known Mersenne primes. It remains an open question as to whether there are an infinite number of Mersenne primes.

Properties of Divisors

(i) $a|b$ and $a|c$ then $a|b + c$
(ii) $a|b$ then $a|bc$
(iii) $a|b$ and $b|c$ then $a|c$

Proof (of i) Suppose $a|b$ and $a|c$ then $b = k_1a$ and $c = k_2a$.
  Then, $b + c = k_1a + k_2a = (k_1 + k_2)a$ and so $a|b + c$.

Proof (of iii) Suppose $a|b$ and $b|c$ then $b = k_1a$ and $c = k_2b$.
  Then, $c = k_2b = (k_2\ k_1)\ a$ and thus $a|c$.

Perfect and Amicable Numbers

Perfect and amicable numbers have been studied for millennia. A positive integer m is said to be *perfect* if it is the sum of its proper divisors. Two positive integers m and n are said to be an *amicable pair* if m is equal to the sum of the proper divisors of n and vice versa.

A *perfect number* is a number whose divisors add up to the number itself. For example, the number 6 is perfect since it has divisors 1, 2, 3 and $1 + 2 + 3 = 6$.

Perfect numbers are quite rare and Euclid showed that $2^{p-1} (2^p - 1)$ is an even perfect number whenever $(2^p - 1)$ is prime. Euler later showed that all even perfect numbers are of this form. It is an open question as to whether there are any odd perfect numbers, and if such an odd perfect number N was to exist then $N > 10^{1500}$.

A prime number of the form $(2^p - 1)$, where p is prime called as *Mersenne prime*. Mersenne primes are quite rare and each Mersenne prime generates an even perfect number and vice versa. That is, there is a one to one correspondence between the number of Mersenne primes and the number of even perfect numbers.

Fig. 3.6 Marin Mersenne

It remains an open question as to whether there are an infinite number of Mersenne primes and perfect numbers.

An *amicable pair* of numbers is a pair of numbers such that each number is the sum of divisors of the other number. For example, the numbers 220 and 284 are an amicable pair since the divisors of 220 are 1, 2, 4, 5, 10, 11, 20, 22, 44, 55, 110, which have sum 284, and the divisors of 284 are 1, 2, 4, 71, 142, which have sum 220.

Theorem 3.1 (Division Algorithm) *For any integer a and any positive integer b there exist unique integers q and r such that*

$$a = bq + r \quad 0 \leq r < b$$

Proof The first part of the proof is to show the existence of integers q and r such that the equality holds, and the second part of the proof is to prove uniqueness of q and r.

Consider ... $-3b, -2b, -b, 0, b, 2b, 3b, \ldots$ then there must be an integer q such that

$$qb \leq a < (q+1)b$$

Then $a - qb = r$ with $0 \leq r < b$ and so $a = bq + r$ and the existence of q and r is proved.

The second part of the proof is to show the uniqueness of q and r. Suppose q_1 and r_1 also satisfy $a = bq_1 + r_1$ with $0 \leq r_1 < b$ and suppose $r < r_1$. Then $bq + r = bq_1 + r_1$ and so $b(q - q_1) = r_1 - r$ and clearly $0 < (r_1 - r) < b$. Therefore, $b|(r_1 - r)$ which is impossible unless $r_1 - r = 0$. Hence, $r = r_1$ and $q = q_1$.

Theorem 3.2 (Irrationality of Square Root of Two) *The square root of two is an irrational number (i.e., it cannot be expressed as the quotient of two integer numbers).*

Proof The Pythagoreans[3] discovered this result and it led to a crisis in their community as number was considered to be the essence of everything in their

---

[3]Pythagoras of Samos (a Greek island in the Aegean sea) was an influential ancient mathematician and philosopher of the sixth century B.C. He gained his mathematical knowledge from his travels throughout the ancient world (especially in Egypt and Babylon). He became convinced that everything is number and he and his followers discovered the relationship between mathematics and the physical world as well as relationships between numbers and music. On his return to Samos he founded a school and he later moved to Croton in southern Italy to set up a school. This school and the Pythagorean brotherhood became a secret society with religious beliefs such as reincarnation and they were focused on the study of mathematics. They maintained secrecy of the mathematical results that they discovered. Pythagoras is remembered today for Pythagoras's Theorem, which states that for a right-angled triangle that the square of the hypotenuse is equal to the sum of the square of the other two sides. The Pythagorean's discovered the irrationality of the square root of two and as this result conflicted in a fundamental way with their philosophy that number is everything, and they suppressed the truth of this mathematical result.

world. The proof is indirect: i.e., the opposite of the desired result is assumed to be correct and it is showed that this assumption leads to a contradiction. Therefore, the assumption must be incorrect and so the result is proved.

Suppose $\sqrt{2}$ is rational then it can be put in the form p/q, where p and q are integers and $q \neq 0$. Therefore, we can choose p, q to be co-prime (i.e., without any common factors) and so

$$(p/q)^2 = 2$$
$$\Rightarrow p^2/q^2 = 2$$
$$\Rightarrow p^2 = 2q^2$$
$$\Rightarrow 2|p^2$$
$$\Rightarrow 2|p$$
$$\Rightarrow p = 2k$$
$$\Rightarrow p^2 = 4k^2$$
$$\Rightarrow 4k^2 = 2q^2$$
$$\Rightarrow 2k^2 = q^2$$
$$\Rightarrow 2|q^2$$
$$\Rightarrow 2|q$$

This is a contradiction as we have chosen p and q to be co-prime, and our assumption that there is a rational number that is the square root of two results in a contradiction. Therefore, this assumption must be false and we conclude that there is no rational number whose square is two.

3.3 Prime Number Theory

A positive integer $n > 1$ is called prime if its only divisors are n and 1. A number that is not a prime is called composite.

Properties of Prime Numbers

(i)   There are an infinite number if primes.
(ii) There is a prime number p between n and $n! + 1$ such that $n < p \leq n! + 1$
(iii) If n is composite then n has a prime divisor p such that $p \leq \sqrt{n}$
(iv) There are arbitrary large gaps in the series of primes (given any $k > 0$ there exist k consecutive composite integers).

Proof (i) Suppose there are a finite number of primes and they are listed as $p_1, p_2, p_3, \ldots, p_k$. Then consider the number N obtained by multiplying all known primes and adding one. That is,

$$N = p_1 p_2 p_3 \ldots p_k + 1.$$

Clearly, N is not divisible by any of p_1, p_2, p_3, ..., p_k since they all leave a remainder of 1. Therefore, N is either a new prime or divisible by a prime q (that is not in the list of p_1, p_2, p_3, ..., p_k.).

This is a contradiction since this was the list of all the prime numbers, and so the assumption that there are a finite number of primes is false, and we deduce that there are an infinite number of primes.

Proof (ii) Consider the integer $N = n! + 1$. If N is prime then we take $p = N$. Otherwise, N is composite and has a prime factor p. We will show that $p > n$.

Suppose, $p \leq n$ then $p|n!$ and since $p|N$ we have $p|n! + 1$ and therefore $p|1$, which is impossible. Therefore, $p > n$ and the result is proved.

Proof (iii) Let p be the smallest prime divisor of n. Since n is composite $n = uv$ and clearly $p \leq u$ and $p \leq v$. Then $p^2 \leq uv = n$ and so $p \leq \sqrt{n}$.

Proof (iv) Consider the k consecutive integers $(k + 1)! + 2$, $(k + 1)! + 3$, ..., $(k + 1)! + k$, $(k + 1)! + k + 1$. Then each of these is composite since $j|(k + 1)! + j$ where $2 \leq j \leq k + 1$.

Algorithm for Determining Primes

The *Sieve of Eratosthenes algorithm* (Fig. 3.7) is a famous algorithm for determining the prime numbers up to a given number n. It was developed by the Hellenistic mathematician, Eratosthenes.

The algorithm involves first listing all of the numbers from 2 to n. The first step is to remove all multiples of two up to n; the second step is to remove all multiples of three up to n; and so on.

The kth step involves removing multiples of the kth prime p_k up to n and the steps in the algorithm continue while $p \leq \sqrt{n}$. The numbers remaining in the list are the prime numbers from 2 to n.

1. List the integers from 2 to n.
2. For each prime p_k up to $\sqrt{n}$ remove all multiples of p_k.
3. The numbers remaining are the prime numbers between 2 and n.

	2	3	4	5	6	7	8	9	10
11	12	13	14	15	16	17	18	19	20
21	22	23	24	25	26	27	28	29	30
31	32	33	34	35	36	37	38	39	40
41	42	43	44	45	46	47	48	49	50

Fig. 3.7 Primes between 1 and 50

The list of primes between 1 and 50 are given in Fig. 3.7. They are 2, 3, 5, 7, 11, 13, 17, 19, 23, 29, 31, 37, 41, 43, and 47.

Theorem 3.3 (Fundamental Theorem of Arithmetic) *Every natural number* n > 1 *may be written uniquely as the product of primes*

$$n = p_1^{\alpha_1} p_2^{\alpha_2} p_3^{\alpha_3} \cdots p_k^{\alpha_k}$$

Proof There are two parts to the proof. The first part shows that there is a factorization and the second part shows that the factorization is unique.

Part(a)

If n is prime then it is a product with a single prime factor. Otherwise, n can be factored into the product of two numbers ab, where $a > 1$ and $b > 1$. The argument can then be applied to each of a and b each of which is either prime or can be factored as the product of two numbers both of which are greater than one. Continue in this way with the numbers involved decreasing with every step in the process until eventually all of the numbers must be prime. (This argument can be made more rigorous using strong induction).

Part(b)

Suppose the factorization is not unique and let $n > 1$ be the smallest number that has more than one factorization of primes. Then n may be expressed as follows:

$$n = p_1 p_2 p_3 \cdots p_k = q_1 q_2 q_3 \cdots q_r$$

Clearly, $k > 1$ and $r > 1$ and $p_i \neq q_j$ for $(i = 1, \ldots k)$ and $(j = 1, \ldots, r)$ as otherwise we could construct a number smaller than n (e.g., n/p_i where $p_i = q_j$) that has two distinct factorizations. Next, without loss of generality take $p_1 < q_1$ and define the number N by

$$
\begin{aligned}
N &= (q_1 - p_1) q_2 q_3 \cdots q_r \\
&= p_1 p_2 p_3 \cdots p_k - p_1 q_2 q_3 \cdots q_r \\
&= p_1 (p_2 p_3 \cdots p_k - q_2 q_3 \cdots q_r)
\end{aligned}
$$

Clearly $1 < N < n$ and so N is uniquely factorizable into primes. However, clearly p_1 is not a divisor of $(q_1 - p_1)$, and so N has two distinct factorizations, which is a contradiction of the choice of n.

3.3.1 Greatest Common Divisors (GCD)

Let a and b be integers not both zero. The *greatest common divisor* d of a and b is a divisor of a and b (i.e., $d|a$ and $d|b$), and it is the largest such divisor (i.e., if $k|a$ and $k|b$ then $k|d$). It is denoted by gcd (a, b).

Properties of greatest common divisors

(i) Let a and b be integers not both zero then exists integers x and y such that:

$$d = \gcd(a,b) = ax + by$$

(ii) Let a and b be integers not both zero then the set S = $\{ax + by$ where $x, y \in \mathbb{Z}\}$ is the set of all multiples of $d = \gcd(a, b)$.

Proof (of i) Consider the set of all linear combinations of a and b forming the set $\{ka + nb: k, n \in \mathbb{Z}\}$. Clearly, this set includes positive and negative numbers. Choose x and y such that $m = ax + by$ is the smallest positive integer in the set. Then we shall show that m is the greatest common divisor.

We know from the division algorithm that $a = mq + r$ where $0 \le r < m$. Thus

$$r = a - mq = a - (ax + by)q = (1 - qx)a + (-yq)b$$

r is a linear combination of a and b and so r must be 0 from the definition of m. Therefore, $m|a$ and similarly $m|b$ and so m is a common divisor of a and b. Since, the greatest common divisor d is such that $d|a$ and $d|b$ and $d \le m$ we must have $d = m$.

Proof (of ii) This follows since $d|a$ and $d|b \Rightarrow d|ax + by$ for all integers x and y and so every element in the set S = $\{ax + by$ where $x, y \in \mathbb{Z}\}$ is a multiple of d.

Relatively Prime

Two integers a, b are relatively prime if $\gcd(a, b) = 1$

Properties

If p is a prime and $p|ab$ then $p|a$ or $p|b$.

Proof Suppose $p \nmid a$ then from the results on the greatest common divisor we have $\gcd(a, p) = 1$. That is,

$$ra + sp = 1$$
$$\Rightarrow rab + spb = b$$
$$\Rightarrow p\,|\,b\,(\text{since } p\,|\,rab \text{ and } p\,|\,spb \text{ and so } p\,|\,rab + spb)$$

3.3.2   Least Common Multiple (LCM)

If m is a multiple of a and m is a multiple of b then it is said to be a *common multiple* of a and b. The least common multiple is the smallest of the common multiples of a and b and it is denoted by lcm (a, b).

Fig. 3.8 Euclid of
Alexandria

Properties

If x is a common multiple of a and b then $m|x$. That is, every common multiple of
a and b is a multiple of the least common multiple m.

Proof We assume that both a and b are nonzero as otherwise the result is trivial
(since all common multiples are 0). Clearly, by the division algorithm we have

$$x = mq + r \qquad \text{where } 0 \le r < m$$

Since x is a common multiple of a and b we have $a|x$ and $b|x$ and also that $a|m$ and $b|
m$. Therefore, $a|r$ and $b|r$. and so r is a common multiple of a and b and since m is
the least common multiple we have r is 0. Therefore, x is a multiple of the least
common multiple m as required,

3.3.3 Euclid's Algorithm

Euclid's[4] algorithm is one of the oldest known algorithms and it provides a pro-
cedure for finding the greatest common divisor of two numbers. It appears in
Book VII of Euclid's Elements, and the algorithm was known prior to Euclid
(Fig. 3.8).

Lemma *Let a, b, q, and r be integers with $b > 0$ and $0 \le r < b$ such that
$a = bq + r$. Then $gcd(a, b) = gcd(b, r)$.*

Proof Let $K = gcd(a, b)$ and let $L = gcd(b, r)$ and we therefore need to show that
$K = L$. Suppose m is a divisor of a and b then as $a = bq + r$ we have m which is a
divisor of r and so any common divisor of a and b is a divisor of r.

Similarly, any common divisor n of b and r is a divisor of a. Therefore, the greatest
common divisor of a and b is equal to the greatest common divisor of b and r.

---

[4]Euclid was a third century B.C. Hellenistic mathematician and is considered the father of
geometry.

Theorem 3.4 (Euclid's Algorithm) *Euclid's algorithm for finding the greatest common divisor of two positive integers a and b involves applying the division algorithm repeatedly as follows:*

$$a = bq_0 + r_1 \qquad 0 < r_1 < b$$
$$b = r_1 q_1 + r_2 \qquad 0 < r_2 < r_1$$
$$r_1 = r_2 q_2 + r_3 \qquad 0 < r_3 < r_2$$
$$\dots\dots\dots\dots$$
$$\dots\dots\dots\dots$$
$$r_{n-2} = r_{n-1} q_{n-1} + r_n \quad 0 < r_n < r_{n-1}$$
$$r_{n-1} = r_n q_n$$

Then r_n (i.e., the last nonzero remainder) is the greatest common divisor of a and b: i.e., gcd(a, b) = r_n.

Proof t is clear from the construction that r_n is a divisor of $r_{n-1}, r_{n-2}, \dots, r_3, r_2, r_1$ and of a and b. Clearly, any common divisor of a and b will also divide r_n. Using the results from the lemma above we have

$$\gcd(a, b)$$
$$= \gcd(b, r_1)$$
$$= \gcd(r_1 \, r_2)$$
$$= \dots$$
$$= \gcd(r_{n-2} \, r_{n-1})$$
$$= \gcd(r_{n-1}, \, r_n)$$
$$= r_n$$

Lemma *Let n be a positive integer greater than one then the positive divisors of n are precisely those integers of the form:*

$$d = p_1^{\beta_1} p_2^{\beta_2} p_3^{\beta_3} \dots p_k^{\beta_k} \qquad (\text{where } 0 \leq \beta_i \leq \alpha_i),$$

where the unique factorization of n is given by

$$n = p_1^{\alpha_1} p_2^{\alpha_2} p_3^{\alpha_3} \dots p_k^{\alpha_k}$$

Proof Suppose d is a divisor of n then $n = dq$. By the unique factorization theorem the prime factorization of n is unique, and so the prime numbers in the factorization of d must appear in the prime factors $p_1, p_2, p_3, \dots, p_k$ of n.

Clearly, the power β_i of p_i must be less than or equal to α_i: i.e., $\beta_i \leq \alpha_i$. Conversely, whenever $\beta_i \leq \alpha_i$ then clearly d divides n.

3.3.4   Distribution of Primes

We already have shown that there are an infinite number of primes. However, most integer numbers are composite and a reasonable question to ask is how many primes are there less than a certain number. The number of primes less than or equal to x is known as the prime distribution function (denoted by $\pi(x)$) and it is defined by

$$\pi(x) = \sum_{p \leq x} 1 \quad \text{(where } p \text{ is prime)}$$

The prime distribution function satisfies the following properties:

(i) $\lim\limits_{x \to \infty} \dfrac{\pi(x)}{x} = 0$

(ii) $\lim\limits_{x \to \infty} \pi(x) = \infty$

The first property expresses the fact that most integer numbers are composite, and the second property expresses the fact that there are an infinite number of prime numbers.

There is an approximation of the prime distribution function in terms of the logarithmic function $(x/\ln x)$ as follows:

$$\lim_{x \to \infty} \frac{\pi(x)}{x/\ln x} = 1 \quad \text{(Prime Number Theorem)}$$

The approximation $x/\ln x$ to $\pi(x)$ gives an easy way to determine the approximate value of $\pi(x)$ for a given value of x. This result is known as the *Prime Number Theorem*, and Gauss originally conjectured this theorem.

Palindromic Primes

A palindromic prime is a prime number that is also a palindrome (i.e., it reads the same left to right as right to left). For example, 11, 101, 353 are all palindromic primes.

All palindromic primes (apart from 11) have an odd number of digits. It is an open question as to whether there are an infinite number of palindromic primes.

Let $\sigma(m)$ denote the sum of all the positive divisors of m (including m):

$$\sigma(m) = \Sigma_{d \mid m} d$$

Let $s(m)$ denote the sum of all the positive divisors of m (excluding m):

$$s(m) = \sigma(m) - m.$$

Clearly, $s(m) = m$ and $\sigma(m) = 2m$ when m is a perfect number.

Theorem 3.5 (Euclid–Euler Theorem) *The positive integer* n *is an even perfect number if and only if* n = $2^{p-1}(2^p - 1)$, *where* $2^p - 1$ *is a Mersenne prime.*

Proof Suppose $n = 2^{p-1}(2^p - 1)$, where $2^p - 1$ is a Mersenne prime then

$$
\begin{aligned}
\sigma(n) &= \sigma\left(2^{p-1}(2^p-1)\right) \\
&= \sigma\left(2^{p-1}\right)\sigma(2^p-1) \\
&= \sigma\left(2^{p-1}\right)2^p \qquad (2^p-1 \text{ is prime with 2 divisors : 1 and itself}) \\
&= (2^p-1)\,2^p \qquad (\text{Sum of arithmetic series}) \\
&= (2^p-1)\,2.2^{p-1} \\
&= 2.2^{p-1}(2^p-1) \\
&= 2n
\end{aligned}
$$

Therefore, n is a perfect number since $\sigma(n) = 2n$.

The next part of the proof is to show that any even perfect number must be of the form above. Let n be an arbitrary even perfect number ($n = 2^{p-1}q$) with q odd and so the gcd $(2^{p-1}, q) = 1$ and so

$$
\begin{aligned}
\sigma(n) \\
&= \sigma\left(2^{p-1}q\right) \\
&= \sigma\left(2^{p-1}\right)\sigma(q) \\
&= (2^p-1)\sigma(q)
\end{aligned}
$$

$$
\begin{aligned}
\sigma(n) \\
&= 2n \qquad (\text{since } n \text{ is perfect}) \\
&= 2.2^{p-1}q \\
&= 2^p q
\end{aligned}
$$

Therefore,

$$
\begin{aligned}
2^p q \\
&= (2^p-1)\sigma(q) \\
&= (2^p-1)\,(s(q)+q) \\
&= (2^p-1)s(q)+(2^p-1)q \\
&= (2^p-1)s(q)+2^p q - q
\end{aligned}
$$

Therefore, $(2^p - 1)\,s(q) = q$

Therefore, $d = s(q)$ is a proper divisor of q. However, $s(q)$ is the sum of all the proper divisors of q including d, and so d is the only proper divisor of q and $d = 1$. Therefore, $q = (2^p - 1)$ is a Mersenne prime.

Fig. 3.9 Leonard Euler

Euler φ Function

The Euler[5] φ function (also known as the *totient function*) is defined for a given positive integer n to be the number of positive integers k less than n that are relatively prime to n (Fig. 3.9). Two integers a, b are relatively prime if $\gcd(a, b) = 1$.

$$\varphi(n) = \sum_{1 \leq k < n} 1 \qquad \text{where } \gcd(k, n) = 1$$

3.4 Theory of Congruences[6]

Let a be an integer and n a positive integer greater than 1 then $(a \bmod n)$ is defined to be the remainder r when a is divided by n. That is,

$$a = kn + r \qquad \text{where } 0 \leq r < n.$$

Definition Suppose a, b are integers and n a positive integer then a is said to be congruent to b modulo n denoted by $a \equiv b \pmod{n}$ if they both have the same remainder when divided by n.

This is equivalent to n being a divisor of $(a - b)$ or $n|(a - b)$ since we have $a = k_1 n + r$ and $b = k_2 n + r$ and so $(a - b) = (k_1 - k_2) n$ and so $n|(a - b)$.

---

[5]Euler was an eighteenth century Swiss mathematician who made important contributions to mathematics and physics. His contributions include graph theory (e.g., the well-known formula $V - E + F = 2$), calculus, infinite series, the exponential function for complex numbers, and the totient function.

[6]The theory of congruences was introduced by the German mathematician, Carl Friedrich Gauss.

Theorem 3.6 *Congruence modulo n is an equivalence relation on the set of integers: i.e., it is a reflexive, symmetric and transitive relation.*

Proof

   (i)   Reflexive
         For any integer a it is clear that $a \equiv a$ (mod n) since $a - a = 0.n$
  (ii)   Symmetric
         Suppose $a \equiv b$ (mod n) then $a - b = kn$. Clearly, $b - a = -kn$ and so $b \equiv a$ (mod n).
 (iii)   Transitive.

$$\text{Suppose } a \equiv b \,(\mathrm{mod}\, n) \text{ and } b \equiv c \,(\mathrm{mod}\, n)$$
$$\Rightarrow a - b = k_1 n \text{ and } b - c = k_2 n$$
$$\Rightarrow a - c = (a - b) + (b - c)$$
$$= k_1 n + k_2 n$$
$$= (k_1 + k_2)n$$
$$\Rightarrow a \equiv c \,(\mathrm{mod}\, n).$$

Therefore, congruence modulo n is an equivalence relation, and an equivalence relation partitions a set S into equivalence classes (Theorem 2.2). The integers are partitioned into n equivalence classes for the congruence modulo n equivalence relation, and these are called *congruence classes* or *residue classes*.   The residue class of a modulo n is denoted by $[a]_n$ or just $[a]$ when n is clear. It is the set of all those integers that are congruent to a modulo n.

$$[a]_n = \{x : x \in \mathbb{Z} \text{ and } x \equiv a(\mathrm{mod}\, n)\} = \{a + kn : k \in \mathbb{Z}\}$$

Any two equivalence classes $[a]$ and $[b]$ are either equal or disjoint: i.e., we have $[a] = [b]$ or $[a] \cap [b] = \varnothing$. The set of all residue classes modulo n is denoted by

$$\mathbb{Z}/n\mathbb{Z} = \mathbb{Z}_n = \{[a]_n : 0 \leq a \leq n - 1\} = \{[0]_n, [1]_n, \ldots, [n-1]_n\}$$

For example, consider $\mathbb{Z}_4$ the residue classes mod 4 then

$$[0]_4 = \{\ldots, -8, -4, 0, 4, 8, \ldots\}$$
$$[1]_4 = \{\ldots, -7, -3, 1, 5, 9, \ldots\}$$
$$[2]_4 = \{\ldots, -6, -2, 2, 6, 10, \ldots\}$$
$$[3]_4 = \{\ldots, -5, -1, 3, 7, 11, \ldots\}$$

The *reduced residue class* is a set of integers r_i such that $(r_i, n) = 1$ and r_i is not congruent to r_j (mod n) for $i \neq j$, and such that every x relatively prime to n is congruent modulo n to for some element r_i of the set. There are $\varphi(n)$ elements $\{r_1, r_2, \ldots, r_{\varphi(n)}\}$ in the reduced residue class set S.

Modular Arithmetic

Addition, subtraction and multiplication may be defined in $\mathbb{Z}/n\mathbb{Z}$ and are similar to these operations in $\mathbb{Z}$. Given a positive integer n and integers a, b, c, d such that $a \equiv b \pmod{n}$ and $c \equiv d \pmod{n}$ then the following are properties of modular arithmetic.

(i) $a + c \equiv b + d \pmod{n}$ and $a - c \equiv b - d \pmod{n}$
(ii) $ac \equiv bd \pmod{n}$
(iii) $a^m \equiv b^m \pmod{n} \; \forall m \in \mathbb{N}$

Proof (of ii) Let $a = kn + b$ and $c = ln + d$ for some $k, l \in \mathbb{Z}$ then

$$
\begin{aligned}
ac &= (kn + b)(ln + d) \\
&= (kn)(ln) + (kn)d + b(ln) + bd \\
&= (knl + kd + bl)n + bd \\
&= sn + bd, \qquad (\text{where } s = knl + kd + bl)
\end{aligned}
$$

and $ac \equiv bd \pmod{n}$

The three properties above may be expressed in the following equivalent formulation:

(i) $[a + c]_n = [b + d]_n$ and $[a - c]_n = [b - d]_n$
(ii) $[ac]_n = [bd]_n$
(iii) $[a^m]_n = [b^m]_n \; \forall m \in \mathbb{N}$

Two integers x, y are said to be multiplicative inverses of each other modulo n if

$$xy \equiv 1 \pmod{n}$$

However, x does not always have an inverse modulo n, and this is clear since, for example, $[3]_6$ is a zero divisor modulo 6, i.e., $[3]_6 \cdot [2]_6 = [0]_6$ and it does not have a multiplicative inverse. However, if n and x are relatively prime then it is easy to see that x has an inverse $\pmod{n}$ since we know that there are integers k, l such that $kx + ln = 1$.

Given $n > 0$ there are $\varphi(n)$ numbers b that are relatively prime to n and there are $\varphi(n)$ numbers that have an inverse modulo n. Therefore, for p prime there are $p - 1$ elements that have an inverse $\pmod{p}$.

Theorem 3.7 (Euler's Theorem) *Let* a *and* n *be positive integers with* gcd(a, n) = *1. Then*

$$a^{\phi(n)} \equiv 1 (\bmod\, n)$$

Proof Let $\{r_1, r_2, \ldots, r_{\varphi(n)}\}$ be the reduced residue system (mod n). Then $\{ar_1, ar_2, \ldots, ar_{\varphi(n)}\}$ is also a reduced residue system (mod n) since $ar_i \equiv ar_j$ (mod n) and $(a, n) = 1$ implies that $r_i \equiv r_j$ (mod n).

For each r_i there is exactly one r_j such that $ar_i \equiv r_j$ (mod n), and different r_i will have different corresponding ar_j Therefore, $\{ar_1, ar_2, \ldots, ar_{\varphi(n)}\}$ are just the residues module n of $\{r_1, r_2, \ldots, r_{\varphi(n)}\}$ but not necessarily in the same order. Multiplying we get

$$\prod_{j=1}^{\varphi(n)} (ar_j) \equiv \prod_{i=1}^{\varphi(n)} r_i \qquad (\bmod\, n)$$

$$a^{\phi(n)} \prod_{j=1}^{\varphi(n)} (r_j) \equiv \prod_{i=1}^{\varphi(n)} r_i \qquad (\bmod\, n)$$

Since $(r_j, n) = 1$ we can deduce that $a^{\phi(n)} \equiv 1$ (mod n) from the result that $ax \equiv ay$ (mod n) and $(a, n) = 1$ then $x \equiv y$ (mod n).

Theorem 3.8 (Fermat's Little Theorem) *Let a be a positive integer and p a prime. If gcd* $(a, p) = 1$ *then*

$$a^{p-1} \equiv 1 \;(\bmod\, p)$$

Proof This result is an immediate corollary to Euler's Theorem as $\varphi(p) = p - 1$.

Theorem 3.9 (Wilson's Theorem) *If p is a prime then* $(p - 1)! \equiv -1$ (mod p).

Proof Each element $a \in 1, 2, \ldots, p - 1$ has an inverse a^{-1} such that $aa^{-1} \equiv 1$ (mod p). Exactly two of these elements 1 and $p - 1$ are their own inverse (i.e., $x^2 \equiv 1$ (mod p) has two solutions 1 and $p - 1$). Therefore, the product 1, 2, ..., $p - 1$ (mod p) = $p - 1$ (mod p) $\equiv -1$ (mod p)

Diphantine equations

The word "*Diophantine*" is derived from the name of the third century mathematician, Diophantus, who lived in the city of Alexandria in Egypt. Diophantus studied various polynomial equations of the form $f(x,y,z,\ldots) = 0$ with integer coefficients to determine which of them had integer solutions.

A Diophantine equation may have no solution, a finite number of solutions or an infinite number of solutions. The integral solutions of a Diophantine equation $f(x, y) = 0$ may be interpreted geometrically as the points on the curve with integral coordinates.

Example A linear Diophantine equation $ax + by = c$ is an algebraic equation with two variables x and y, and the problem is to find integer solutions for x and y.

Table 3.1 Binary number system

Binary	Dec.	Binary	Dec.	Binary	Dec.	Binary	Dec.
0000	0	0100	4	1000	8	1100	12
0001	1	0101	5	1001	9	1101	13
0010	2	0110	6	1010	10	1110	14
0011	3	0111	7	1011	11	1111	15

3.5 Binary System and Computer Representation of Numbers

Arithmetic has traditionally been done using the decimal notation,[7] and this positional number system involves using the digits 0, 1, 2, ..., 9. Leibniz[8] was one of the earliest people to recognize the potential of the binary number system, and this base 2 system uses just two digits namely "0" and "1". Leibniz described the binary system in *Explication de l'Arithmétique Binaire* [1], which was published in 1703. His 1703 paper describes how binary numbers may be added, subtracted, multiplied and divided, and Leibniz was an advocate of their use.

The number two is represented by 10; the number four by 100; and so on. A table of values for the first fifteen binary numbers is given in Table 3.1.

The binary number system (base 2) is a positional number system, which uses two binary digits 0 and 1, and an example binary number is 1001.01_2 which represents $1 \times 2^3 + 1 + 1 \times 2^{-2} = 8 + 1 + 0.25 = 9.25$.

The binary system is ideally suited to the digital world of computers, as a binary digit may be implemented by an *on off switch*. In the digital world devices that store information or data on permanent storage media such as disks, and CDs, or temporary storage media such as random access memory (RAM) consist of a large number of memory elements that may be in one of two states (i.e., on or off).

The digit 1 represents that the switch is on, and the digit 0 represents that the switch is off. Claude Shannon showed in his Master's thesis [2] that the binary digits (i.e., 0 and 1) can be represented by electrical switches. This allows binary arithmetic and more complex mathematical operations to be performed by relay circuits, and provided the foundation of digital computing.

---

[7]Other bases have been employed such as the segadecimal (or base-60) system employed by the Babylonians. The decimal system was developed by Indian and Arabic mathematicians between 800–900AD, and it was introduced to Europe in the late twelfth/early thirteenth century. It is known as the *Hindu–Arabic system*.

[8]Wilhelm Gottfried Leibniz was a German philosopher, mathematician and inventor in the field of mechanical calculators. He developed the binary number system used in digital computers, and invented the Calculus independently of Sir Issac Newton. He was embroiled in a bitter dispute towards the end of his life with Newton, as to who developed the calculus first.

The decimal system (base 10) is more familiar for everyday use, and there are algorithms to convert numbers from decimal to binary and vice versa. For example, to convert the decimal number 25 to its binary representation we proceed as follows:

2	25
	12	1
	6	0
	3	0
	1	1
	0	1

The base 2 is written on the left and the number to be converted to binary is placed in the first column. At each stage in the conversion the number in the first column is divided by 2 to form the quotient and remainder, which are then placed on the next row. For the first step the quotient when 25 is divided by 2 is 12 and the remainder is 1. The process continues until the quotient is 0, and the binary representation result is then obtained by reading the second column from the bottom up. Thus, we see that the binary representation of 25 is 11001_2.

Similarly, there are algorithms to convert decimal fractions to binary representation (to a defined number of binary digits as the representation may not terminate), and the conversion of a number that contains an integer part and a fractional part involves converting each part separately and then combining them.

The octal (base 8) and hexadecimal (base 16) are often used in computing, as the bases 2, 8 and 16 are related bases and easy to convert between, as to convert between binary and octal involves grouping the bits into groups of three on either side of the point. Each set of 3-bits corresponds to one digit in the octal representation. Similarly, the conversion between binary and hexadecimal involves grouping into sets of 4 digits on either side of the point. The conversion the other way from octal to binary or hexadecimal to binary is equally simple, and involves replacing the octal (or hexadecimal) digit with the 3-bit (or 4-bit) binary representation.

Numbers are represented in a digital computer as sequences of bits of fixed length (e.g., 16-bits, 32-bits). There is a difference in the way in which integers and real numbers are represented, with the representation of real numbers being more complicated.

An integer number is represented by a sequence (usually 2 or 4) bytes where each byte is 8-bits. For example, a 2-byte integer has 16 bits with the first bit used as the sign bit (the sign is 1 for negative numbers and 0 for positive integers), and the remaining 15 bits represent the number. This means that two bytes may be used to represent all integer numbers between $-32,768$ and $32,767$. A positive number is represented by the normal binary representation discussed earlier, whereas a negative number is represented using 2's complement of the original number (i.e., 0 changes to 1 and 1 changes to 0 and the sign bit is 1). All of the standard arithmetic operations may then be carried out (using modulo 2 arithmetic).

The representation of floating point real numbers is more complicated, and a real number is represented to a fixed number of significant digits (the significand) and scaled using an exponent in some base (usually 2). That is, the number is represented (approximated as):

$$\text{significand} \times \text{base}^{\text{exponent}}$$

The significand (also called mantissa) and exponent have a sign bit. For example, in simple floating point representation (4 bytes) the mantissa is generally 24-bits and the exponent 8-bits, whereas for double precision (8 bytes) the mantissa is generally 53 bits and the exponent 11 bits. There is an IEEE standard for floating point numbers (IEEE 754).

3.6 Review Questions

1. Show that
   (i) if $a|b$ then $a|bc$
   (ii) If $a|b$ and $c|d$ then $ac|bd$
2. Show that 1184 and 1210 are an amicable pair.
3. Use the Euclidean Algorithm to find $g = \gcd(b, c)$ where $b = 42,823$ and $c = 6409$, and find integers x and y such that $bx + cy = g$
4. List all integers x in the range $1 \le x \le 100$ such that $x \equiv 7 \pmod{17}$.
5. Evaluate $\phi(m)$ for $m = 1, 2, 3, \ldots. 12$.
6. Determine a complete and reduced residue system modulo 12.
7. Convert 767 to binary, octal and hexadecimal.
8. Convert (you may need to investigate) 0.32_{10} to binary (to 5 places).
9. Explain the difference between binary, octal and hexadecimal.
10. Find the 16-bit integer representation of -4961.

3.7   Summary

Number theory is concerned with the mathematical properties of the natural numbers and integers. These include properties such as, whether a number is prime or composite, the prime factors of a number, the greatest common divisor and least common multiple of two numbers and so on.

The natural numbers $\mathbb{N}$ consist of the numbers $\{1, 2, 3, \ldots\}$. The integer numbers $\mathbb{Z}$ consist of $\{\ldots -2, -1, 0, 1, 2, \ldots\}$. The rational numbers $\mathbb{Q}$ consist of all numbers of the form $\{p/q$ where p and q are integers and $q \neq 0\}$. Number theory has been applied to cryptography in the computing field.

Prime numbers have no factors apart from themselves and one, and there are an infinite number of primes. The Sieve of Eratosthene's algorithm may be employed to determine prime numbers, and the approximation to the distribution of prime numbers less than a number n is given by the prime distribution function $\pi(n) = n/\ln n$. Prime numbers are the key building blocks in number theory, and the fundamental theorem of arithmetic states that every number may be written uniquely as the product of factors of prime numbers.

Mersenne primes and perfect numbers were considered and it was shown that there is a one to one correspondence between the Mersenne primes and the even perfect numbers.

Modulo arithmetic including addition, subtraction and multiplication were defined, and the residue classes and reduced residue classes discussed. There are unsolved problems in number theory such as Goldbach's conjecture that states that every even integer is the sum of two primes. Other open questions include whether there are an infinite number of Mersenne primes and palindromic primes.

We discussed the binary number system, which is ideally suited for digital computers. We discussed the conversion between binary and decimal systems, as well as the octal and hexadecimal systems. Finally, we discussed the representation of integers and real numbers on a computer. For more detailed information on number theory see [3].

References

1. *Explication de l'Arithmétique Binaire* Wilhelm Gottfried Leibniz. *Memoires de l'Academie Royale des Sciences.* 1703.
2. A Symbolic Analysis of Relay and Switching Circuits. Claude Shannon. Masters Thesis. Massachusetts Institute of Technology. 1937.
3. Number Theory for Computing. Song Y. Yan 2nd Edition. Springer. 1998.

Mathematical Induction and Recursion

4

Key Topics

Mathematical Induction
Strong and weak Induction
Base Case
Inductive Step
Recursion
Recursive Definition
Structural Induction

4.1 Introduction

Mathematical induction is an important proof technique used in mathematics, and it is often used to establish the truth of a statement for all natural numbers. There are two parts to a proof by induction, and these are the base step and the inductive step. The first step is termed the *base case*, and it involves showing that the statement is true for some natural number (usually the number 1). The second step is termed the *inductive step*, and it involves showing that if the statement is true for some natural number $n = k$, then the statement is true for its successor $n = k + 1$. This is often written as $P(k) \rightarrow P(k + 1)$.

The statement $P(k)$ that is assumed to be true when $n = k$ is termed the *inductive hypothesis*. From the base step and the inductive step, we infer that the statement is true for all natural numbers (that are greater than or equal to the number specified in

© Springer International Publishing Switzerland 2016
G. O'Regan, *Guide to Discrete Mathematics*, Texts in Computer Science,
DOI 10.1007/978-3-319-44561-8_4

the base case). Formally, the proof technique used in mathematical induction is of the form[1]

$$(P(1) \wedge \forall k(P(k) \rightarrow P(k+1))) \rightarrow \forall n \, P(n).$$

Mathematical induction (weak induction) may be used to prove a wide variety of theorems, and especially theorems of the form $\forall n \, P(n)$. It may be used to provide a proof of theorems about summation formulae, inequalities, set theory, and the correctness of algorithms and computer programs. One of the earliest inductive proofs was the sixteenth century proof that the sum of the first n odd integers is n^2, which was proved by Francesco Maurolico in 1575. Later mathematicians made the method of mathematical induction more precise.

We distinguish between *strong induction* and *weak induction*, where strong induction also has a base case and an inductive step, but the inductive step is a little different. It involves showing that if the statement is true for all natural numbers less than or equal to an arbitrary number k, then the statement is true for its successor $k + 1$. *Structural induction* is another form of induction and this mathematical technique is used to prove properties about recursively defined sets and structures.

Recursion is often used in mathematics to define functions, sequences and sets. However, care is required with a recursive definition to ensure that it actually defined something, and that what is defined makes sense. Recursion defines a concept in terms of itself, and we need to ensure that the definition is not circular (i.e. that it does not lead to a vicious circle).

Recursion and induction are closely related and are often used together. Recursion is extremely useful in developing algorithms for solving complex problems, and induction is a useful technique in verifying the correctness of such algorithms.

Example 4.1 Show that the sum of the first n natural numbers is given by the formula

$$1 + 2 + 3 + \cdots + n = \frac{n(n+1)}{2}$$

Proof
Base Case

We consider the case where $n = 1$ and clearly $1 = \frac{1(1+1)}{2}$ and so the base case P (1) is true.

$$1 + 2 + 3 + \cdots + k = \frac{k(k+1)}{2}$$

---

[1]This definition of mathematical induction covers the base case of $n = 1$, and would need to be adjusted if the number specified in the base case is higher.

Inductive Step

Suppose the result is true for some number k then we have $P(k)$

Then consider the sum of the first $k + 1$ natural numbers, and we use the inductive hypothesis to show that its sum is given by the formula.

$$1 + 2 + 3 + \cdots + k + (k+1)$$
$$= \frac{k(k+1)}{2} + (k+1) \qquad \text{(by inductive hypothesis)}$$
$$= \frac{k^2 + k}{2} + \frac{(2k+2)}{2}$$
$$= \frac{k^2 + 3k + 2)}{2}$$
$$= \frac{(k+1)(k+2)}{2}$$

Thus, we have shown that if the formula is true for an arbitrary natural number k, then it is true for its successor $k + 1$. That is, $P(k) \rightarrow P(k + 1)$. We have shown that $P(1)$ is true, and so it follows from mathematical induction that $P(2)$, $P(3)$, ... are true, and so $P(n)$ is true, for all natural numbers and the theorem is established.

Note 4.1

There are opportunities to make errors in proofs with induction, and the most common mistakes are not to complete the base case or inductive step correctly. These errors can lead to strange results and so care is required. It is important to be precise in the statements of the base case and inductive step.

Example 4.2 (**Binomial Theorem**) Prove the binomial theorem using induction (permutations and combinations are discussed in Chap. 5). That is,

$$(1+x)^n = 1 + \binom{n}{1}x + \binom{n}{2}x^2 + \ldots + \binom{n}{r}x^r + \ldots + \binom{n}{n}x^n$$

Proof
Base Case

We consider the case where $n = 1$ and clearly $(1+x)^1 = (1+x) = 1 + \binom{1}{1}x^1$

and so the base case $P(1)$ is true.

Inductive Step

Suppose the result is true for some number k then we have $P(k)$

$$(1+x)^k = 1 + \binom{k}{1}x + \binom{k}{2}x^2 + \cdots + \binom{k}{r}x^r + \cdots + \binom{k}{k}x^k$$

Then consider $(1 + x)^{k+1}$ and we use the inductive hypothesis to show that it is given by the formula.

$(1+x)^{k+1}$

$$= (1+x)^k (1+x)$$

$$= \left(1 + \binom{k}{1}x + \binom{k}{2}x^2 + \cdots + \binom{k}{r}x^r + \cdots + \binom{k}{k}x^k\right)(1+x)$$

$$= \left(1 + \binom{k}{1}x + \binom{k}{2}x^2 + \cdots + \binom{k}{r}x^r + \cdots + \binom{k}{k}x^k\right)$$

$$+ x + \binom{k}{1}x^2 + \cdots + \binom{k}{r}x^{r+1} + \cdots + \binom{k}{k}x^{k+1}$$

$$= 1 + \binom{k}{1}x + \binom{k}{2}x^2 + \cdots + \binom{k}{r}x^r + \cdots + \binom{k}{k}x^k$$

$$+ \binom{k}{0}x + \binom{k}{1}x^2 + \cdots + \binom{k}{r-1}x^r + \cdots + \binom{k}{k-1}x^k + \binom{k}{k}x^{k+1}$$

$$= 1 + \binom{k+1}{1}x + \cdots + \binom{k+1}{r}x^r + \cdots + \binom{k+1}{k}x^k + \binom{k+1}{k+1}x^{k+1}$$

(which follows from Exercise 7 below)

Thus, we have shown that if the binomial theorem is true for an arbitrary natural number k, then it is true for its successor $k + 1$. That is, $P(k) \rightarrow P(k + 1)$. We have shown that $P(1)$ is true, and so it follows from mathematical induction that $P(n)$ is true, for all natural numbers, and so the theorem is established.

The standard formula of the binomial theorem $(x + y)^n$ follows immediately from the formula for $(1 + x)^n$, by noting that $(x + y)^n = \{x(1 + {}^y/_x)\}^n = x^n(1 + {}^y/_x)^n$.

4.2   Strong Induction

Strong induction is another form of mathematical induction, which is often employed when we cannot prove a result with (weak) mathematical induction. It is similar to weak induction in that there is a base step and an inductive step. The base step is identical to weak mathematical induction, and it involves showing that the statement is true for some natural number (usually the number 1). The inductive step is a little different, and it involves showing that if the statement is true for all natural numbers less than or equal to an arbitrary number k, then the statement is true for its successor $k + 1$. This is often written as $(P(1) \wedge P(2) \wedge \ldots \wedge P(k)) \rightarrow P(k + 1)$.

From the base step and the inductive step, we infer that the statement is true for all natural numbers (that are greater than or equal to the number specified in the base case). Formally, the proof technique used in mathematical induction is of the form[2]

$$(P(1) \wedge \forall k[(P(1) \wedge P(2) \wedge \ldots \wedge P(k)) \rightarrow P(k+1)]) \rightarrow \forall n\, P(n).$$

Strong and weak mathematical induction are equivalent in that any proof done by weak mathematical induction may also be considered a proof using strong induction, and a proof conducted with strong induction may also be converted into a proof using weak induction.

Weak mathematical induction is generally employed when it is reasonably clear how to prove $P(k + 1)$ from $P(k)$, with strong mathematical typically employed where it is not so obvious. The validity of both forms of mathematical induction follows from the *well-ordering property* of the Natural Numbers, which states that every non-empty set has a least element.

Well-Ordering Principle

Every nonempty set of natural numbers has a least element. The well-ordering principle is equivalent to the principle of mathematical induction.

Example 4.3 Show that every natural number greater than one is divisible by a prime number.

Proof
Base Case
We consider the case of $n = 2$ which is trivially true, since 2 is a prime number and is divisible by itself.
Inductive Step (strong induction)
Suppose that the result is true for every number less than or equal to k. Then we consider $k + 1$, and there are there are two cases to consider. If $k + 1$ is prime then it is divisible by itself. Otherwise it is composite and it may be factored as the product of two numbers each of which is less than or equal to k. Each of these numbers is divisible by a prime number by the strong inductive hypothesis, and so $k + 1$ is divisible by a prime number.

Thus, we have shown that if all natural numbers less than or equal to k are divisible by a prime number, then $k + 1$ is divisible by a prime number. We have shown that the base case $P(2)$ is true, and so it follows from strong mathematical induction that every natural numbers greater than one is divisible by some prime number.

---

[2]As before this definition covers the base case of $n = 1$ and would need to be adjusted if the number specified in the base case is higher.

4.3 Recursion

Some functions (or objects) used in mathematics (e.g. the Fibonacci sequence) are difficult to define explicitly, and are best defined by a *recurrence relation*: (i.e. an equation that recursively defines a sequence of values, once one or more initial values are defined). Recursion may be employed to define functions, sequences and sets.

There are two parts to a recursive definition namely the *base case*, and the *recursive (inductive) step*. The base case usually defines the value of the function at $n = 0$ or $n = 1$, whereas the recursive step specifies how the application of the function to a number may be obtained from its application to one or more smaller numbers.

It is important that care is taken with the recursive definition, to ensure that that it is not circular, and does not lead to an infinite regress. The argument of the function on the right-hand side of the definition in the recursive step is usually smaller than the argument on the left-hand side to ensure termination (there are some unusual recursively defined functions such as the *McCarthy* 91 *function* where this is not the case).

It is natural to ask when presented with a recursive definition whether it means anything at all, and in some cases the answer is negative. Fixed-point theory provides the mathematical foundations for recursion, and ensures that the functions/objects are well defined.

Chapter 12 (Sect. 12.6) discusses various mathematical structures such as partial orders, complete partial orders and lattices, which may be employed to give a secure foundation for recursion. A precise mathematical meaning is given to recursively defined functions in terms of domains and fixed-point theory, and it is essential that the conditions in which recursion maybe used safely be understood. The reader is referred to [1] for more detailed information.

A recursive definition will include at least one non-recursive branch with every recursive branch occurring in a context that is different from the original, and brings it closer to the non-recursive case. Recursive definitions are a powerful and elegant way of giving the denotational semantics of language constructs.

Next, we present examples of the recursive definition of the factorial function and Fibonacci numbers.

Example 4.4 (**Recursive Definition of Functions**) The factorial function $n!$ is very common in mathematics and its well-known definition is $n! = n(n - 1)(n - 2)...$ $3.2.1$ and $0! = 1$. The formal definition in terms of a base case and inductive step is given as follows:

$$
\begin{aligned}
&\text{Base Step} && \mathrm{fac}\,(0) = 1 \\
&\text{Recursive Step} && \mathrm{fac}\,(n) = n * \mathrm{fac}\,(n - 1)
\end{aligned}
$$

This recursive definition defines the procedure by which the factorial of a number is determined from the base case, or by the product of the number by the factorial of its predecessor. The definition of the factorial function is built up in a sequence: fac(0), fac(1), fac(2), ...

The Fibonacci sequence[3] is named after the Italian mathematician Fibonacci, who introduced it in the thirteenth century. It had been previously described in Indian mathematics, and the Fibonacci numbers are the numbers in the following integer sequence:

$$1, 1, 2, 3, 5, 8, 13, 21, 34$$

Each Fibonacci number (apart from the first two in the sequence) is obtained by adding the two previous Fibonacci numbers in the sequence together. Formally, the definition is given by

Base Step $\quad F_1 = 1, F_2 = 1$
Recursive Step $\quad F_n = F_{n-1} + F_{n-2}$ (Definition for when $n > 2$)

Example 4.5 (**Recursive Definition of Sets and Structures**) Sets and sequences may also be defined recursively, and there are two parts to the recursive definition (as before). The base case specifies an initial collection of elements in the set, whereas the inductive step provides rules for adding new elements to the set based on those already there. Properties of recursively defined sets may often be proved by a technique called structural induction.

Consider the subset S of the Natural Numbers defined by

Base Step $\quad 5 \in S$
Recursive Step $\quad$ For $x \in S$ then $x + 5 \in S$

Then the elements in S are given by the set of all multiples of 5, as clearly $5 \in S$; therefore by the recursive step $5 + 5 = 10 \in S$; $5 + 10 = 15 \in S$; and so on.

The recursive definition of the set of strings Σ^* over an alphabet Σ is given by

Base Step $\quad \Lambda \in \Sigma^*$ (Λ is the empty string)
Recursive Step $\quad$ For $\sigma \in \Sigma^*$ and $v \in \Sigma$ then $\sigma v \in \Sigma^*$

Clearly, the empty string is created from the base step. The recursive step states that a new string is obtained by adding a letter from the alphabet to the end of an existing string in Σ^*. Each application of the inductive step produces a new string that contains one additional character. For example, if $\Sigma = \{0, 1\}$ then the strings in Σ^* are the set of bit strings Λ, 0, 1, 00, 01, 10, 11, 000, 001, 010, etc.

---

[3]We are taking the Fibonacci sequence as starting at 1, whereas others take it as starting at 0.

We can define an operation to determine the length of a string (len: $\Sigma^* \to \mathbb{N}$) recursively.

Base Step   len $(\Lambda) = 0$
Recursive Step   len $(\sigma v) = $ len$(\sigma) + 1$ (where $\sigma \in \Sigma^*$ and $v \in \Sigma$)

A binary tree[4] is a well-known data structure in computer science, and it consists of a root node together with a left and right binary tree. A binary tree is defined as a finite set of nodes (starting with the root node), where each node consists of a data value and a link to a left subtree and a right subtree. Recursion is often used to define the structure of a binary tree.

Base Step   A single node is a binary tree (root)
Recursive Step

(i) Suppose X and Y are binary trees and x is a node then XxY is a binary tree, where X is the left subtree, Y the right subtree, and x is the new root node.
(ii) Suppose X is a binary tree and x is a node then xX and Xx are binary trees, which consist of the root node x and a single child left or right subtree.

That is, a binary tree has a root node and it may have no subtrees; it may consist of a root node with a left subtree only; a root node with a right subtree only; or a root node with both a left and right subtree.

4.4 Structural Induction

Structural induction is a mathematical technique that is used to prove properties about recursively defined sets and structures. It may be used to show that all members of a recursively defined set have a certain property, and there are two parts to the proof (as before) namely the base case and the recursive (inductive) step.

The first part of the proof is to show that the property holds for all elements specified in the base case of the recursive definition. The second part of the proof involves showing that if the property is true for all elements used to construct the new elements in the recursive definition then the property holds for the new elements. From the base case and the recursive step we deduce that the property holds for all elements of the set (structure).

---

[4]We will give an alternate definition of a tree in terms of a connected acyclic graph in Chap. 9 on graph theory.

Example 4.6 (**Structural Induction**) We gave a recursive definition of the subset S of the natural numbers that consists of all multiples of 5. We did not prove that all elements of the set S is divisible by 5, and we use structural induction to prove this.

Base Step $5 \in S$ (and clearly the base case is divisible by 5)

Inductive Step Suppose $q \in S$ then $q = 5k$ for some k. From the inductive hypothesis $q + 5 \in S$ and $q + 5 = 5k + 5 = 5(k + 1)$ and so $q + 5$ is divisible by 5.

Therefore, all elements of S are divisible by 5.

4.5 Review Questions

1. Show that $9^n + 7$ is always divisible by 8.
2. Show that the sum of $1^2 + 2^2 + \cdots + n^2 = n(n + 1)(2n + 1)/6$
3. Explain the difference between strong and weak induction.
4. What is structural induction?
5. Explain how recursion is used in mathematics.
6. Investigate the recursive definition of the Mc Carthy 91 function, and explain how it differs from usual recursive definitions.
7. Show that $\binom{r}{r} + \binom{n}{r-1} = \binom{n+1}{r}$
8. Determine the standard formula for the binomial theorem $(x + y)^n$ from the formula for $(1 + x)^n$.

4.6 Summary

Mathematical induction is an important proof technique that is used to establish the truth of a statement for all natural numbers. There are two parts to a proof by induction, and these are the base case and the inductive step. The base case involves showing that the statement is true for some natural number (usually for the number $n = 1$). The inductive step involves showing that if the statement is true for some natural number $n = k$, then the statement is true for its successor $n = k + 1$.

From the base step and the inductive step, we infer that the statement is true for all natural numbers (that are greater than or equal to the number specified in the base case). Mathematical induction may be used to prove a wide variety of theorems, such as theorems about summation formulae, inequalities, set theory, and the correctness of algorithms and computer programs.

Strong induction is often employed when we cannot prove a result with (weak) mathematical induction. It also has a base case and an inductive step, where the inductive step is a little different, and it involves showing that if the statement is true for all natural numbers less than or equal to an arbitrary number k, then the statement is true for its successor $k + 1$.

Recursion may be employed to define functions, sequences and sets in mathematics, and there ar two parts to a recursive definition namely the base case and the recursive step. The base case usually defines the value of the function at $n = 0$ or $n = 1$, whereas the recursive step specifies how the application of the function to a number may be obtained from its application to one or more smaller numbers. It is important that care is taken with the recursive definition, to ensure that that it is not circular, and does not lead to an infinite regress.

Structural induction is a mathematical technique that is used to prove properties about recursively defined sets and structures. It may be used to show that all members of a recursively defined set have a certain property, and there are two parts to the proof namely the base case and the recursive (inductive) step.

Reference

1. Introduction to the Theory of Programming Languages. Bertrand Meyer. Prentice Hall. 1990.

Sequences, Series and Permutations and Combinations

5

Key Topics

Arithmetic sequence
Arithmetic series
Geometric Sequence
Geometric Series
Simple and compound interest
Annuities
Present Value
Permutations and Combinations
Counting Principle

5.1 Introduction

The goal of this chapter is to provide an introduction to sequences and series, including arithmetic and geometric sequences, and arithmetic and geometric series. We derive formulae for the sum of an arithmetic series and geometric series, and we discuss the convergence of a geometric series when $|r| < 1$, and the limit of its sum as n gets larger and larger.

We discuss the calculation of simple and compound interest, and the concept of the time value of money, and its application to determine the present value of a payment to be made in the future. We then discuss annuities, which are a series of payments made at regular intervals over a period of time, and we determine the present value of an annuity.

© Springer International Publishing Switzerland 2016
G. O'Regan, *Guide to Discrete Mathematics*, Texts in Computer Science,
DOI 10.1007/978-3-319-44561-8_5

We consider the counting principle where one operation has m possible outcomes and a second operation has n possible outcomes. We determine that the total number of outcomes after performing the first operation is followed by the second operation to be $m \times n$. A permutation is an arrangement of a given number of objects, by taking some or all of them at a time. The order of the arrangement is important, as the arrangement 'abc' is different from 'cba'. A combination is a selection of a number of objects in any order, where the order of the selection is unimportant. That is, the selection 'abc' is the same as the selection 'cba'.

5.2 Sequences and Series

A sequence a_1, a_2, ... a_n ... is any succession of terms (usually numbers), and we discussed the Fibonacci sequence earlier in Chap. 4. Each term in the Fibonacci sequence (apart from the first two terms) is obtained from the sum of the previous two terms in the sequence.

$$1, 1, 2, 3, 5, 8, 13, 21, \ldots.$$

A sequence may be finite (with a fixed number of terms) or infinite. The Fibonacci sequence is infinite whereas the sequence 2, 4, 6, 8, 10 is finite. We distinguish between convergent and divergent sequences, where a *convergent* sequence approaches a certain value as n gets larger and larger (technically we say that $\lim_{n \to \infty} a_n$ exists (i.e. the limit of a_n exists). Otherwise, the sequence is said to be *divergent*.

Often, there is a mathematical expression for the nth term in a sequence (e.g. for the sequence of even integers 2, 4, 6, 8, ... the general expression for a_n is given by $a_n = 2n$). Clearly, the sequence of the even integers is divergent, as it does not approach a particular value and as n gets larger and larger. Consider the following sequence:

$$1, -1, 1, -1, 1, -1$$

Then this sequence is divergent since it does not approach a certain value, as n gets larger and larger and since it continues to alternate between 1 and -1. The formula for the nth term in the sequence may be given by

$$(-1)^{n.+1}$$

The sequence 1, 1/2, 1/3. 1/4, ...1/n ... is convergent and it converges to 0. The nth term in the sequence is given by $1/n$, and as n gets larger and larger it gets closer and closer to 0.

A series is the sum of the terms in a sequence, and the sum of the first n terms of the sequence $a_1, a_2, \ldots, a_n \ldots$ is given by $a_1 + a_2 + \cdots + a_n$ which is denoted by

$$\sum_{k=1}^{n} a_k$$

A series is convergent if its sum approaches a certain value S as n gets larger and larger, and this is written formally as

$$\lim_{n \to \infty} \sum_{k=1}^{n} a_k = S$$

Otherwise, the series is said to be divergent.

5.3 Arithmetic and Geometric Sequences

Consider the sequence 1, 4, 7, 10,... where each term is obtained from the previous term by adding the constant value 3. This is an example of an arithmetic sequence, and there is a difference of 3 between any term and the previous one. The general form of a term in this sequence is $a_n = 3n - 2$.

The general form of an *arithmetic sequence* is given by

$$a, a+d, a+2d, a+3d, \cdots a+(n-1)d, \cdots$$

The value a is the initial term in the sequence, and the value d is the constant difference between a term and its successor. For the sequence, 1, 4, 7, ..., we have $a = 1$ and $d = 3$, and the sequence is not convergent. In fact, all arithmetic sequences (apart from the constant sequence $a, a, \ldots a$ which converges to a) are divergent.

Consider, the sequence 1, 3, 9, 27, 81,... where each term is achieved from the previous term by multiplying by the constant value 3. This is an example of a geometric sequence, and the general form of a geometric sequence is given by

$$a, ar, ar^2, ar^3, \cdots, ar^{n-1}$$

The first term in the geometric sequence is a and r is the common ratio. Each term is obtained from the previous one by multiplying by the common ratio r. For the sequence 1, 3, 9, 27 the value of a is 1 and r is 3.

A geometric sequence is convergent if $r < 1$, and for this case it converges to 0. It is also convergent if $r = 1$, as for this case it is simply the constant sequence $a, a, a, \ldots$, which converges to a. For the case where $r > 1$ the sequence is divergent.

5.4 Arithmetic and Geometric Series

An arithmetic series is the sum of the terms in an arithmetic sequence, and a geometric sequence is the sum of the terms in a geometric sequence. It is possible to derive a simple formula for the sum of the first n terms in an arithmetic and geometric series.

Arithmetic Series
We write the series two ways: first the normal left to right addition, and then the reverse, and then we add both series together.

$$Sn = a \quad + (a+d) + (a+2d) + (a+3d) + \ldots + (a+(n-1)d)$$
$$Sn = a + (n-1)d + a + (n-2)d + \ldots + \quad + (a+d) + a$$

$$2Sn = [2a+(n-1)d] + [2a+(n-1)d] + \ldots + [2a+(n-1)d] \qquad (n\,\text{times})$$
$$2Sn = n \times [2a+(n-1)d]$$

Therefore, we conclude that

$$S_n = \frac{n}{2}[2a+(n-1)d]$$

Example (**Arithmetic Series**) Find the sum of the first n terms in the following arithmetic series 1, 3, 5, 7, 9.

Solution
Clearly, $a = 1$ and $d = 2$. Therefore, applying the formula we get

$$S_n = \frac{n}{2}[2.1+(n-1)2] = \frac{2n^2}{2} = n^2$$

Geometric Series
For a geometric series we have

$$S_n = a + ar + ar^2 + ar^3 + \cdots + ar^{n-1}$$
$$\Rightarrow rS_n = \quad ar + ar^2 + ar^3 + \cdots + ar^{n-1} + ar^n$$

$$\Rightarrow rS_n - S_n = ar^n - a = a(r^n - 1)$$
$$\Rightarrow (r-1)S_n = a(r^n - 1)$$

Therefore, we conclude that (where $r \neq 1$) that

$$S_n = a\frac{(r^n - 1)}{r - 1} = a\frac{(1 - r^n)}{1 - r}$$

The case of when $r = 1$ corresponds to the arithmetic series $a + a + \cdots + a$, and the sum of this series is simply na. The geometric series converges when $|r| < 1$ as $r^n \to 0$ as $n \to \infty$, and so

$$S_n \to \frac{a}{1 - r} \qquad as\ n \to \infty$$

Example (**Geometric Series**) Find the sum of the first n terms in the following geometric series 1, 1/2, 1/4, 1/8, ... What is the sum of the series?

Solution
Clearly, $a = 1$ and $r = {}^1/_2$. Therefore, applying the formula we get

$$S_n = 1\frac{(1 - 1/2^n)}{1 - 1/2} = \frac{(1 - 1/2^n)}{1 - 1/2} = 2(1 - 1/2^n)$$

The sum of the series is the limit of the sum of the first n terms as n approaches infinity. This is given by

$$\lim_{n \to \infty} S_n = \lim_{n \to \infty} 2(1 - 1/2^n) = 2$$

5.5 Simple and Compound Interest

Savers receive interest on placing deposits at the bank for a period of time, whereas lenders pay interests on their loans to the bank. We distinguish between simple and compound interest, where *simple interest* is always calculated on the original principal, whereas for *compound interest*, the interest is added to the principal sum, so that interest is also earned on the added interest for the next compounding period.

For example, if Euro 1000 is placed on deposit at a bank with an interest rate of 10 % per annum for 2 years, it would earn a total of Euro 200 in simple interest. The interest amount is calculated by

$$\frac{1000 * 10 * 2}{100} = \text{Euro } 200$$

The general formula for calculating simple interest on principal P, at a rate of interest I, and for time T (in years:) is

$$A = \frac{P \times I \times T}{100}$$

The calculation of compound interest is more complicated as may be seen from the following example:

Example (**Compound Interest**) Calculate the interest earned and what the new principal will be on Euro 1000, which is placed on deposit at a bank, with an interest rate of 10 % per annum (compound) for 3 years.

Solution
At the end of year 1, Euro 100 of interest is earned, and this is capitalized making the new principal at the start of year 2 Euro 1100. At the end of year 2, Euro 110 is earned in interest, making the new principal at the start of year 3 Euro 1210. Finally, at the end of year 3 a further Euro 121 is earned in interest, and so the new principal is Euro 1331 and the total interest earned for the 3 years is the sum of the interest earned for each year (i.e. Euro 331). This may be seen from Table 5.1.

The new principal each year is given by the geometric sequence with $a = 1000$ and $r = 10/100 = 0.1$.

$$1000, \ 1000(1.1), \ 1000(1.1)^2, \ 1000(1.1)^3, \ldots \ldots$$

In general, if a principal amount P is invested for T years at a rate R of interest (r is expressed as a proportion, i.e. $r = R/100$) then it will amount to

$$A = P(1 + r)^T$$

For our example above, $A = 1000$, $T = 3$ and $r = 0.1$. Therefore,

$$A = 1000(1.1)^3$$
$$= 1331 \,(\text{as before})$$

There are variants of the compound interest formula to cover situations where there are m-compounding periods per year, and so the reader may consult the available texts.

Table 5.1 Calculation of compound interest

Year	Principal	Interest earned
1	1000	100
2	1100	110
3	1210	121

5.6 Time Value of Money and Annuities

The time value of money discusses the concept that the earlier that cash is received the greater value it has to the recipient. Similarly, the later that a cash payment is made, the lower its value to the recipient, and the lower its cost to the payer.

This is clear if we consider the example of a person who receives $1000 now and a person who receives $1000 five years from now. The person who receives $1000 now is able to invest it and to receive annual interest on the principal, whereas the other person who receives $1000 in 5 years earns no interest during the period. Further, the inflation during the period means that the purchasing power of $1000 is less in 5 years time and is less than it is today.

We presented the general formula for what the future value of a principal P invested for n years at a compound rate r of interest as $A = P (1 + r)^n$

We can determine the present value of an amount A received in n years time at a discount rate r by

$$P = \frac{A}{(1+r)^n}$$

An annuity is a series of equal cash payments made at regular intervals over a period of time, and so there is a need to calculate the present value of the series of payments made over the period. The actual method of calculation is clear from Table 5.2.

Example (Annuities) Calculate the present value of a series of payments of $1000 (made at the end of each year) with the payments made for 5 years at a discount rate of 10 %.

Solution
The regular payment A is 1000, the rate r is 0.1 and $n = 5$. The present value of the payment received at the end of year of year 1 is $1000/1.1 = 909.91$; at the end of year 2 it is $1000/(1.1)^2 = 826.45$; and so on. The total present value of the payments over the 5 years is given by the sum of the individual present values and is $3791 (Table 5.2).

We may easily derive a formula for the present value of a series of payments A over a period of n years at a discount rate of r as follows: Clearly, the present value is given by

Table 5.2 Calculation of present value of annuity

Year	Amount	Present value ($r = 0.1$)
1	1000	909.91
2	1000	826.44
3	1000	751.31
4	1000	683.01
5	1000	620.92

$$\frac{A}{(1+r)} + \frac{A}{(1+r)^2} + \cdots + \frac{A}{(1+r)^n}$$

This is a geometric series where the constant ratio is $\frac{1}{1+r}$ and the present value of the annuity is given by its sum

$$PV = \frac{A}{r}[1 - \frac{1}{(1+r)^n}]$$

$$PV = \frac{1000}{0.1}[1 - \frac{1}{(1.1)^5}]$$

For the example above we apply the formula and get

$$= 10{,}000(0.3791)$$
$$= \$3791$$

5.7   Permutations and Combinations

A permutation is an arrangement of a given number of objects, by taking some or all of them at a time. A combination is a selection of a number of objects where the order of the selection is unimportant. Permutations and combinations are defined in terms of the factorial function, which was defined in Chap. 4. Recall that $n! = n(n-1)\cdots 3.2.1$.

Principles of Counting

(a) Suppose one operation has m possible outcomes and a second operation has n possible outcomes, then the total number of possible outcomes when performing the first operation followed by the second operation is $m \times n$. (**Product Rule**).

(b) Suppose one operation has m possible outcomes and a second operation has n possible outcomes then the possible outcomes of the first operation **or** the second operation is given by $m + n$. (**Sum Rule**)

Example (**Counting Principle (a)**) Suppose a dice is thrown and a coin is then tossed. How many different outcomes are there and what are they?

Solution
There are six possible outcomes from a throw of the dice: 1, 2, 3, 4, 5 or 6, and there are two possible outcomes from the toss of a coin: H or T. Therefore, the total number of outcomes is determined from the product rule as $6 \times 2 = 12$. The outcomes are given by

$(1, H), (2, H), (3, H), (4, H), (5, H), (6, H), (1, T), (2, T), (3, T), (4, T), (5, T), (6, T)$

Example (**Counting Principle (b)**) Suppose a dice is thrown and if the number is even a coin is tossed and if it is odd then there is a second throw of the dice. How many different outcomes are there?

Solution
There are two experiments involved with the first experiment involving an even number and a toss of a coin. There are three possible outcomes that result in an even number and two outcomes from the toss of a coin. Therefore, there are $3 \times 2 = 6$ outcomes from the first experiment.

The second experiment involves an odd number from the throw of a dice and the further throw of the dice. There are three possible outcomes that result in an odd number and six outcomes from the throw of a dice. Therefore, there are $3 \times 6 = 18$ outcomes from the second experiment.

Finally, there are six outcomes from the first experiment and 18 outcomes from the second experiment, and so from the sum rule there are a total of $6 + 18 = 24$ outcomes.

Pigeonhole Principle
The pigeonhole principle states that if n items are placed into m containers (with $n > m$) then at least one container must contain more than one item.

Examples (**Pigeonhole Principle**)

(a) Suppose there is a group of 367 people then there must be at least two people with the same birthday.
    This is clear as there are 365 days in a year (with 366 days in a leap year), and so as there are at most 366 possible birthdays in a year. The group size is 367 people, and so there must be at least two people with the same birthday.
(b) Suppose that a class of 102 students are assessed in an examination (the outcome from the exam is a mark between 0 and 100). Then, there are at least two students who receive the same mark.
    This is clear as there are 101 possible outcomes from the test (as the mark that a student may achieve is between is between 0 and 100), and as there are 102 students in the class and 101 possible outcomes from the test, then there must be at least two students who receive the same mark.

Permutations
A permutation is an arrangement of a number of objects in a definite order.

Consider the three letters A, B and C. If these letters are written in a row then there are six possible arrangements:

ABC or ACB or BAC or BCA or CAB or CBA

There is a choice of three letters for the first place, then there is a choice of two letters for the second place, and there is only one choice for the third place. Therefore, there are $3 \times 2 \times 1 = 6$ arrangements.

If there are n different objects to arrange then the total number of arrangements (permutations) of n objects is given by $n! = n(n - 1)(n - 2) \ldots 3.2.1$.

Consider the four letters A, B, C and D. How many arrangements (taking two letters at a time with no repetition) of these letters can be made?

There are four choices for the first letter and three choices for the second letter, and so there are 12 possible arrangements. These are given by

AB or AC or AD or BA or BC or BD or CA or CB or CD or DA or DB or DC

The total number of arrangements of n different objects taking r at a time ($r \le n$) is given by $^nP_r = n(n - 1)(n - 2) \ldots (n - r + 1)$. It may also be written as

$$nP_r = \frac{n!}{(n - r)!}$$

Example (**Permutations**) Suppose A, B, C, D, E and F are six students. How many ways can they be seated in a row if

(i)   There is no restriction on the seating.
(ii) A and B must sit next to one another
(iii) A and B must not sit next to one another

Solution
For unrestricted seating the number of arrangements is given by $6.5.4.3.2.1 = 6! = 720$.

For the case where A and B must be seated next to one another, then consider A and B as one person, and then the five people may be arranged in $5! = 120$ ways. There are $2! = 2$ ways in which AB may be arranged, and so there are $2! \times 5! = 240$ arrangements.

AB	C	D	E	F

For the case where A and B must not be seated next to one another, then this is given by the difference between the total number of arrangements and the number of arrangements with A and B together: i.e. $720 - 240 = 480$.

Combinations

A combination is a selection of a number of objects in any order, and the order of the selection is unimportant, in that both AB and BA represent the same selection. The total number of arrangements of n different objects taking r at a time is given by ^nP_r, and we can determine that the number of ways that r objects can be selected from n different objects from this, as each selection may be permuted $r!$ times, and so the total number of selections is $r! \times$ total number of combinations. That is, $^nP_r = r! \times {}^nC_r$, and we may also write this as

$$\binom{n}{r} = \frac{n!}{r!(n-r)!} = \frac{n(n-1)\ldots(n-r+1)}{r!}$$

It is clear from the definition that

$$\binom{n}{r} = \binom{n}{n-r}$$

Example 1 (**Combinations**) How many ways are there to choose a team of 11 players from a panel of 15 players?

Solution

Clearly, the number of ways is given by $\binom{15}{11} = \binom{15}{4}$

That is, $15.14.13.12/4.3.2.1 = 1365$.

Example 2 (**Combinations**) How many ways can a committee of four people be chosen from a panel of 10 people where

(i)   There is no restriction on membership of the panel.
(ii) A certain person must be a member.
(iii) A certain person must not be a member.

Solution

For (i) with no restrictions on membership the number of selections of a committee of four people from a panel of 10 people is given by $\binom{10}{4} = 210$

For (ii) where one person must be a member of the committee then this involves choosing three people from a panel of nine people and is given by $\binom{9}{3} = 84$

For (iii) where one person must not be a member of the committee then this involves choosing four people from a panel of nine people, and is given by $\binom{9}{4} = 126$

5.8   Review Questions

1. Determine the formula for the general term and the sum of the following arithmetic sequence:

$$1, 4, 7, 10, \ldots.$$

2. Write down the formula for the nth term in the following sequence:

$$1/4, 1/12, 1/36, 1/108, \ldots.$$

3. Find the sum of the following geometric sequence:

$$1/3, 1/6, 1/12, 1/24, \ldots.$$

4. How many years will it take a principal of $5000 to exceed $10,000 at a constant annual growth rate of 6 % compound interest?
5. What is the present value of $5000 to be receive in 5 years time at a discount rate of 7 %?
6. Determine the present value of a 20-year annuity of an annual payment of $5000 per year at a discount rate of 5 %.
7. How many different five-digit numbers can be formed from the digits 1, 2, 3, 4, 5 where

   (i)   No restrictions on digits and repetitions allowed.
   (ii) The number is odd and no repetitions are allowed.
   (iii) The number is even and repetitions are allowed.

8. (i)   How many ways can a group of five people be selected from nine people?
   (ii) How many ways can a group be selected if two particular people are always included?
   (iii) How many ways can a group be selected if two particular people are always excluded?

5.9 Summary

This chapter provided a brief introduction to sequences and series, including arithmetic and geometric sequences, and arithmetic series and geometric series. We derived formulae for the sum of an arithmetic series and geometric series, and we discussed the convergence of a geometric series when $|r| < 1$.

We discussed the calculation of simple and compound interest, and the concept of the time value of money, and its application to determine the present value of a payment to be made in the future. We discussed annuities, which are a series of payments made at regular intervals over a period of time, and we calculated the present value of an annuity.

We considered counting principles including the product and sum rules. The product rule is concerned with where one operation has m possible outcomes and a second operation has n possible outcomes then the total number of possible outcomes when performing the first operation followed by the second operation is $m \times n$.

We discussed the pigeonhole principle, which states that if n items are placed into m containers (with $n > m$) then at least one container must contain more than one item. We discussed permutations and combinations where permutations are an arrangement of a given number of objects, by taking some or all of them at a time. A combination is a selection of a number of objects in any order, and the order of the selection is unimportant.

Algebra

6

Key Topics

Simultaneous equations
Quadratic equations
Polynomials
Indices
Logs
Abstract Algebra
Groups
Rings
Fields
Vector Spaces

6.1 Introduction

Algebra is the branch of mathematics that uses letters in the place of numbers, where the letters stand for variables or constants that are used in mathematical expressions. Algebra is the study of such mathematical symbols and the rules for manipulating them, and it is a powerful tool for problem solving in science and engineering.

The origins of algebra are in work done by Islamic mathematicians during the Golden age in Islamic civilization, and the word '*algebra*' comes from the Arabic '*al-jabr*', which appears as part of the title of a book by the Islamic mathematician,

© Springer International Publishing Switzerland 2016
G. O'Regan, *Guide to Discrete Mathematics*, Texts in Computer Science,
DOI 10.1007/978-3-319-44561-8_6

Al Khwarizmi, in the ninth century A.D. The third century A.D. Hellenistic mathematician, Diophantus, also did early work on algebra.

Algebra covers many areas such as elementary algebra, linear algebra and abstract algebra. Elementary algebra includes the study of symbols and rules for manipulating them to form valid mathematical expressions; simultaneous equations; quadratic equations; polynomials; indices and logarithms. Linear algebra is concerned with the solution of a set of linear equations, and the study of matrices (see Chap. 8) and vectors. Abstract algebra is concerned with the study of abstract algebraic structures such as monoids, groups, rings, integral domains, fields and vector spaces.

6.2   Simple and Simultaneous Equations

A simple equation is an equation with one unknown, and the unknown may be on both the left-hand side and right-hand side of the equation. The method of solving such equations is to bring the unknowns to one side of the equation, and the values to the other side.

Simultaneous equations are equations with two (or more) unknowns. There are a number of methods to find a solution to two simultaneous equations such as elimination, substitution and graphical techniques. The solution of n linear equations with n unknowns may be done using Gaussian elimination and matrix theory (see Chap. 8).

Example (**Simple Equation**) Solve the simple equation $4 - 3x = 2x - 11$

Solution (**Simple Equation**)

$$4 - 3x = 2x - 11$$
$$4 - (-11) = 2x - (3x)$$
$$4 + 11 = 2x + 3x$$
$$15 = 5x$$
$$3 = x$$

Example (**Simultaneous Equation—Substitution Method**) Solve the following simultaneous equations by the method of substitution.

$$x + 2y = -1$$
$$4x - 3y = 18$$

Solution

(Simultaneous Equation—Substitution Method) The method of substitution involves expressing x in terms of y and substituting it in the other equation (or vice versa expressing y in terms of x and substituting it in the other equation). For this example, we use the first equation to *express* x in terms of y.

$$x + 2y = -1$$
$$x = -1 - 2y$$

We then substitute for x $(-1 - 2y)$ in the second equation, and we get a simple equation involving just the unknown y.

$$4(-1 - 2y) - 3y = 18$$
$$\Rightarrow -4 - 8y - 3y = 18$$
$$\Rightarrow -11y = 18 + 4$$
$$\Rightarrow -11y = 22$$
$$\Rightarrow y = -2$$

We then obtain the value of x from the substitution

$$x = -1 - 2y$$
$$\Rightarrow x = -1 - 2(-2)$$
$$\Rightarrow x = -1 + 4$$
$$\Rightarrow x = 3$$

We can then verify that our solution is correct by checking our answer for both equations.

$$3 + 2(-2) = -1 \quad ✔$$
$$4(3) - 3(-2) = 18 \quad ✔$$

Example (**Simultaneous Equation—Method of Elimination**) Solve the following simultaneous equations by the method of elimination.

$$3x + 4y = 5$$
$$2x - 5y = -12$$

Solution

(Simultaneous Equation—Method of Elimination) The approach is to manipulate both equations so that we may eliminate either x or y, and so reduce to a simple equation with just x or y. For this example, we are going to eliminate x, and so we multiply equation (1) by 2 and equation (2) by -3 and this yields two equations with the opposite coefficient of x.

$$6x + 8y = 10$$
$$-6x + 15y = 36$$
$$- - - - - - - -$$
$$0x + 23y = 46$$
$$y = 2$$

We then add both equations together and conclude that $y = 2$. We then determine the value of x by replacing y with 2 in equation (1).

$$3x + 4(2) = 5$$
$$3x + 8 = 5$$
$$3x = 5 - 8$$
$$3x = -3$$
$$x = -1$$

We can then verify that our solution is correct as before by checking our answer for both equations

Example (**Simultaneous Equation—Graphical Techniques**) Find the solution to the following simultaneous equations using graphical techniques:

$$x + 2y = -1$$
$$4x - 3y = 18$$

Solution
(**Simultaneous Equation—Graphical Techniques**) Each simultaneous equation represents a straight line, and so the solution to the two simultaneous equations is the point of intersection of both lines (if there is such a point). Therefore, the solution involves drawing each line and finding the point of intersection of both lines (Fig. 6.1).

First we find two points on line 1: e.g. (0, −0.5) and (−1, 0) are on line 1, since when $x = 0$ we have $2y = -1$ and so $y = -0.5$. Similarly, when $y = 0$ we have $x = -1$. Next we find two points on line 2 in a similar way: e.g. when x is 0 y is –6 and when y is 0 we have $x = 4.5$ and so the points (0–6) and (4.5, 0) are on line 2.

Fig. 6.1 Graphical solution
to simultaneous equations

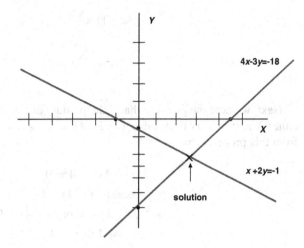

We then draw the X axis and the Y axis, draw the scales on the axes, label the axes, plot the points and draw both lines. Finally, we find the point of intersection of both lines (if there is such a point), and this is our solution to the simultaneous equations.

For this example, there is a point of intersection for the lines, and so we determine the x and y coordinate and the solution is then given by $x = 3$ and $y = -2$. The solution using graphical techniques requires care (as inaccuracies may be introduced from poor drawing) and graph paper is required for accuracy.

6.3   Quadratic Equations

A quadratic equation is an equation of the form $ax^2 + bx + c = 0$, and solving the quadratic equation is concerned with finding the unknown value x (roots of the quadratic equation). There are several techniques to solve quadratic equations such as factorization; completing the square; the quadratic formula; and graphical techniques.

Example (**Quadratic Equations—Factorization**) Solve the quadratic equation $3x^2 - 11x - 4 = 0$ by factorization.

Solution
(**Quadratic Equations—Factorization**) The approach taken is to find the factors of the quadratic equation. Sometimes this is easy, but often other techniques will need to be employed. For the above quadratic equation we note immediately that its factors are $(3x + 1)(x - 4)$ since

$$(3x + 1)(x - 4)$$
$$= 3x^2 - 12x + x - 4$$
$$= 3x^2 - 11x - 4$$

Next, we note the property that if the product of two numbers A and B is 0 then either A is 0 or B is 0. Another words, $AB = 0 \Rightarrow A = 0$ or $B = 0$. We conclude from this property that as

$$3x^2 - 11x - 4 = 0$$
$$\Rightarrow \quad (3x + 1)(x - 4) = 0$$
$$\Rightarrow \quad (3x + 1) = 0 \text{ or } (x - 4) = 0$$
$$\Rightarrow \quad 3x = -1 \text{ or } x = 4$$
$$\Rightarrow \quad x = -0.33 \text{ or } x = 4$$

Therefore, the solution (or roots) of the quadratic equation $3x^2 - 11x - 4 = 0$ are $x = -0.33$ or $x = 4$.

Example (**Quadratic Equations—Completing the Square**) Solve the quadratic equation $2x^2 + 5x - 3 = 0$ by completing the square.

Solution
(**Quadratic Equations—Completing the Square**) First we convert the quadratic equation to an equivalent quadratic with a unary coefficient of x^2. This involves division by 2. Next, we examine the coefficient of x (in this case 5/2) and we add the square of half the coefficient of x to both sides. This allows us to complete the square, and we then to take the square root of both sides. Finally, we solve for x.

$$2x^2 + 5x - 3 = 0$$
$$\Rightarrow \quad x^2 + 5/2x - 3/2 = 0$$
$$\Rightarrow \quad x^2 + 5/2x = 3/2$$
$$\Rightarrow \quad x^2 + 5/2x + (5/4)^2 = 3/2 + (5/4)^2$$
$$\Rightarrow \quad (x + 5/4)^2 = 3/2 + (25/16)$$
$$\Rightarrow \quad (x + 5/4)^2 = 29/16 + (25/16)$$
$$\Rightarrow \quad (x + 5/4)^2 = 49/16$$
$$\Rightarrow \quad (x + 5/4) = \pm 7/4$$
$$\Rightarrow \quad x = -5/4 \pm 7/4$$
$$\Rightarrow \quad x = -5/4 - 7/4 \text{ or } x = -5/4 + 7/4$$
$$\Rightarrow \quad x = -12/4 \text{ or } x = 2/4$$
$$\Rightarrow \quad x = -3 \text{ or } x = 0.5$$

Example 1 (**Quadratic Equations—Quadratic Formula**) Establish the quadratic formula for solving quadratic equations.

Solution
(**Quadratic Equations—Quadratic Formula**) We complete the square and the result will follow.

$$ax^2 + bx + c = 0$$
$$\Rightarrow x^2 + b/ax + c/a = 0$$
$$\Rightarrow x^2 + b/ax = -c/a$$
$$\Rightarrow x^2 + b/ax + (b/2a)^2 = -c/a + (b/2a)^2$$
$$\Rightarrow (x + b/2a)^2 = -c/a + (b/2a)^2$$
$$\Rightarrow (x + b/2a)^2 = \frac{-4ac}{4a^2} + \frac{b^2}{4a^2}$$
$$\Rightarrow (x + b/2a)^2 = \frac{b^2 - 4ac}{4a^2}$$
$$\Rightarrow (x + b/2a) = \pm \frac{\sqrt{b^2 - 4ac}}{2a}$$
$$\Rightarrow x = \frac{-b \pm \sqrt{b^2 - 4ac}}{2a}$$

Example 2 (**Quadratic Equations—Quadratic Formula**) Solve the quadratic equation $2x^2 + 5x - 3 = 0$ using the quadratic formula.

Solution
(**Quadratic Equations—Quadratic Formula**) For this example $a = 2$; $b = 5$; and $c = -3$, and we put these values into the quadratic formula.

$$x = \frac{-5 \pm \sqrt{5^2 - 4.2.(-3)}}{2.2} = \frac{-5 \pm \sqrt{25 + 24}}{4}$$
$$x = \frac{-5 \pm \sqrt{49}}{4} = \frac{-5 \pm 7}{4}$$
$$x = 0.5 \text{ or } x = -3.$$

Example (**Quadratic Equations—Graphical Techniques**) Solve the quadratic equation $2x^2 - x - 6 = 0$ using graphical techniques given that the roots of the quadratic equation lie between $x = -3$ and $x = 3$

Solution
(**Quadratic Equations—Graphical Techniques**) The approach is first to create a table of values (Table 6.1) for the curve $y = 2x^2 - x - 6$, and to draw the X and

Fig. 6.2 Graphical solution
to quadratic equation

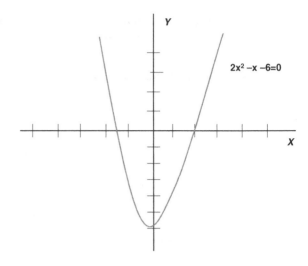

$2x^2 - x - 6 = 0$

Table 6.1 Table of values
for quadratic equation

x	−3	−2	−1	0	1	2	3
$y = 2x^2 - x - 6$	15	4	−3	−6	−5	0	9

Y axis and scales, and then to plot the points from the table of values, and to join the points together to form the curve (Fig. 6.2).

The graphical solution is to the quadratic equation is then given by the points where the curve intersects the X axis (i.e. $y = 0$ on the X axis). There may be no solution (i.e. the curve does not intersect the X axis), one solution (a double root), or two solutions.

The graph for the curve $y = 2x^2 - x - 6$ is given below, and so the points where the curve intersects the X axis are determined. We note from the graph that the curve intersects the X axis at two distinct points, and we see from the graph that the roots of the quadratic equation are given by $x = -1.5$ and $x = 2$.

The solution to quadratic equations using graphical techniques requires care (as with the solution to simultaneous equations using graphical techniques), and graph paper is required for accuracy.

6.4 Indices and Logarithms

The product $a.a.a.....a$ (n times) is denoted by a^n, and the number n is the index of a. The following are properties of indices:

$$a^0 = 1$$
$$a^{m+n} = a^m.a^n$$
$$a^{mn} = (a^m)^n$$
$$a^{-n} = \frac{1}{a^n}$$
$$a^{\frac{1}{n}} = \sqrt[n]{a}$$

Logarithms are closely related to indices, and if the number b can be written in the form $b = a^x$, then we say that log to the base a of b is x: i.e. $\log_a b = x \Leftrightarrow a^x = b$. Clearly, $\log_{10} 100 = 2$ since $10^2 = 100$. The following are properties of logarithms:

$$\log_a AB = \log_a A + \log_a B$$
$$\log_a A^n = n\log_a A$$
$$\log \frac{A}{B} = \log A - \log B$$

We will prove the first property of logarithms. Suppose $\log_a A = x$ and $\log_a B = y$. Then $A = a^x$ and $B = a^y$ and so $AB = a^x a^y = a^{x+y}$ and so $\log_a AB = x + y = \log_a A + \log_a B$.

The law of logarithms may be used to solve certain equations involving powers (called indicial equations). We illustrate this by an example

Example (**Indicial Equations**) Solve the equation $2^x = 3$, correct to 4 significant places.

Solution
(Indicial Equations)

$$2^x = 3$$
$$\Rightarrow \log_{10} 2^x = \log_{10} 3$$
$$\Rightarrow x\log_{10} 2 = \log_{10} 3$$
$$\Rightarrow x = \frac{\log_{10} 3}{\log_{10} 2}$$
$$= \frac{0.4771}{0.3010}$$
$$\Rightarrow x = 1.585$$

6.5 Horner's Method for Polynomials

Horner's Method is a computationally efficient way to evaluate a polynomial function. It is named after William Horner who was a nineteenth century British mathematician and schoolmaster. Chinese mathematicians were familiar with the method in the third century A.D.

The normal method for the evaluation of a polynomial involves computing exponentials, and this is computationally expensive. Horner's method has the advantage that fewer calculations are required, and it eliminates all exponentials using nested multiplication and addition. It also provides a computationally efficient way to determine the derivative of the polynomial.

Horner's Method and Algorithm
Consider a polynomial $P(x)$ of degree n defined by

$$P(x) = a_n x^n + a_{n-1} x^{n-1} + a_{n-2} x^{n-2} + \cdots + a_1 x + a_0$$

The Horner method to evaluate $P(x_0)$ essentially involves writing $P(x)$ as

$$P(x) = (((a_n x + a_{n-1}) x + a_{n-2}) x + \cdots + a_1) x + a_0$$

The computation of $P(x_0)$ involves defining a set of coefficients b_k such that

$$b_n = a_n$$
$$b_{n-1} = a_{n-1} + b_n x_0$$
$$\ldots\ldots$$
$$b_k = a_k + b_{k+1} x_0$$
$$\ldots\ldots$$
$$b_1 = a_1 + b_2 x_0$$
$$b_0 = a_0 + b_1 x_0$$

Then the computation of $P(x_0)$ is given by

$$P(x_0) = b_0$$

Further, if $Q(x) = b_n x^{n-1} + b_{n-1} x^{n-2} + b_{n-2} x^{n-3} + \cdots + b_1$ then it is easy to verify that

$$P(x) = (x - x_0) Q(x) + b_0$$

This also allows the derivative of $P(x)$ to be easily computed for x_0 since

$$P'(x) = Q(x) + (x - x_0)Q'(x)$$
$$P'(x_0) = Q(x_0)$$

Algorithm (*To evaluate polynomial and its derivative*)

(i)   Initialize y to a_n and z to a_n (Compute b_n for P and b_{n-1} for Q)
(ii)   For each j from $n - 1$, $n - 2$ to 1 compute b_j for P and b_{j-1} for Q by
      Set y to $x_0 y + a_j$ (i.e. b_j for P) and z to $x_0 z + y$ (i.e. b_{j-1} for Q)
(iii) Compute b_0 by setting y to $x_0 y + a_0$

Then $P(x_0) = y$ and $P'(x_0) = z$.

6.6   Abstract Algebra

One of the important features of modern mathematics is the power of the abstract approach. This has opened up whole new areas of mathematics, and it has led to a large body of new results and problems. The term '*abstract*' is subjective, as what is abstract to one person may be quite concrete to another. We shall introduce some important algebraic structures in this section including monoids, groups, rings, fields, and vector spaces.

6.6.1   Monoids and Groups

A non-empty set M together with a binary operation '*' is called a *monoid* if for all elements $a, b, c \in M$ the following properties hold

(1) $a * b \in M$	(Closure property)
(2) $a * (b * c) = (a * b) * c$	(Associative property)
(3) $\exists u \in M$ such that: $a * u = u * a = a$ $(\forall a \in M)$	(Identity element)

A monoid is commutative if $a * b = b * a$ for all $a, b \in M$. A *semi-group* $(M, *)$ is a set with a binary operation '*' such that the closure and associativity properties hold (but it may not have an identity element).

Example 6.1 (**Monoids**)

(i) The set of sequences $\Sigma*$ under concatenation with the empty sequence Λ the identity element.

(ii) The set of integers under addition forms an infinite monoid in which 0 is the identity element.

A non-empty set G together with a binary operation '$*$' is called a *group* if for all elements $a,b,c \in G$ the following properties hold

(1) $a * b \in G$	(Closure property)
(2) $a * (b * c) = (a * b) * c$	(Associative property)
(3) $\exists e \in G$ such that: $a * e = e * a = a$ ($\forall a \in G$)	(Identity element)
(4) For every $a \in G$, $\exists a^{-1} \in G$, such that: $a * a^{-1} - a^{-1} * a = e$	(Inverse element)

The identity element is unique, and the inverse a^{-1} of an element a is unique (see Exercise 5). A *commutative group* has the additional property that $a * b = b * a$ for all $a, b \in G$. The order of a group G is the number of elements in G, and is denoted by $o(G)$. If the order of G is finite then G is said to be a finite group.

Example 6.1 (**Groups**)

(i) The set of integers under addition $(\mathbb{Z}, +)$ forms an infinite group in which 0 is the identity element.

(ii) The set of integer 2×2 matrices under addition, where the identity element is $\begin{pmatrix} 0 & 0 \\ 0 & 0 \end{pmatrix}$

(iii) The set of integers under multiplication $(\mathbb{Z}, \times)$ forms an infinite monoid with 1 as the identity element.

A *cyclic group* is a group where all elements $g \in G$ are obtained from the powers a^i of one element $a \in G$, with $a^0 = e$. The element 'a' is termed the generator of the cyclic group G. A finite cyclic group with n elements is of the form $\{a^0, a^1, a^2, \ldots, a^{n-1}\}$.

A non-empty subset H of a group G is said to be a *subgroup* of G if for all $a, b \in H$ then $a * b \in H$, and for any $a \in H$ then $a^{-1} \in H$. A subgroup N is termed a *normal subgroup* of G if $gng^{-1} \in G$ for all $g \in G$ and all $n \in N$. Further, if G is a group and N is a normal subgroup of G, then the *quotient group* G/N may be formed.

Lagrange's theorem states the relationship between the order of a subgroup H of G, and the order of G. The theorem states that if G is a finite group, and H is a subgroup of G, then $o(H)$ is a divisor of $o(G)$.

We may also define mapping between similar algebraic structures termed *homomorphism*, and these mapping preserve structure. If the homomorphism is one to one and onto it is termed an *isomorphism*, which means that the two structures are identical in some sense (apart from a relabelling of elements).

6.6.2 Rings

A *ring* is a non-empty set R together with two binary operations '+' and '×' where $(R, +)$ is a commutative group; $(R, \times)$ is a semi-group; and the left and right distributive laws hold. Specifically, for all elements a, b, $c \in R$ the following properties hold:

(1) $a + b \in R$	(Closure property)
(2) $a + (b + c) = (a + b) + c$	(Associative property)
(3) $\exists 0 \in R$ such that $\forall a \in R$: $a + 0 = 0 + a = a$	(Identity element)
(4) $\forall a \in R$: $\exists(-a) \in R$: $a + (-a) = (-a) + a = 0$	(Inverse element)
(5) $a + b = b + a$	(Commutativity)
(6) $a \times b \in R$	(Closure property)
(7) $a \times (b \times c) = (a \times b) \times c$	(Associative property)
(8) $a \times (b + c) = a \times b + a \times c$	(Distributive law)
(9) $(b + c) \times a = b \times a + c \times a$	(Distributive law)

The element 0 is the identity element under addition, and the additive inverse of an element a is given by $-a$. If a ring $(R, \times, +)$ has a multiplicative identity 1 where $a \times 1 = 1 \times a = a$ for all $a \in R$ then R is termed a ring with a unit element. If $a \times b = b \times a$ for all $a, b \in R$ then R is termed a *commutative ring*.

An element $a \neq 0$ in a ring R is said to be a *zero divisor* if there exists $b \in R$, with $b \neq 0$ such that $ab = 0$. A commutative ring is an *integral domain* if it has no zero divisors. A ring is said to be a *division ring* if its non-zero elements form a group under multiplication.

Example 6.2 (**Rings**)

(i)   The set of integers $(\mathbb{Z}, +, \times)$ forms an infinite commutative ring with multiplicative unit element 1. Further, since it has no zero divisors it is an integral domain.

(ii) The set of integers mod 4 (i.e. $\mathbb{Z}_4$ where addition and multiplication is performed modulo 4)[1] is a finite commutative ring with unit element $[1]_4$. Its elements are $\{[0]_4, [1]_4, [2]_4, [3]_4\}$. It has zero divisors since $[2]_4[2]_4 = [0]_4$ and so it is not an integral domain.

---

[1] Recall from Chap. 3 that $\mathbb{Z}/n\mathbb{Z} = \mathbb{Z}_n = \{[a]_n: 0 \leq a \leq n - 1\} = \{[0]_n, [1]_n, ..., [n - 1]_n\}$.

(iii)   The Quaternions (discussed in [1]) are an example of a non-commutative ring (they form a division ring).
(iv)   The set of integers mod 5 (i.e. $\mathbb{Z}_5$ where addition and multiplication is performed modulo 5) is a finite commutative division ring[2] and it has no zero divisors.

6.6.3   Fields

A *field* is a non-empty set F together with two binary operation '+' and '×' where $(F, +)$ is a commutative group; $(F \backslash \{0\}, \times)$ is a commutative group; and the distributive properties hold. The properties of a field are

(1) $a + b \in F$	(Closure property)
(2) $a + (b + c) = (a + b) + c$	(Associative property)
(3) $\exists 0 \in F$ such that $\forall a \in F$ $a + 0 = 0 + a = a$	(Identity Element)
(4) $\forall a \in F$ $\exists (-a) \in F$ $a + (-a) = (-a) + a = 0$	(Inverse Element)
(5) $a + b = b + a$	(Commutativity)
(6) $a \times b \in F$	(Closure property)
(7) $a \times (b \times c) = (a \times b) \times c$	(Associative property)
(8) $\exists 1 \in F$ such that $\forall a \in F$ $a \times 1 = 1 \times a = a$	(Identity Element)
(10) $\forall a \in F \backslash \{0\}$ $\exists a^{-1} \in F$ $a \times a^{-1} = a^{-1} \times a = 1$	(Inverse Element)
(11) $a \times b = b \times a$	(Commutativity)
(12) $a \times (b + c) = a \times b + a \times c$	(Distributive Law)
(13) $(b + c) \times a = b \times a + c \times a$	(Distributive Law)

The following are examples of fields:

Example 6.3 (**Fields**)

(i)   The set of rational numbers $(\mathbb{Q}, +, \times)$ forms an infinite commutative field. The additive identity is 0, and the multiplicative identity is 1.
(ii)   The set of real numbers $(\mathbb{R}, +, \times)$ forms an infinite commutative field. The additive identity is 0, and the multiplicative identity is 1.
(iii) The set of complex numbers $(\mathbb{C}, +, \times)$ forms an infinite commutative field. The additive identity is 0, and the multiplicative identity is 1.
(iv)   The set of integers mod 7 (i.e. $\mathbb{Z}_7$ where addition and multiplication is performed mod 7) is a finite field.

---

[2]A finite division ring is actually a field (i.e. it is commutative under multiplication), and this classic result was proved by Wedderburn.

(v) The set of integers mod p where p is a prime (i.e. $\mathbb{Z}_p$ where addition and multiplication is performed mod p) is a finite field with p elements. The additive identity is $[0]$ and the multiplicative identity is $[1]$.

A field is a commutative division ring but not every division ring is a field. For example, the quaternions (discovered by Hamilton) are an example of a division ring, which is not a field. If the number of elements in the field F is finite then F is called a finite field, and F is written as F_q where q is the number of elements in F. In fact, every finite field has $q = p^k$ elements for some prime p, and some $k \in \mathbb{N}$ and $k > 0$.

6.6.4 Vector Spaces

A non-empty set V is said to be a *vector space* over a field F if V is a commutative group under vector addition $+$, and if for every $\alpha \in F$, $v \in V$ there is an element αv in V such that the following properties hold for $v, w \in V$ and $\alpha, \beta \in F$:

1. $u + v \in V$
2. $u + (v + w) = (u + v) + w$
3. $\exists 0 \in V$ such that $\forall v \in V\ v + 0 = 0 + v = v$
4. $\forall v \in V\ \exists (-v) \in V$ such that $v + (-v) = (-v) + v = 0$
5. $v + w = w + v$
6. $\alpha(v + w) = \alpha v + \alpha w$
7. $(\alpha + \beta)v = \alpha v + \beta v$
8. $\alpha(\beta v) = (\alpha \beta)v$
9. $1v = v$

The elements in V are referred to as *vectors* and the elements in F are referred to as *scalars*. The element 1 refers to the identity element of the field F under multiplication.

Application of Vector Spaces to Coding Theory

The representation of codewords in coding theory (which is discussed in Chap. 11), is by n-dimensional vectors over the finite field F_q. A codeword vector v is represented as the n-tuple

$$v = (a_0, a_1, \ldots a_{n-1})$$

where each $a_i \in F_q$. The set of all n-dimensional vectors is the n-dimensional vector space F_q^n with q^n elements. The addition of two vectors v and w, where $v = (a_0, a_1, \ldots a_{n-1})$ and $w = (b_0, b_1, \ldots b_{n-1})$ is given by

$$v + w = (a_0 + b_0, a_1 + b_1, \ldots a_{n-1} + b_{n-1})$$

The scalar multiplication of a vector $v = (a_0, a_1, \ldots a_{n-1}) \in F_q^n$ by a scalar $\beta \in F_q$ is given by

$$\beta v = (\beta a_0, \beta a_1, \ldots \beta a_{n-1})$$

The set F_q^n is called the vector space over the finite field F_q, if the vector space properties above hold. A finite set of vectors v_1, v_2, ... v_k is said to be *linearly independent* if

$$\beta_1 v_1 + \beta_2 v_2 + \ldots + \beta_k v_k = 0 \Rightarrow \beta_1 = \beta_2 = \ldots \beta_k = 0$$

Otherwise, the set of vectors v_1, v_2, ... v_k is said to be *linearly dependent*.

A non-empty subset W of a vector space $V (W \subseteq V)$ is said to be a *subspace* of V, if W forms a vector space over F under the operations of V. This is equivalent to W being closed under vector addition and scalar multiplication: i.e. w_1, $w_2 \in W$, α, $\beta \in F$ then $\alpha w_1 + \beta w_2 \in W$.

The *dimension* (dim W) of a subspace $W \subseteq V$ is k if there are k linearly independent vectors in W but every $k + 1$ vectors are linearly dependent. A subset of a vector space is a *basis* for V if it consists of linearly independent vectors, and its linear span is V (i.e. the basis generates V). We shall employ the basis of the vector space of codewords (see Chap. 11) to create the generator matrix to simplify the encoding of the information words. The linear span of a set of vectors v_1, v_2, ..., v_k is defined as $\beta_1 v_1 + \beta_2 v_2 + \cdots + \beta_k v_k$.

Example 6.4 (**Vector Spaces**)

(i) The Real coordinate space $\mathbb{R}^n$ forms an n-dimensional vector space over $\mathbb{R}$. The elements of $\mathbb{R}^n$ are the set of all n tuples of elements of $\mathbb{R}$, where an element x in $\mathbb{R}^n$ is written as

$$x = (x_1, x_2, \ldots x_n)$$

where each $x_i \in \mathbb{R}$ and vector addition and scalar multiplication are given by

$$\alpha x = (\alpha x_1, \alpha x_2, \ldots \alpha x_n)$$
$$x + y = (x_1 + y_1, x_2 + y_2 \ldots x_n + y_n)$$

(ii) The set of $m \times n$ matrices over the real numbers forms a vector space, with vector addition given by matrix addition, and the multiplication of a matrix by a scalar given by the multiplication of each entry in the matrix by the scalar.

6.7 Review Questions

1. Solve the simple equation: $4(3x + 1) = 7(x + 4) - 2(x + 5)$

2. Solve the following simultaneous equations by

$$x + 2y = -1$$
$$4x - 3y = 18$$

   (a) Graphical techniques
   (b) Method of substitution
   (c) Method of Elimination

3. Solve the quadratic equation $3x^2 + 5x - 2 = 0$ given that the solution is between $x = -3$ and $x = 3$ by:

   (a) Graphical techniques
   (b) Factorization
   (c) Quadratic Formula

4. Solve the following indicial equation using logarithms

$$2^{x=1} = 3^{2x-1}$$

5. Explain the differences between semigroups, monoids and groups.
6. Show that the following properties are true for groups.

   (i) The identity element is unique in a group.
   (ii) The inverse of an element is unique in a group.

7. Explain the differences between rings, commutative rings, integral domains, division rings and fields.
8. What is a vector space?
9. Explain how vector spaces may be applied to coding theory (see Chap. 11 for more details).

6.8 Summary

This chapter provided a brief introduction to algebra, which is the branch of mathematics that studies mathematical symbols and the rules for manipulating them. Algebra is a powerful tool for problem solving in science and engineering.

Elementary algebra includes the study of simultaneous equations (i.e. two or more equations with two or more unknowns); the solution of quadratic equations $ax^2 + bx + c = 0$; and the study of polynomials, indices and logarithms. Linear algebra is concerned with the solution of a set of linear equations, and the study of matrices and vector spaces.

Abstract algebra is concerned with the study of abstract algebraic structures such as monoids, groups, rings, integral domains, fields and vector spaces. The abstract approach in modern mathematics has opened up whole new areas of mathematics as well as applications in areas such as coding theory in the computing field.

Reference

1. Mathematics in Computing. Second Edition, Gerard O' Regan. Springer. 2012.

Automata Theory

Key Topics

Finite State Automata
State transition
Deterministic FSA
Non-deterministic FSA
Pushdown automata
Turing Machine

7.1 Introduction

Automata Theory is the branch of computer science that is concerned with the study of abstract machines and automata. These include finite-state machines, pushdown automata, and Turing machines. Finite-state machines are abstract machines that may be in one of a finite number of states. These machines are in only one state at a time (current state), and the input symbol causes a transition from the current state to the next state. Finite state machines have limited computational power due to memory and state constraints, but they have been applied to a number of fields including communication protocols, neurological systems and linguistics.

Pushdown automata have greater computational power than finite-state machines, and they contain extra memory in the form of a stack from which symbols may be pushed or popped. The state transition is determined from the current state of the machine, the input symbol and the element on the top of the stack. The action may be to change the state and/or push/pop an element from the stack.

© Springer International Publishing Switzerland 2016
G. O'Regan, *Guide to Discrete Mathematics*, Texts in Computer Science,
DOI 10.1007/978-3-319-44561-8_7

The Turing machine is the most powerful model for computation, and this theoretical machine is equivalent to an actual computer in the sense that it can compute exactly the same set of functions. The memory of the Turing machine is a tape that consists of a potentially infinite number of one-dimensional cells. The Turing machine provides a mathematical abstraction of computer execution and storage, as well as providing a mathematical definition of an algorithm. However, Turing machines are not suitable for programming, and therefore they do not provide a good basis for studying programming and programming languages.

7.2 Finite-State Machines

The neurophysiologists Warren McCulloch and Walter Pitts published early work on finite state automata in 1943. They were interested in modelling the thought process for humans and machines. Moore and Mealy developed this work further, and their finite-state machines are referred to as the 'Mealy machine' and the 'Moore machine'. The Mealy machine determines its outputs through the current state and the input, whereas the output of Moore's machine is based upon the current state alone.

Definition 7.1 (*Finite State Machine*) A finite state machine (FSM) is an abstract mathematical machine that consists of a finite number of states. It includes a start state q_0 in which the machine is in initially; a finite set of states Q; an input alphabet Σ; a state transition function δ; and a set of final accepting states F (where $F \subseteq Q$).

The state transition function δ takes the current state and an input symbol, and returns the next state. That is, the transition function is of the form

$$\delta : Q \times \Sigma \rightarrow Q$$

The transition function provides rules that define the action of the machine for each input symbol, and its definition may be extended to provide output as well as a transition of the state. State diagrams are used to represent finite state machines, and each state accepts a finite number of inputs. A finite-state machine (Fig. 7.1) may be deterministic or non-deterministic, and a *deterministic machine* changes to exactly (or at most)[1] one state for each input transition, whereas a *non-deterministic machine* may have a choice of states to move to for a particular input symbol.

Finite state automata can compute only very primitive functions, and so they are not adequate as a model for computing. There are more powerful automata such as the Turing machine that is essentially a finite automaton with a potentially infinite storage (memory). Anything that is computable is computable by a Turing machine.

---

[1]The transition function may be undefined for a particular input symbol and state.

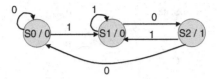

Fig. 7.1 Finite state machine

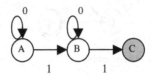

Fig. 7.2 Deterministic FSM

A finite-state machine can model a system that has a finite number of states, and a finite number of inputs/events that can trigger transitions between states. The behaviour of the system at a point in time is determined from the current state and input, with behaviour defined for the possible input to that state. The system starts in a particular initial state.

A finite-state machine (also known as finite-state automata) is a quintuple (Σ, Q, δ, q_0, F). The alphabet of the FSM is given by Σ; the set of states is given by Q; the transition function is defined by $\delta: Q \times \Sigma \to Q$; the initial state is given by q_0; and the set of accepting states is given by F where F is a subset of Q. A string is given by a sequence of alphabet symbols: i.e. $s \in \Sigma^*$, and the transition function δ can be extended to $\delta^*: Q \times \Sigma^* \to Q$.

A string $s \in \Sigma^*$ is accepted by the finite-state machine if $\delta^*(q_0, s) = q_f$ where $q_f \in F$, and the set of all strings accepted by a finite-state machine is the language generated by the machine. A finite-state machine is termed *deterministic* (Fig. 7.2) if the transition function δ is a function,[2] and otherwise (where it is a relation) it is said to be *non-deterministic*. A non-deterministic automata is one for which the next state is not uniquely determined from the present state and input symbol, and the transition may be to a set of states rather than a single state.

For the example above the input alphabet is given by $\Sigma = \{0, 1\}$; the set of states by $\{A, B, C\}$; the start state by A; the final state by $\{C\}$; and the transition function is given by the state transition table below (Table 7.1). The language accepted by the automata is the set of all binary strings that end with a one that contain exactly two ones.

---

[2]It may be a total or a partial function (as discussed in Chap. 2).

Table 7.1 State transition
table

State	0	1
A	A	B
B	B	C
C	–	–

Fig. 7.3 Non-deterministic finite state machine

A *non-deterministic* automaton (NFA) or non-deterministic finite state machine is a finite state machine where from each state of the machine and any given input, the machine may jump to several possible next states. However, a non-deterministic automaton (Fig. 7.3) is equivalent to a deterministic automaton, in that they both recognize the same formal language (i.e. regular languages as defined in Chomsky's classification). For any non-deterministic automaton, it is possible to construct the equivalent deterministic automaton using power set construction.

NFAs were introduced by Scott and Rabin in 1959, and a NFA is defined formally as a 5-tuple $(Q, \Sigma, \delta, q_0, F)$ as in the definition of a deterministic automaton, and the only difference is in the transition function δ.

$$\delta : Q \times \Sigma \to \mathbb{P}Q$$

The non-deterministic finite state machine $M_1 = (Q, \Sigma, \delta, q_0, F)$ may be converted to the equivalent deterministic machine $M_2 = (Q', \Sigma, \delta', q_0', F')$ where

$Q' = \mathbb{P}Q$ (the set of all subsets of Q)
$q_0' = \{q_0\}$
$F' = \{q \in Q' \text{ and } q \cap F \neq \varnothing\}$
$\delta' (q, \sigma) = \cup_{p \in q} \delta(p, \sigma)$ for each state $q \in Q'$ and $\sigma \in \Sigma$.

The set of strings (or language) accepted by an automaton M is denoted $L(M)$. That is, $L(M) = \{s : |\delta^*(q_0, s) = q_f \text{ for some } q_f \in F\}$. A language is termed regular if it is accepted by some finite-state machine. Regular sets are closed under union, intersection, concatenation, complement, and transitive closure. That is, for regular sets A, B $\subseteq \Sigma^*$ then

- A $\cup$ B and A $\cap$ B are regular.
- $\Sigma^*\backslash$A (i.e. A^c) is regular.
- AB and A* is regular.

Fig. 7.4 Components of
pushdown automata

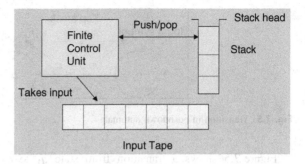

The proof of these properties is demonstrated by constructing finite-state
machines to accept these languages. The proof for A $\cap$ B is to construct a machine
$M_{A \cap B}$ that mimics the execution of M_A and M_B and is in a final state if and only if
both M_A and M_B are in a final state. Finite-state machines are useful in designing
systems that process sequences of data.

7.3 Pushdown Automata

A pushdown automaton (PDA) is essentially a finite-state machine with a stack, and
its three components (Fig. 7.4) are an input tape; a control unit; and a potentially
infinite stack. The stack head scans the top symbol of the stack, and two operations
(push or pop) may be performed on the stack. The *push* operation adds a new
symbol to the top of the stack, whereas the *pop* operation reads and removes an
element from the top of the stack.

A pushdown automaton may remember a potentially infinite amount of information, whereas a finite state automaton remembers only a finite amount of
information. A PDA also differs from a FSM in that it may use the top of the stack
to decide on which transition to take, and it may manipulate the stack as part of
performing a transition. The input and current state determine the transition in a
finite-state machine, and a FSM has no stack to work with.

A pushdown automaton is defined formally as a 7-tuple (Σ, Q, Γ, δ, q_0, Z, F).
The set Σ is a finite set which is called the input alphabet; the set Q is a finite set of
states; Γ is the set of stack symbols; δ, is the transition function which maps
$Q \times \{\Sigma \cup \{\varepsilon\}\}^3 \times \Gamma$ into finite subsets of $Q \times \Gamma^{*4}$; q_0 is the initial state; Z is the
initial stack top symbol on the stack (i.e. $Z \in \Gamma$); and F is the set of accepting states
(i.e. $F \subseteq Q$).

---

[3]The use of$\{\Sigma \cup \{\varepsilon\}\}$is to formalize that the PDA can either read a letter from the input, or
proceed leaving the input untouched.

[4]This could also be written as $\delta : Q \times \{\Sigma \cup \{\varepsilon\}\} \times \Gamma \rightarrow \mathbb{P}(Q \times \Gamma^*)$. It may also be described as a
transition relation.

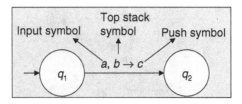

Fig. 7.5 Transition in pushdown automata

Figure 7.5 shows a transition from state q_1 to q_2, which is labelled as a, $b \rightarrow c$. This means that at if the input symbol a occurs in state q_1, and the symbol on the top of the stack is b, then b is popped from the stack and c is pushed onto the stack. The new state is q_2.

In general, a pushdown automaton has several transitions for a given input symbol, and so pushdown automata are mainly *non-deterministic*. If a pushdown automaton has at most one transition for the same combination of state, input symbol, and top of stack symbol it is said to be a *deterministic PDA* (DPDA). The set of strings (or language) accepted by a pushdown automaton M is denoted $L(M)$.

The class of languages accepted by pushdown automata is the context free languages, and every context free grammar can be transformed into an equivalent non-deterministic pushdown automaton. Chapter 12 has more detailed information on the classification of languages,

Example (Pushdown Automata)
Construct a non-deterministic pushdown automaton which recognizes the language $\{0^n 1^n | n \geq 0\}$.

Solution
We construct a pushdown automaton $M = (\Sigma, Q, \Gamma, \delta, q_0, Z, F)$ where $\Sigma = \{0,1\}$; $Q = \{q_0, q_1, q_f\}$; $\Gamma = \{A, Z\}$; q_0 is the start state; the start stack symbol is Z; and the set of accepting states is given by $\{q_f\}$:. The transition function (relation) δ is defined by

$$1. \quad (q_0, 0, Z) \rightarrow (q_0, AZ)$$
$$2. \quad (q_0, 0, A) \rightarrow (q_0, AA)$$
$$3. \quad (q_0, \varepsilon, Z) \rightarrow (q_1, Z)$$
$$4. \quad (q_0, \varepsilon, A) \rightarrow (q_1, A)$$
$$5. \quad (q_1, 1, A) \rightarrow (q_1, \varepsilon)$$
$$6. \quad (q_1, \varepsilon, Z) \rightarrow (q_f, Z)$$

The transition function (Fig. 7.6) essentially says that whenever the value 0 occurs in state q_0 then A is pushed onto the stack. Parts (3) and (4) of the transition function essentially states that the automaton may move from state q_0 to state q_1 at any moment. Part (5) states when the input symbol is 1 in state q_1 then one symbol

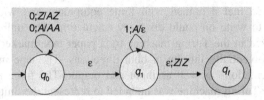

Fig. 7.6 Transition function for pushdown automata M

A is popped from the stack. Finally, part (6) states the automaton may move from state q_1 to the accepting state q_f only when the stack consists of the single stack symbol Z.

For example, it is easy to see that the string 0011 is accepted by the automaton, and the sequence of transitions is given by

$$(q_0, \ 0011, Z) \vdash (q_0, \ 011, AZ) \vdash (q_0, \ 11, AAZ) \vdash (q_1, \ 11, AAZ)$$
$$\vdash (q_1, \ 1, AZ) \vdash (q_1, \varepsilon, Z) \vdash (q_f, Z).$$

7.4 Turing Machines

Turing introduced the theoretical Turing Machine in 1936, and this abstract mathematical machine consists of a head and a potentially infinite tape that is divided into frames (Fig. 7.7). Each frame may be either blank or printed with a symbol from a finite alphabet of symbols. The input tape may initially be blank or have a finite number of frames containing symbols. At any step, the head can read the contents of a frame; the head may erase a symbol on the tape, leave it unchanged, or replace it with another symbol. It may then move one position to the right, one position to the left, or not at all. If the frame is blank, the head can either leave the frame blank or print one of the symbols.

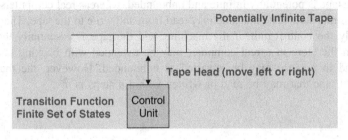

Fig. 7.7 Turing machine

Turing believed that a human with finite equipment and with an unlimited supply of paper to write on could do every calculation. The unlimited supply of paper is formalized in the Turing machine by a paper tape marked off in squares, and the tape is potentially infinite in both directions. The tape may be used for intermediate calculations as well as input and output. The finite number of configurations of the Turing machine was intended to represent the finite states of mind of a human calculator.

The transition function determines for each state and the tape symbol what the next state to move to and what should be written on the tape, and where to move the tape head.

Definition 7.2 (*Turing Machine*) A Turing machine $M = (Q, \Gamma, b, \Sigma, \delta, q_0, F)$ is a 7-tuple as defined formally in [1] as:

- Q is a finite set of *states*
- Γ is a finite set of the *tape alphabet/symbols*
- $b \in \Gamma$ is the *blank symbol* (This is the only symbol that is allowed to occur infinitely often on the tape during each step of the computation)
- Σ is the set of input symbols and is a subset of Γ (i.e. $\Gamma = \Sigma \cup \{b\}$).
- $\delta: Q \times \Gamma \rightarrow Q \times \Gamma \times \{L, R\}^5$ is the transition function. This is a partial function where L is left shift and R is right shift
- $q_0 \in Q$ is the initial state.
- $F \subseteq Q$ is the set of final or accepting states.

The Turing machine is a simple machine that is equivalent to an actual physical computer in the sense that it can compute exactly the same set of functions. It is much easier to analyse and prove things about than a real computer, but it is not suitable for programming and therefore does not provide a good basis for studying programming and programming languages.

Figure 7.8 illustrates the behaviour when the machine is in state q_1 and the symbol under the tape head is a, where b is written to the tape and the tape head moves to the left and the state changes to q_2.

A Turing machine is essentially a finite-state machine (FSM) with an unbounded tape. The tape is potentially infinite and unbounded, whereas real computers have a large but finite store. The machine may read from and write to the tape. The FSM is essentially the control unit of the machine, and the tape is essentially the store. However, the store in a real computer may be extended with backing tapes and disks, and in a sense may be regarded as unbounded. However, the maximum amount of tape that may be read or written within n steps is n.

---

[5]We may also allow no movement of the tape head to be represented by adding the symbol 'N' to the set.

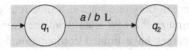

Fig. 7.8 Transition on turing machine

A Turing machine has an associated set of rules that defines its behaviour. Its actions are defined by the transition function. It may be programmed to solve any problem for which there is an algorithm. However, if the problem is unsolvable then the machine will either stop or compute forever. The solvability of a problem may not be determined beforehand. There is, of course, some answer (i.e. either the machine halts or it computes forever). The applications of the Turing machine to computability and decidability are discussed in Chap. 13.

Turing also introduced the concept of a Universal Turing Machine and this machine is able to simulate any other Turing machine.

7.5 Review Questions

1. What is a finite state machine?
2. Explain the difference between a deterministic and non-deterministic finite state machine.
3. Show how to convert the non-deterministic finite state automaton in Fig. 7.3 to a deterministic automaton.
4. What is a pushdown automaton?
5. What is a Turing machine?
6. Explain what is meant by the language accepted by an automaton.
7. Give an example of a language accepted by a pushdown automaton but not by a finite state machine.
8. Describe the applications of the Turing machine to computability and decidability.

7.6 Summary

Automata Theory is concerned with the study of abstract machines and automata. These include finite-state machines, pushdown automata and Turing machines. Finite-state machines are abstract machines that may be in one of a finite number of states. These machines are in only one state at a time (current state), and the state transition function determines the new state from the current state and the input symbol. Finite-state machines have limited computational power due to memory and state constraints, but they have been applied to a number of fields including communication protocols and linguistics.

Pushdown automata have greater computational power than finite-state machines, and they contain extra memory in the form of a stack from which symbols may be pushed or popped. The state transition is determined from the current state of the machine, the input symbol and the element on the top of the stack. The action may be to change the state and/or push/pop an element from the stack.

The Turing machine is the most powerful model for computation, and it is equivalent to an actual computer in the sense that it can compute exactly the same set of functions. The Turing machine provides a mathematical abstraction of computer execution and storage, as well as providing a mathematical definition of an algorithm

Reference

1. Introduction to Automata Theory, Languages and Computation. Hopcroft, J.E., Ullman, J.D.: Addison-Wesley, Boston (1979).

Matrix Theory

8

Key Topics

Matrix
Matrix Operations
Inverse of a Matrix
Determinant
Eigen Vectors and Values
Cayley–Hamilton Theorem
Cramer's Rule

8.1 Introduction

A *matrix* is a rectangular array of numbers that consists of horizontal rows and vertical columns. A matrix with m rows and n columns is termed an $m \times n$ matrix, where m and n are its dimensions. A matrix with an equal number of rows and columns (e.g. n rows and n columns) is termed a *square* matrix. Figure 8.1 is an example of a square matrix with four rows and four columns.

The entry in the ith row and the jth column of a matrix A is denoted by $A[i, j]$, $A_{i, j}$, or a_{ij}, and the matrix A may be denoted by the formula for its (i, j)th entry: i.e. (a_{ij}) where i ranges from 1 to m and j ranges from 1 to n.

An $m \times 1$ matrix is termed a *column vector*, and a $1 \times n$ matrix is termed a *row vector*. Any row or column of a $m \times n$ matrix determines a row or column vector which is obtained by removing the other rows (respectively, columns) from the

© Springer International Publishing Switzerland 2016
G. O'Regan, *Guide to Discrete Mathematics*, Texts in Computer Science,
DOI 10.1007/978-3-319-44561-8_8

$$\begin{pmatrix} 6 & 0 & -2 & 3 \\ 4 & 2 & 3 & 7 \\ 11 & -5 & 5 & 3 \\ 3 & -5 & -8 & 1 \end{pmatrix}$$

Fig. 8.1 Example of a 4 × 4 square matrix

matrix. For example, the row vector (11, −5, 5, 3) is obtained from the matrix example by removing rows 1, 2, and 4 of the matrix.

Two matrices A and B are equal if they are both of the same dimensions, and if $a_{ij} = b_{ij}$ for each $i = 1, 2, …, m$ and each $j = 1, 2, …, n$.

Matrices be added or multiplied (provided certain conditions are satisfied). There are identity matrices under the addition and multiplication binary operations such that the addition of the (additive) identity matrix to any matrix A yields A and similarly for the multiplicative identity. Square matrices have inverses (provided that their determinant is non-zero), and every square matrix satisfies its characteristic polynomial.

It is possible to consider matrices with infinite rows and columns, and although it is not possible to write down such matrices explicitly it is still possible to add, subtract and multiply by a scalar provided there is a well-defined entry in each (i, j)th element of the matrix.

Matrices are an example of an algebraic structure known as *algebra*. Chapter 6 discussed several algebraic structures such as groups, rings, fields and vector spaces. The matrix algebra for $m \times n$ matrices A, B, C and scalars λ, μ satisfies the following properties (there are additional multiplicative properties for square matrices).

1. $A + B = B + A$
2. $A + (B + C) = (A + B) + C$
3. $A + 0 = 0 + A = A$
4. $A + (−A) = (−A) + A = 0$
5. $\lambda(A + B) = \lambda A + \lambda B$
6. $(\lambda + \mu)A = \lambda A + \mu B$
7. $\lambda (\mu A) = (\lambda \mu) A$
8. $1A = A$

Matrices have many applications including their use in graph theory to keep track of the distance between pairs of vertices in the graph; a rotation matrix may be employed to represent the rotation of a vector in three-dimensional space. The product of two matrices represents the composition of two linear transformations, and matrices may be employed to determine the solution to a set of linear equations.

They also arise in computer graphics and may be employed to project a three-dimensional image onto a two-dimensional screen. It is essential to employ efficient algorithms for matrix computation, and this is an active area of research in the field of numerical analysis.

8.2 Two × Two Matrices

Matrices arose in practice as a means of solving a set of linear equations. One of the earliest examples of their use is in a Chinese text dating from between 300 B.C. and 200 A.D. The Chinese text showed how matrices could be employed to solve simultaneous equations. Consider the set of equations:

$$ax + by = r$$
$$cx + dy = s$$

Then the coefficients of the linear equations in x and y above may be represented by the matrix A, where A is given by:

$$A = \begin{pmatrix} a & b \\ c & d \end{pmatrix}$$

The linear equations may be represented as the multiplication of the matrix A and a vector $\underline{x}$ resulting in a vector $\underline{v}$:

$$A\underline{x} = \underline{v}.$$

The matrix representation of the linear equations and its solution are as follows:

$$\begin{pmatrix} a & b \\ c & d \end{pmatrix} \begin{pmatrix} x \\ y \end{pmatrix} = \begin{pmatrix} r \\ s \end{pmatrix}$$

The vector $\underline{x}$ may be calculated by determining the inverse of the matrix A (provided that its inverse exists). The vector $\underline{x}$ is then given by:

$$\underline{x} = A^{-1}\underline{v}$$

The solution to the set of linear equations is then given by:

$$\begin{pmatrix} x \\ y \end{pmatrix} = \begin{pmatrix} a & b \\ c & d \end{pmatrix}^{-1} \begin{pmatrix} r \\ s \end{pmatrix}$$

The inverse of a matrix A exists if and only if its *determinant* is non-zero, and if this is the case the vector $\underline{x}$ is given by:

$$\begin{pmatrix} x \\ y \end{pmatrix} = \frac{1}{\det A} \begin{pmatrix} d & -b \\ -c & a \end{pmatrix} \begin{pmatrix} r \\ s \end{pmatrix}$$

The determinant of a 2×2 matrix A is given by:

$$\det A = ad - cb.$$

The determinant of a 2×2 matrix is denoted by:

$$\begin{vmatrix} a & b \\ c & d \end{vmatrix}$$

A key property of determinants is that

$$\det(AB) = \det(A) \cdot \det(B)$$

The transpose of a 2×2 matrix A (denoted by A^T) involves exchanging rows and columns, and is given by:

$$A^T = \begin{pmatrix} a & c \\ b & d \end{pmatrix}$$

The inverse of the matrix A (denoted by A^{-1}) is given by:

$$A^{-1} = \frac{1}{\det A} \begin{pmatrix} d & -b \\ -c & a \end{pmatrix}$$

Further, $A \cdot A^{-1} = A^{-1} \cdot A = I$ where I is the identity matrix of the algebra of 2×2 matrices under multiplication. That is:

$$AA^{-1} = A^{-1}A = \begin{pmatrix} 1 & 0 \\ 0 & 1 \end{pmatrix}$$

The addition of two 2×2 matrices A and B is given by a matrix whose entries are the addition of the individual components of A and B. The addition of two matrices is commutative and we have:

$$A + B = B + A = \begin{pmatrix} a+p & b+q \\ c+r & d+s \end{pmatrix}$$

where A, B are given as

$$A = \begin{pmatrix} a & b \\ c & d \end{pmatrix} \quad B = \begin{pmatrix} p & q \\ r & s \end{pmatrix}$$

The identity matrix under addition is given by the matrix whose entries are all 0, and it has the property that $A + 0 = 0 + A = A$.

$$\begin{pmatrix} 0 & 0 \\ 0 & 0 \end{pmatrix}$$

The multiplication of two 2×2 matrices is given as

$$AB = \begin{pmatrix} ap + br & aq + bs \\ cp + dr & cq + ds \end{pmatrix}$$

The multiplication of matrices is not commutative: i.e. $AB \neq BA$. The multiplicative identity matrix I has the property that $A \cdot I = I \cdot A = A$, and it is given as

$$I = \begin{pmatrix} 1 & 0 \\ 0 & 1 \end{pmatrix}$$

A matrix A may be multiplied by a scalar λ, and this yields the matrix λA where each entry in A is multiplied by the scalar λ. That is the entries in the matrix λA are λa_{ij}.

8.3 Matrix Operations

More general sets of linear equations may be solved with $m \times n$ matrices (i.e. a matrix with m rows and n columns) or square $n \times n$ matrices. In this section we consider several matrix operations including addition, subtraction, multiplication of matrices, scalar multiplication and the transpose of a matrix.

The addition and subtraction of two matrices A, B is meaningful if and only if A and B have the same dimensions: i.e. they are both $m \times n$ matrices. In this case, $A + B$ is defined by adding the corresponding entries:

$$(A + B)_{ij} = A_{ij} + B_{ij}$$

The additive identity matrix for the square $n \times n$ matrices is denoted by 0, where 0 is a $n \times n$ matrix whose entries are zero: i.e. $r_{ij} = 0$ for all i, j where $1 \leq i \leq n$ and $1 \leq j \leq n$.

The scalar multiplication of a matrix A by a scalar k is meaningful and the resulting matrix kA is given by:

$$\begin{pmatrix} a_{11} & a_{12} & a_{13} & \cdots & a_{1n} \\ a_{21} & a_{22} & a_{23} & \cdots & a_{2n} \\ a_{31} & a_{32} & a_{33} & \cdots & a_{3n} \\ \cdots & \cdots & \cdots & \cdots & \cdots \\ \cdots & \cdots & \cdots & \cdots & \cdots \\ a_{m1} & a_{m2} & a_{m3} & \cdots & a_{mn} \end{pmatrix} \begin{pmatrix} b_{11} & b_{12} & \cdots & b_{1p} \\ b_{21} & b_{22} & \cdots & b_{2p} \\ b_{31} & b_{32} & \cdots & b_{3p} \\ \cdots & \cdots & \cdots & \cdots \\ \cdots & \cdots & \cdots & \cdots \\ \cdots & \cdots & \cdots & \cdots \\ b_{n1} & b_{n2} & \cdots & b_{np} \end{pmatrix} = \begin{pmatrix} c_{11} & c_{12} & \cdots & c_{1p} \\ c_{21} & c_{22} & \cdots & c_{2p} \\ c_{31} & c_{32} & \cdots & c_{3p} \\ \cdots & \cdots & \cdots & \cdots \\ \cdots & \cdots & \cdots & \cdots \\ c_{m1} & c_{m2} & \cdots & c_{mp} \end{pmatrix}$$

  m rows, n columns   n rows, p columns   m rows, p columns

Fig. 8.2 Multiplication of two matrices

$$(kA)_{ij} = kA_{ij}$$

The multiplication of two matrices A and B is meaningful if and only if the number of columns of A is equal to the number of rows of B (Fig. 8.2): i.e. A is an $m \times n$ matrix and B is a $n \times p$ matrix and the resulting matrix AB is a $m \times p$ matrix.

Let $A = (a_{ij})$ where i ranges from 1 to m and j ranges from 1 to n, and let $B = (b_{jl})$ where j ranges from 1 to n and l ranges from 1 to p. Then AB is given by (c_{il}) where i ranges from 1 to m and l ranges from 1 to p with c_{il} given as

$$c_{il} = \sum_{k=1}^{n} a_{ik} b_{kl}.$$

That is, the entry (c_{il}) is given by multiplying the ith row in A by the lth column in B followed by a summation. Matrix multiplication is not commutative: i.e. $AB \neq BA$.

The identity matrix I is a $n \times n$ matrix and the entries are given by r_{ij} where $r_{ii} = 1$ and $r_{ij} = 0$ where $i \neq j$ (Fig. 8.3). A matrix that has non-zero entries only on the diagonal is termed a *diagonal matrix*. A triangular matrix is a square matrix in which all the entries above or below the main diagonal are zero. A matrix is an *upper triangular* matrix if all entries below the main diagonal are zero, and *lower triangular* if all of the entries above the main diagonal are zero. Upper triangular and lower triangular matrices form a sub algebra of the algebra of square matrices.

A key property of the identity matrix is that for all $n \times n$ matrices A we have:

$$AI = IA = A$$

The inverse of a $n \times n$ matrix A is a matrix A^{-1} such that:

$$AA^{-1} = A^{-1}A = I$$

The inverse A^{-1} exists if and only if the determinant of A is non-zero.

$$\begin{pmatrix} 1 & 0 & 0 & .... & 0 \\ 0 & 1 & 0 & ... & 0 \\ 0 & 0 & 1 & .... & 0 \\ ... & .... & .... & .... & .... \\ ... & .... & .... & .... & .... \\ ... & .... & .... & .... & .... \\ 0 & 0 & 0 & .... & 1 \end{pmatrix}$$

Fig. 8.3 Identity matrix I_n

$$\begin{pmatrix} a_{11} & a_{12} & a_{13} & .... & a_{1n} \\ a_{21} & a_{22} & a_{23} & ... & a_{2n} \\ a_{31} & a_{32} & a_{33} & .... & a_{3n} \\ ... & .... & .... & .... & .... \\ ... & .... & .... & .... & .... \\ a_{m1} & a_{m2} & a_{m3} & .... & a_{mn} \end{pmatrix}^T = \begin{pmatrix} a_{11} & a_{21} & a_{31} & .... & a_{m1} \\ a_{12} & a_{22} & a_{32} & ... & a_{m2} \\ a_{13} & a_{23} & a_{33} & .... & a_{m3} \\ ... & ... & ... & .... & ... \\ ... & ... & ... & .... & ... \\ ... & ... & ... & .... & ... \\ a_{1n} & a_{2n} & a_{3n} & .... & a_{mn} \end{pmatrix}$$

        m rows, n columns   *n rows, m columns*

Fig. 8.4 Transpose of a matrix

The *transpose* of a matrix $A = (a_{ij})$ involves changing the rows to columns and vice versa to form the transpose matrix A^T. The result of the operation is that the $m \times n$ matrix A is converted to the $n \times m$ matrix A^T (Fig. 8.4). It is defined by:

$$\left(A^T\right)_{ij} = \left(A_{ji}\right) \qquad 1 \leq j \leq n. \text{ and } 1 \leq i \leq m$$

A matrix is *symmetric* if it is equal to its transpose: i.e. $A = A^T$.

8.4 Determinants

The determinant is a function defined on square matrices and its value is a scalar. A key property of determinants is that a matrix is invertible if and only if its determinant is non-zero. The determinant of a 2×2 matrix is given by:

$$\begin{vmatrix} a & b \\ c & d \end{vmatrix} = ad - bc$$

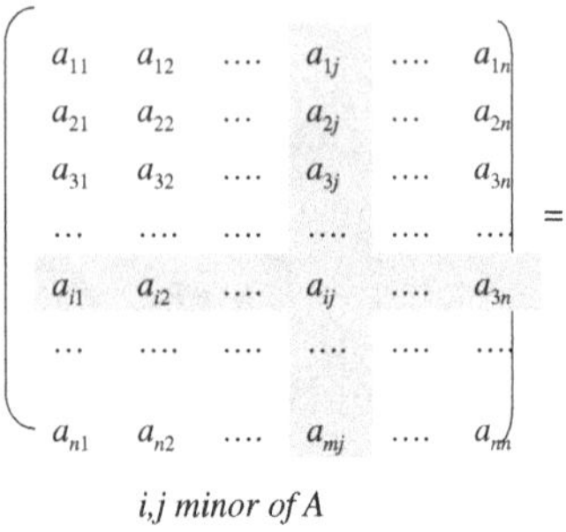

Fig. 8.5 Determining the
(i, j) minor of A

$$\begin{pmatrix}
a_{11} & a_{12} & \cdots & a_{1j} & \cdots & a_{1n} \\
a_{21} & a_{22} & \cdots & a_{2j} & \cdots & a_{2n} \\
a_{31} & a_{32} & \cdots & a_{3j} & \cdots & a_{3n} \\
\cdots & \cdots & \cdots & \cdots & \cdots & \cdots \\
a_{i1} & a_{i2} & \cdots & a_{ij} & \cdots & a_{3n} \\
\cdots & \cdots & \cdots & \cdots & \cdots & \cdots \\
a_{n1} & a_{n2} & \cdots & a_{mj} & \cdots & a_{nn}
\end{pmatrix} =$$

i,j minor of A

The determinant of a 3×3 matrix is given by:

$$\begin{vmatrix} a & b & c \\ d & e & f \\ g & h & i \end{vmatrix} = aei + bfg + cdh - afh - bdi - ceg$$

Cofactors

Let A be an $n \times n$ matrix. For $1 \leq i, j \leq n$, the (i, j) *minor* of A is defined to be the $(n - 1) \times (n - 1)$ matrix obtained by deleting the ith row and jth column of A (Fig. 8.5).

The shaded row is the ith row and the shaded column is the jth column. These are both deleted from A to form the (i, j) minor of A, and this is a $(n - 1) \times (n - 1)$ matrix.

The (i, j) *cofactor* of A is defined to be $(-1)^{i+j}$ times the determinant of the (i, j) minor. The (i, j) cofactor of A is denoted by $K_{ij}(A)$.

The cofactor matrix *Cof* A is formed in this way where the (i, j)th element in the cofactor matrix is the (i, j) cofactor of A.

Definition of Determinant

The determinant of a matrix is defined as

$$\det A = \sum_{j=1}^{n} A_{ij} K_{ij}$$

In other words, the determinant of A is determined by taking any row of A and multiplying each element by the corresponding cofactor and adding the results. The determinant of the product of two matrices is the product of their determinants.

$$\det(AB) = \det A \times \det B$$

Definition
The *adjugate* of A is the $n \times n$ matrix $Adj(A)$ whose (i, j) entry is the (j, i) cofactor $K_{ji}(A)$ of A. That is, the adjugate of A is the transpose of the cofactor matrix of A.

Inverse of A
The inverse of A is determined from the determinant of A and the adjugate of A. That is,

$$A^{-1} = \frac{1}{\det A} Adj\, A = \frac{1}{\det A} (Cof\, A)^{\mathrm{T}}$$

A matrix is invertible if and only if its determinant is non-zero: i.e. A is invertible if and only if $\det(A) \neq 0$.

Cramer's Rule
Cramer's rule is a theorem that expresses the solution to a system of linear equations with several unknowns using the determinant of a matrix. There is a unique solution if the determinant of the matrix is non-zero.

For a system of linear equations of the $A\underline{x} = \underline{v}$ where $\underline{x}$ and $\underline{v}$ are n-dimensional column vectors, then if $\det A \neq 0$ then the unique solution for each x_i is

$$x_i = \frac{\det U_i}{\det A}$$

where U_i is the matrix obtained from A by replacing the ith column in A by the v-column.

Characteristic Equation
For every $n \times n$ matrix A there is a polynomial equation of degree n satisfied by A. The *characteristic polynomial* of A is a polynomial in x of degree n. It is given as

$$cA(x) = \det(xI - A).$$

Cayley-Hamilton Theorem
Every matrix A satisfies its characteristic polynomial: i.e. $p(A) = 0$ where $p(x)$ is the characteristic polynomial of A.

8.5 Eigen Vectors and Values

A number λ is an eigenvalue of a $n \times n$ matrix A if there is a non-zero vector v such that the following equation holds:

$$Av = \lambda v$$

The vector v is termed an eigenvector and the equation is equivalent to:

$$(A - \lambda I)v = 0$$

This means that $(A - \lambda I)$ is a zero divisor and hence it is not an invertible matrix. Therefore,

$$\det (A - \lambda I) = 0$$

The polynomial function $p(\lambda) = \det (A - \lambda I)$ is called the characteristic polynomial of A, and it is of degree n. The characteristic equation is $p(\lambda) = 0$ and as the polynomial is of degree n there are at most n roots of the characteristic equation, and so there at most n eigenvalues.

The *Cayley–Hamilton theorem* states that every matrix satisfies its characteristic equation: i.e. the application of the characteristic polynomial to the matrix A yields the zero matrix.

$$p(A) = 0$$

8.6   Gaussian Elimination

Gaussian elimination with backward substitution is an important method used in solving a set of linear equations. A matrix is used to represent the set of linear equations, and Gaussian elimination reduces the matrix to a *triangular* or *reduced form*, which may then be solved by backward substitution.

This allows the set of n linear equations (E_1 to E_n) defined below to be solved by applying operations to the equations to reduce the matrix to triangular form. This reduced form is easier to solve and it provides exactly the same solution as the original set of equations. The set of equations is defined as

$$
\begin{aligned}
E_1 : \quad & a_{11}x_1 + a_{12}x_2 + \cdots + a_{1n}x_n = b_1 \\
E_2 : \quad & a_{21}x_1 + a_{22}x_2 + \cdots + a_{2n}x_n = b_2 \\
& \vdots \qquad \vdots \qquad \vdots \qquad\qquad \vdots \qquad \vdots \\
E_n : \quad & a_{n1}x_1 + a_{n2}x_2 + \cdots + a_{nn}x_n = b_n
\end{aligned}
$$

Three operations are permitted on the equations and these operations transform the linear system into a reduced form. They are

(a) Any equation may be multiplied by a non-zero constant.
(b) An equation E_i may be multiplied by a constant and added to another equation E_j, with the resulting equation replacing E_j
(c) Equations E_i and E_j may be transposed with E_j replacing E_i and vice versa.

This method for solving a set of linear equations is best illustrated by an example, and we consider an example taken from [1]. Then, the solution to a set of linear equations with four unknowns may be determined as follows:

$$
\begin{aligned}
E_1 : \quad & x_1 + x_2 && + 3x_4 = 4 \\
E_2 : \quad & 2x_1 + x_2 - x_3 + x_4 = 1 \\
E_3 : \quad & 3x_1 - x_2 - x_3 + 2x_4 = -3 \\
E_4 : \quad & -x_1 + 2x_2 + 3x_3 - x_4 = 4
\end{aligned}
$$

First, the unknown x_1 is eliminated from E_2, E_3, and E_4 and this is done by replacing E_2 with E_2-2E_1; replacing E_3 with E_3-3E_1; and replacing E_4 with $E_4 + E_1$. The resulting system is

$$
\begin{aligned}
E_1 : \quad & x_1 + x_2 && + 3x_4 = 4 \\
E_2 : \quad & -x_2 - x_3 - 5x_4 = -7 \\
E_3 : \quad & -4x_2 - x_3 - 7x_4 = -15 \\
E_4 : \quad & 3x_2 + 3x_3 + 2x_4 = 8
\end{aligned}
$$

The next step is then to eliminate x_2 from E_3 and E_4. This is done by replacing E_3 with E_3-4E_2 and replacing E_4 with $E_4 + 3E_2$. The resulting system is now in triangular form and the unknown variable may be solved easily by backward substitution. That is, we first use equation E_4 to find the solution to x_4 and then we use equation E_3 to find the solution to x_3. We then use equations E_2 and E_1 to find the solutions to x_2 and x_1.

$$
\begin{aligned}
E_1 : \quad & x_1 + x_2 && + 3x_4 = 4 \\
E_2 : \quad & -x_2 - x_3 - 5x_4 = -7 \\
E_3 : \quad & 3x_3 + 13x_4 = 13 \\
E_4 : \quad & -13x_4 = -13
\end{aligned}
$$

The usual approach to Gaussian elimination is to do it with an augmented matrix. That is, the set of equations is a $n \times n$ matrix and it is augmented by the column vector to form the augmented $n \times n + 1$ matrix. Gaussian elimination is then applied to the matrix to put it into triangular form, and it is then easy to solve the unknowns.

The other common approach to solving a set of linear equation is to employ Cramer's rule, which was discussed in Sect. 13.4. Finally, another possible (but computationally expensive) approach to solving the set of linear equations $A\underline{x} = \underline{v}$ is to compute the determinant and inverse of A, and to then compute $\underline{x} = A^{-1}\underline{v}$.

8.7   Review Questions

1. Show how 2×2 matrices may be added and multiplied.
2. What is the additive identity for 2×2 matrices? The multiplicative identity?
3. What is the determinant of a 2×2 matrix?
4. Show that a 2×2 matrix is invertible if its determinant is non-zero.
5. Describe general matrix algebra including addition and multiplication, determining the determinant and inverse of a matrix.
6. What is Cramer's rule?
7. Show how Gaussian elimination may be used to solve a set of linear equations.
8. Write a program to find the inverse of a 3×3 and then a $(n \times n)$ matrix.

8.8   Summary

A matrix is a rectangular array of numbers that consists of horizontal rows and vertical columns. A matrix with m rows and n columns is termed an $m \times n$ matrix, where m and n are its dimensions. A matrix with an equal number of rows and columns (e.g. n rows and n columns) is termed a square matrix.

Matrices arose in practice as a means of solving a set of linear equations, and one of the earliest examples of their use is from a Chinese text dating from between 300 B.C. and 200 A.D.

Matrices of the same dimensions may be added, subtracted, and multiplied by a scalar. Two matrices A and B may be multiplied provided that the number of columns of A equals the number of rows in B.

Matrices have an identity matrix under addition and multiplication, and a square matrix has an inverse provided that its determinant is non-zero. The inverse of a matrix involves determining its determinant, constructing the cofactor matrix, and transposing the cofactor matrix.

The solution to a set of linear equations may be determined by Gaussian elimination to convert the matrix to upper triangular form, and then employing backward substitution. Another approach is to use Cramer's rule.

Eigenvalues and eigenvectors lead to the characteristic polynomial and every matrix satisfies its characteristic polynomial. The characteristic polynomial is of degree n, and a square $n \times n$ matrix has at most n eigenvalues.

Reference

1. Numerical Analysis. 4th Edition. Richard L. Burden and J. Douglas Faires. PWS Kent. 1989.

Graph Theory

9

9.1 Introduction

Graph theory is a practical branch of mathematics that deals with the arrangements of certain objects known as vertices (or nodes) and the relationships between them. It has been applied to practical problems such as the modelling of computer networks, determining the shortest driving route between two cities, the link structure of a website, the travelling salesman problem and the four-colour problem.[1]

Consider a map of the London underground, which is issued to users of the underground transport system in London. Then, this map does not represent every feature of the city of London, as it includes only material that is relevant to the users of the London underground transport system. In this map the exact geographical location of the stations is unimportant, and the essential information is how the stations are interconnected to one another, as this allows a passenger to plan a route

---

[1]The 4-colour theorem states that given any map it is possible to colour the regions of the map with no more than four colours such that no two adjacent regions have the same colour. This result was finally proved in the mid-1970s.

© Springer International Publishing Switzerland 2016
G. O'Regan, *Guide to Discrete Mathematics*, Texts in Computer Science,
DOI 10.1007/978-3-319-44561-8_9

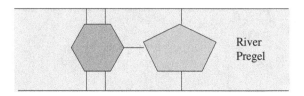

Fig. 9.1 Königsberg seven bridges problem

Fig. 9.2 Königsberg graph

from one station to another. That is, the map of the London underground is essentially a model of the transport system that shows how the stations are interconnected.

The seven bridges of Königsberg[2] (Fig. 9.1) is one of the earliest problems in graph theory. The city was set on both sides of the Pregel River in the early eighteenth century, and it consisted of two large islands that were connected to each other and the mainland by seven bridges. The problem was to find a walk through the city that would cross each bridge only once.

Euler showed that the problem had no solution, and his analysis helped to lay the foundations for graph theory as a discipline. This problem in graph theory is concerned with the question as to whether it is possible to travel along the edges of a graph starting from a vertex and returning to it and travelling along each edge exactly once. An Euler Path in a graph G is a simple path containing every edge of G.

Euler noted, in effect, that for a walk through a graph traversing each edge exactly once depends on the *degree* of the nodes (i.e. the number of edges touching it). He showed that a necessary and sufficient condition for the walk is that the graph is connected and has zero or two nodes of odd degree. For the Köningberg graph, the four nodes (i.e. the land masses) have odd degree (Fig. 9.2).

A *graph* is a collection of objects that are interconnected in some way. The objects are typically represented by vertices (or nodes), and the interconnections between them are represented by edges (or lines). We distinguish between directed

---

[2]Königsberg was founded in the thirteenth century by Teutonic knights and was one of the cities of the Hanseatic League. It was the historical capital of East Prussia (part of Germany), and it was annexed by Russia at the end of the Second World War. The German population either fled the advancing Red army or were expelled by the Russians in 1949. The city is now called Kaliningrad. The famous German philosopher, Immanuel Kant, spent all his life in the city, and is buried there.

and adirected graphs, where a *directed graph* is mathematically equivalent to a binary relation, and an *adirected (undirected) graph* is equivalent to a symmetric binary relation.

9.2 Undirected Graphs

An *undirected graph* (*adirected graph*) (Fig. 9.3) G is a pair of finite sets (V, E) such that E is a binary symmetric relation on V. The set of vertices (or nodes) is denoted by $V(G)$ and the set of edges is denoted by $E(G)$.

A *directed graph* (Fig. 9.4) is a pair of finite sets (V, E) where E is a binary relation (that may not be symmetric) on V. A *directed acyclic graph* (*dag*) is a directed graph that has no cycles. The example below is of a directed graph with three edges and four vertices.

An edge $e \in E$ consists of a pair $\langle x, y \rangle$ where x, y are adjacent nodes in the graph. The *degree* of x is the number of nodes that are adjacent to x. The set of edges is denoted by $E(G)$, and the set of vertices is denoted by $V(G)$.

A *weighted graph* is a graph $G = (V, E)$ together with a weighting function $w : E \rightarrow \mathbb{N}$, *which* associates a weight with every edge in the graph. A weighting function may be employed in modelling computer networks: for example, the weight of an edge may be applied to model the bandwidth of a telecommunications link between two nodes. Another application of the weighting function is in determining the distance (or shortest path) between two nodes in the graph (where such a path exists).

Fig. 9.3 Undirected graph

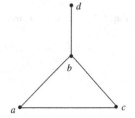

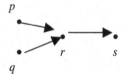

Fig. 9.4 Directed graph

For an adirected graph the weight of the edge is the same in both directions: i.e. $w(v_i, v_j) = w(v_j, v_i)$ for all edges $\langle v_i, v_j \rangle$ in the graph G, whereas the weights may be different for a directed graph.

Two vertices x, y are adjacent if $xy \in E$, and x and y are said to be incident to the edge xy. A matrix may be employed to represent the adjacency relationship.

Example 9.1

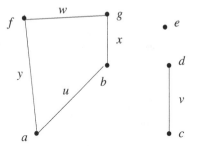

Consider the graph $G = (V, E)$ where $E = \{u = ab, v = cd, w = fg, x = bg, y = af\}$

An adjacency matrix (Fig. 9.5) may be employed to represent the relationship of adjacency in the graph. Its construction involves listing the vertices in the rows and columns, and an entry of 1 is made in the table if the two vertices are adjacent and 0 otherwise.

Similarly, we can construct a table describing the incidence of edges and vertices by constructing an incidence matrix (Fig. 9.6). This matrix lists the vertices and edges in the rows and columns, and an entry of 1 is made if the vertex is one of the nodes of the edge and 0 otherwise.

Fig. 9.5 Adjacency matrix

	a	b	c	d	e	f	g
a	0	1	0	0	0	1	0
b	1	0	0	0	0	0	1
c	0	0	0	1	0	0	0
d	0	0	1	0	0	0	0
e	0	0	0	0	0	0	0
f	1	0	0	0	0	0	1
g	0	1	0	0	0	1	0

Fig. 9.6 Incidence matrix

	u	v	w	x	y
a	1	0	0	0	1
b	1	0	0	1	0
c	0	1	0	0	0
d	0	1	0	0	0
e	0	0	0	0	0
f	0	0	1	0	1
g	0	0	1	1	0

Two graphs $G = (V, E)$ and $G' = (V', E')$ are said to be isomorphic if there exists a bijection $f : V \rightarrow V'$ such that for any $u, v \in V$, $uv \in E$, $f(u) f(v) \in E'$. The mapping f is called an isomorphism. Two graphs that are isomorphic are essentially equivalent apart from a re-labelling of the nodes and edges.

Let $G = (V, E)$ and $G' = (V', E')$ be two graphs then G' is a *subgraph* of G if $V' \subseteq V$ and $E' \subseteq E$. Given $G = (V, E)$ and $V' \subseteq V$ then we can induce a subgraph $G' = (V', E')$ by restricting G to V' (denoted by $G |_{V'}$). The set of edges in E' is defined as:

$$E' = \{e \in E : e = uv \text{ and } u, v \in V'\}$$

The *degree* of a vertex v is the number of distinct edges incident to v. It is denoted by $\deg v$ where

$$\deg v = |\{e \in E : e = vx \text{ for some } x \in V\}|$$
$$= |\{x \in V : vx \in E\}|$$

A vertex of degree 0 is called an isolated vertex.

Theorem 9.1 *Let $G = (V, E)$ be a graph then*

$$\sum_{v \in V} \deg v = 2|E|$$

Proof This result is clear since each edge contributes one to each of the vertex degrees. The formal proof is by induction based on the number of edges in the graph, and the basis case is for a graph with no edges (i.e. where every vertex is isolated), and the result is immediate for this case.

The inductive step (strong induction) is to assume that the result is true for all graphs with k or fewer edges. We then consider a graph $G = (V, E)$ with $k + 1$ edges.

Choose an edge $e = xy \in E$ and consider the graph $G' = (V, E')$ where $E' = E \backslash \{e\}$. Then G' is a graph with k edges and therefore letting $\deg' v$ represents the degree of a vertex in G' we have:

$$\sum_{v \in V} \deg' v = 2|E'| = 2(|E| - 1) = 2|E| - 2$$

The degree of x and y are one less in G' than they are in G. That is,

$$\sum_{v \in V} \deg v - 2 = \sum_{v \in V} \deg' v = 2|E| - 2$$
$$\Rightarrow \sum_{v \in V} \deg v = 2|E|$$

A graph $G = (V, E)$ is said to be *complete* if all the vertices are adjacent: i.e. $E = V \times V$. A graph $G = (V, E)$ is said to be *simple graph* if each edge connects two different vertices, and no two edges connect the same pair of vertices. Similarly, a graph that may have multiple edges between two vertices is termed a *multigraph*.

A common problem encountered in graph theory is determining whether or not there is a route from one vertex to another. Often, once a route has been identified the problem then becomes that of finding the shortest or most efficient route to the destination vertex. A graph is said to be *connected* if for any two given vertices v_1, v_2 in V there is a path from v_1 to v_2.

Consider a person walking in a forest from A to B where the person does not know the way to B. Often, the route taken will involve the person wandering around aimlessly, and often retracing parts of the route until eventually the destination B is reached. This is an example of a *walk* from v_1 to v_k where there may be repetition of edges.

If all of the edges of a walk are distinct then it is called a *trail*. A *path* $v_1, v_2, \dots, v_k$ from vertex v_1 to v_k is of length $k - 1$ and consists of the sequence of edges $\langle v_1, v_2 \rangle$, $\langle v_2, v_3 \rangle$, $\dots$, $\langle v_{k-1}, v_k \rangle$ where each $\langle v_i, v_{i+1} \rangle$ is an edge in E. The vertices in the path are all distinct apart from possibly v_1 and v_k. The path is said to be a cycle if $v_1 = v_k$. A graph is said to be *acyclic* if it contains no cycles.

Theorem 9.2 *Let $G = (V, E)$ be a graph and $W = v_1, v_2, \dots, v_k$ be a walk from v_1 to, v_k. Then there is a path from v_1 to, v_k using only edges of W.*

Proof The walk W may be reduced to a path by successively replacing redundant parts in the walk of the form $v_i \, v_{i+1} \dots, v_j$ where $v_i = v_j$ with v_i. That is, we successively remove cycles from the walk and this clearly leads to a path (not necessarily the shortest path) from v_1 to, v_k.

Theorem 9.3 *Let $G = (V, E)$ be a graph and let $u, v \in V$ with $u \neq v$. Suppose that there exists two different paths from u to v in G, then G contains a cycle.*

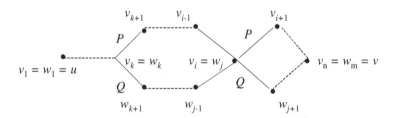

Suppose that $P = v_1, v_2, \dots, v_n$ and $Q = w_1, w_2, \dots, w_m$ are two distinct paths from u to v (where $u \neq v$), and $u = v_1 = w_1$ and $v = v_n = w_m$. Suppose P and Q are identical for the first k vertices (k could be 1), and then differ (i.e., $v_{k+1} \neq w_{k+1}$). Then Q crosses P again at $v_n = w_m$, and possibly several times before then. Suppose

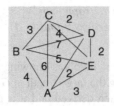

Fig. 9.7 Travelling salesman problem

the first occurrence is at $v_i = w_j$ with $k < i \leq n$. Then $w_k, w_{k+1}, w_{k+2}, \ldots, w_j v_{i-1},$ $v_{i-2}, \ldots, v_k$ is a closed path (i.e. a cycle) since the vertices are all distinct.

If there is a path from v_1 to v_2 then it is possible to define the *distance* between v_1 and v_2. This is defined to be the total length (number of edges) of the shortest path between v_1 and v_2.

9.2.1 Hamiltonian Paths

A *Hamiltonian path*[3] in a graph $G = (V, E)$ is a path that visits every vertex once and once only. In other words, the length of a Hamiltonian path is $|V| - 1$. A graph is Hamiltonian-connected if for every pair of vertices there is a Hamiltonian path between the two vertices.

Hamiltonian paths are applicable to the travelling salesman problem, where a salesman[4] wishes to travel to k cities in the country without visiting any city more than once. In principle, this problem may be solved by looking at all of the possible routes between the various cities, and choosing the route with the minimal distance.

For example, Fig. 9.7 shows five cities and the connections (including distance) between them. Then, a travelling salesman starting at A would visit the cities in the order AEDCBA (or in reverse order ABCDEA) covering a total distance of 14.

However, the problem becomes much more difficult to solve as the number of cities increase, and there is no general algorithm for its solution. For example, for the case of ten cities, the total number of possible routes is given by $9! = 362{,}880$, and an exhaustive search by a computer is feasible and the solution may be determined quite quickly. However, for 20 cities, the total number of routes is given by $19! = 1.2 \times 10^{17}$, and in this case it is no longer feasible to do an exhaustive search by a computer.

There are several sufficient conditions for the existence of a Hamiltonian path, and Theorem 9.4 describes a condition that is sufficient for the existence of a Hamiltonian path.

---

[3]These are named after Sir William Rowan Hamilton, a nineteenth century Irish mathematician and astronomer, who is famous for discovering quaternions [1].

[4]We use the term "salesman" to stand for "salesman" or "saleswoman".

Theorem 9.4 *Let G = (V, E) be a graph with |V| = n and such that deg v + deg w ≥ n − 1 for all non-adjacent vertices v and w. Then G possesses a Hamiltonian path.*

Proof The first part of the proof involves showing that G is connected, and the second part involves considering the largest path in G of length $k − 1$ and assuming that $k < n$. A contradiction is then derived and it is deduced that $k = n$.

We assume that $G' = (V', E')$ and $G'' = (V'', E'')$ are two connected components of G, then $|V'| + |V''| \leq n$ and so if $v \in V'$ and $w \in V''$ then $n − 1 \leq$ deg $v +$ deg $w \leq |V'| − 1 + |V''| − 1 = |V'| + |V''| − 2 \leq n − 2$ which is a contradiction, and so G must be connected.

Let $P = v_1, v_2, ..., v_k$ be the largest path in G and suppose $k < n$. From this a contradiction is derived, and the details for are in [2].

9.3   Trees

An acylic graph is termed a *forest* and a connected forest is termed a *tree*. A graph G is a tree if and only if for each pair of vertices in G there exists a unique path in G joining these vertices. This is since G is connected and acyclic, with the connected property giving the existence of at least one path and the acylic property giving uniqueness.

A *spanning tree* $T = (V, E')$ for the connected graph $G = (V, E)$ is a tree with the same vertex set V. It is formed from the graph by removing edges from it until it is acyclic (while ensuring that the graph remains connected).

Theorem 9.5 *Let G = (V, E) be a tree and let e ∈ E then G' = (V, E\{e}) is disconnected and has two components.*

Proof Let $e = uv$ then since G is connected and acyclic uv is the unique path from u to v, and thus G' is disconnected since there is no path from u to v in G'.

It is thus clear that there are at least two components in G' with u and v in different components. We show that any other vertex w is connected to u or to v in G'.

Since G is connected there is a path from w to u in G, and if this path does not use e then it is in G' as well, and therefore u and w are in the same component of G'.

If it does use e then e is the last edge of the graph since u cannot appear twice in the path, and so the path is of the form $w, ..., v, u$ in G. Therefore, there is a path from w to v in G', and so w and v are in the same component in G'. Therefore, there are only two components in G'

Theorem 9.6 *Any connected graph G = (V, E) possesses a spanning tree.*

Proof This result is proved by considering all connected subgraphs of $(G = V, E)$ and choosing a subgraph T with $|E'|$ as small as possible. The final step is to show

Fig. 9.8 Binary tree

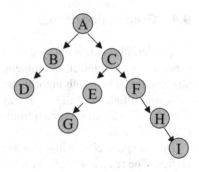

that T is the desired spanning tree, and this involves showing that T is acyclic. The details of the proof are left to the reader.

Theorem 9.7 *Let $G = (V, E)$ be a connected graph, then G is a tree if and only if* $|E| = |V| - 1$.

Proof This result may be proved by induction on the number of vertices $|V|$ and the applications of Theorems 9.5 and 9.6.

9.3.1 Binary Trees

A *binary tree* (Fig. 9.8) is a tree in which each node has at most two child nodes (termed left and right child nodes). A node with children is termed a *parent node*, and the top node of the tree is termed the root node. Any node in the tree can be reached by starting from the root node, and by repeatedly taking either the left branch (left child) or right branch (right child) until the node is reached. Binary trees are used in computing to implement efficient searching algorithms. (We gave an alternative recursive definition of a binary tree in Chap. 4).

The *depth* of a node is the length of the path (i.e. the number of edges) from the root to the node. The depth of a tree is the length of the path from the root to the deepest node in the tree. A *balanced* binary tree is a binary tree in which the depth of the two subtrees of any node never differs by more than one. The root of the binary tree in Fig. 9.8 is A and its depth is 4. The tree is unbalanced and unsorted.

Tree traversal is a systematic way of visiting each node in the tree exactly once, and we distinguish between *breadth first search* in which every node on a particular level is visited before going to a lower level, and *depth first search* where one starts at the root and explores as far as possible along each branch before backtracking. The traversal in depth first search may be in preorder, inorder or postorder.

9.4 Graph Algorithms

Graph algorithms are employed to solve various problems in graph theory including network cost minimization problems; construction of spanning trees; shortest path algorithms; longest path algorithms; and timetable construction problems.

A length function $l : E \to \mathbb{R}$ may be defined on the edges of a connected graph $G = (V, E)$, and a shortest path from u to v in G is a path P with edge set E' such that $l(E')$ is minimal.

Due to space constraints it is not possible to describe graph algorithms in this section. The reader should consult the many texts on graph theory to explore many well-known graph algorithms such as Dijkstra's shortest path algorithm and longest path algorithm (e.g. as described in [2]). Kruskal's minimal spanning tree algorithm and Prim's minimal spanning tree algorithms are described in [2]. Next, we briefly discuss graph colouring in the next section.

9.5 Graph Colouring and Four-Colour Problem

It is very common for maps to be coloured in such a way that neighbouring states or countries are coloured differently. This allows different states or countries to be easily distinguished as well as the borders between them. The question naturally arises as to how many colours are needed (or determining the least number of colours needed) to colour the entire map, as it might be expected that a large number of colours would be needed to colour a large complicated map.

However, it may come as a surprise that in fact very few colours are required to colour any map. A former student of the British logician, Augustus De Morgan, had noticed this in the mid-1800s, and he proposed the conjecture of the four-colour theorem. There were various attempts to prove that four colours were sufficient from the mid-1800s onwards, and it remained a famous unsolved problem in mathematics until the late twentieth century.

Kempe gave an erroneous proof of the four-colour problem in 1879, but his attempt led to the proof that five colours are sufficient (which was proved by Heawod in the late 1800s). Appel and Haken of the University of Illinois finally provided the proof that 4 colours are sufficient in the mid-1970s (using over 1000 h of computer time in their proof).

Each map in the plane can be represented by a graph, with each region of the graph represented by a vertex. Edges connect two vertices if the regions have a common border. The colouring of a graph is the assignment of a colour to each vertex of the graph so that no two adjacent vertices in this graph have the same colour.

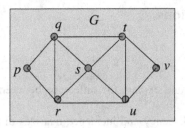

Fig. 9.9 Determining the chromatic colour of *G*

Fig. 9.10 Chromatic
colouring of *G*

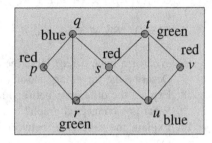

Definition

Let $G = (V, E)$ be a graph and let C be a finite set called the colours. Then, a colouring of *G* is a mapping $\kappa : V \rightarrow C$ such that if $uv \in E$ then $\kappa(u) \neq \kappa(v)$.

That is, the colouring of a simple graph is the assignment of a colour to each vertex of the graph such that if two vertices are adjacent then they are assigned a different colour. The chromatic number of a graph is the least number of colours needed for a colouring of the graph. It is denoted by $\chi(G)$.

Example 9.2 Show that the chromatic colour of the following graph *G* is 3 (this example is adapted from [3]) (Fig. 9.9).

Solution

The chromatic colour of *G* must be at least three since vertices *p*, *q* and *r* must have different colours, and so we need to show that three colours are in fact sufficient to colour *G*. We assign the colours red, blue and green to *p*, *q* and *r*, respectively. We immediately deduce that the colour of *s* must be red (as adjacent to *q* and *r*). From this, we deduce that *t* is coloured green (as adjacent to *q* and *s*) and *u* is coloured blue (as adjacent to *s* and *t*). Finally, *v* must be coloured red (as adjacent to *u* and *t*). This leads to the colouring of the graph *G* in Fig. 9.10.

Theorem 9.8 (Four-Colour Theorem) *The chromatic number of a planar graph G is less than or equal to* 4.

9.6 Review Questions

1. What is a graph and explain the difference between an adirected graph and a directed graph.
2. Determine the adjacency and incidence matrices of the following graph where $V = \{a, b, c, d, e\}$ and $E = \{ab, bc, ae, cd, bd\}$
3. Determine if the two graphs G and G' defined below are isomorphic.
4. $G = (V, E)$, $V = \{a, b, c, d, e, f, g\}$ and $E = \{ab, ad, ae, bd, ce, cf, dg, fg, bf\}$
5. $G' = (V', E')$, $V' = \{a, b, c, d, e, f, g\}$ and $E' = \{ab, bc, cd, de, ef, fg, ga, ac, be\}$
6. What is a binary tree? Describe applications of binary trees.
7. Describe the travelling salesman problem and its applications.
8. Explain the difference between a walk, trail and path.
9. What is a connected graph?
10. Explain the difference between an incidence matrix and an adjacency matrix.
11. Complete the details of Theorems 9.6 and 9.7.
12. Describe the four-colour problem and its applications.

9.7 Summary

This chapter provided a brief introduction to graph theory, which is a practical branch of mathematics that deals with the arrangements of vertices and the edges between them. It has been applied to practical problems such as the modelling of computer networks, determining the shortest driving route between two cities, and the travelling salesman problem.

The seven bridges of Königsberg is one of the earliest problems in graph theory, and it was concerned with the problem was of finding a walk through the city that would cross each bridge once and once only. Euler showed that the problem had no solution, and his analysis helped to lay the foundations for graph theory.

An undirected graph G is a pair of finite sets (V, E) such that E is a binary symmetric relation on V, whereas a directed graph is a binary relation that is not symmetric. An adjacency matrix is used to represent whether two vertices are adjacent to each other, whereas an incidence matrix indicates whether a vertex is part of a particular edge.

A Hamiltonian path in a graph is a path that visits every vertex once and once only. Hamiltonian paths are applicable to the travelling salesman problem, where a salesman wishes to travel to k cities in the country without visiting any city more than once.

Graph colouring arose to answer the question as to how many colours are needed to colour an entire map. It may be expected that many colours would be required, but the four-colour theorem demonstrates that in fact four colours are sufficient to colour a planar graph.

A tree is a connected and acylic graph, and a binary tree is a tree in which each node has at most two child nodes.

References

1. Mathematics in Computing. Second Edition, Gerard O' Regan. Springer. 2012.
2. Discrete Mathematics. An Introduction for Software Engineers. Mike Piff. Cambridge University Press. 1991.
3. Discrete Mathematics and its Applications. 7th Edition. Kenneth H. Rosen Mc Graw Hill. 2012.

Cryptography

10

Key Topics

Caesar Cipher
Enigma Codes
Bletchley Park
Turing
Public and Private Keys
Symmetric Keys
Block Ciphers
RSA

10.1 Introduction

Cryptography was originally employed to protect communication of private information between individuals. Today, it consists of mathematical techniques that provide secrecy in the transmission of messages between computers, and its objective is to solve security problems such as privacy and authentication over a communications channel.

It involves enciphering and deciphering messages, and it employs theoretical results from number theory to convert the original message (or plaintext) into cipher text that is then transmitted over a secure channel to the intended recipient. The cipher text is meaningless to anyone other than the intended recipient, and the recipient uses a key to decrypt the received cipher text and to read the original message.

© Springer International Publishing Switzerland 2016
G. O'Regan, *Guide to Discrete Mathematics*, Texts in Computer Science,
DOI 10.1007/978-3-319-44561-8_10

Alphabet Symbol	abcde fghij klmno pqrst uvwxyz
Cipher Symbol	dfegh ijklm nopqr stuvw xyzabc

Fig. 10.1 Caesar Cipher

The origin of the word "cryptography" is from the Greek "kryptos" meaning hidden, and "graphein" meaning to write. The field of cryptography is concerned with techniques by which information may be concealed in cipher texts and made unintelligible to all but the intended recipient. This ensures the privacy of the information sent, as any information intercepted will be meaningless to anyone other than the recipient.

Julius Caesar developed one of the earliest ciphers on his military campaigns in Gaul. His objective was to communicate important messages safely to his generals. His solution is one of the simplest and widely known encryption techniques, and it involves the substitution of each letter in the plaintext (i.e., the original message) by a letter a fixed number of positions down in the alphabet. The Caesar cipher involves a shift of three positions and this leads to the letter B being replaced by E, the letter C by F, and so on.

The Caesar cipher (Fig. 10.1) is easily broken, as the frequency distribution of letters may be employed to determine the mapping. However, the Gaulish tribes who were mainly illiterate, and it is likely that the cipher provided good security. The translation of the Roman letters by the Caesar cipher (with a shift key of 3) can be seen by the following table.

The process of enciphering a message (i.e., the plaintext) simply involves going through each letter in the plaintext and writing down the corresponding cipher letter. The enciphering of the plaintext message "summer solstice" involves the following:

Plaintext: Summer Solstice
Cipher Text   vxpphu vrovwleh

The process of deciphering a cipher message involves doing the reverse operation: i.e., for each cipher letter the corresponding plaintext letter is identified from the table.

Cipher Text   vxpphu vrovwleh
Plaintext: Summer Solstice

The encryption may also be represented using modular arithmetic. This involves using the numbers 0–25 to represent the alphabet letters, and the encryption of a letter is given by a shift transformation of three (modulo 26). This is simply addition (modula 26): i.e., the encoding of the plaintext letter x is given by

$$x + 3(\text{mod } 26) = a$$

Similarly, the decoding of the cipher letter a is given by

$$a - 3(\text{mod } 26) = x$$

The Caesar cipher was still in use up to the early twentieth century. However, by then frequency analysis techniques were available to break the cipher. The Vignère cipher uses a Caesar cipher with a different shift at each position in the text. The value of the shift to be employed with each plaintext letter is defined using a repeating keyword.

10.2 Breaking the Enigma Codes

The Enigma codes were used by the Germans during the second world war for the secure transmission of naval messages to their submarines. These messages contained top-secret information on German submarine and naval activities in the Atlantic, and the threat that they posed to British and Allied shipping.

The codes allowed messages to be passed secretly using encryption, and this meant that any unauthorized interception was meaningless to the Allies. The plaintext (i.e., the original message) was converted by the Enigma machine (Fig. 10.2) into the encrypted text, and these messages were then transmitted by the German military to their submarines in the Atlantic, or to their bases throughout Europe.

The Enigma cipher was invented in 1918 and the Germans believed it to be unbreakable. A letter was typed in German into the machine, and electrical impulses through a series of rotating wheels and wires produced the encrypted letter which was lit up on a panel above the keyboard. The recipient typed the received message into his machine and the decrypted message was lit up letter by letter above the keyboard. The rotors and wires of the machine could be configured in

Fig. 10.2 The Enigma machine

Fig. 10.3 Bletchley park

Fig. 10.4 Alan Turing

many different ways, and during the war the cipher settings were changed at least once a day. The odds against anyone breaking the Enigma machine without knowing the setting were 150×10^{18} to 1.

The British code and cipher school relocated from London to Bletchley Park (Fig. 10.3) at the start of the second world war. It was located in the town of Bletchley near Milton Keynes (about 50 miles North West of London). It was commanded by Alistair Dennison and was known as Station X, and several thousands were working there during the second world war. The team at Bletchley Park broke the Enigma codes, and therefore made vital contributions to the British and Allied war effort.

Polish cryptanalysts did important work in breaking the Enigma machine in the early 1930s, and they constructed a replica of the machine. They passed their knowledge on to the British and gave them the replica just prior to the German invasion of Poland. The team at Bletchley built upon the Polish work, and the team included Alan Turing[1] (Fig. 10.4) and other mathematicians.

The code-breaking teams worked in various huts in Bletchley park. Hut 6 focused on air force and army ciphers, and hut 8 focused on naval ciphers. The deciphered messages were then converted into intelligence reports, with air force

---

[1]Turing made fundamental contributions to computing, including the theoretical Turing machine.

Fig. 10.5 Replica of bombe

and army intelligence reports produced by the team in hut 3, and naval intelligence reports produced by the team in hut 4. The raw material (i.e., the encrypted messages) to be deciphered came from wireless intercept stations dotted around Britain, and from various countries overseas. These stations listened to German radio messages, and sent them to Bletchley park to be deciphered and analyzed.

Turing devised a machine to assist with breaking the codes (an idea that was originally proposed by the Polish cryptanalysts). This electromechanical machine was known as the bombe (Fig. 10.5), and its goal was to find the right settings of the Enigma machine for that particular day. The machine greatly reduced the odds and the time required to determine the settings on the Enigma machine, and it became the main tool for reading the Enigma traffic during the war. The bombe was first installed in early 1940 and it weighed over a ton. It was named after a cryptological device designed in 1938 by the Polish cryptologist, Marian Rejewski.

A standard Enigma machine employed a set of rotors, and each rotor could be in any of 26 positions. The bombe tried each possible rotor position and applied a test. The test eliminated almost all of the positions and left a smaller number of cases to be dealt with. The test required the cryptologist to have a suitable "crib": i.e., a section of ciphertext for which he could guess the corresponding plaintext.

For each possible setting of the rotors, the bombe employed the crib to perform a chain of logical deductions. The bombe detected when a contradiction had occurred and it then ruled out that setting and moved onto the next. Most of the possible settings would lead to contradictions and could then be discarded. This would leave only a few settings to be investigated in detail.

The Government Communication Headquarters (GCHQ) was the successor of Bletchley Park, and it opened after the war. The site at Bletchley park was then used for training purposes.

The codebreakers who worked at Bletchley Park were required to remain silent about their achievements until the mid-1970s when the wartime information was declassified. The link between British Intelligence and Bletchley Park came to an end in the mid-1980s.

It was decided in the mid-1990s to restore Bletchley Park, and today it is run as a museum by the Bletchley Park Trust.

10.3   Cryptographic Systems

A cryptographic system is a computer system that is concerned with the secure transmission of messages. The message is encrypted prior to its transmission, which ensures that any unauthorized interception and viewing of the message is meaningless to anyone other than the intended recipient. The recipient uses a key to decrypt the cipher text, and to retrieve the original message.

There are essentially two different types of cryptographic systems employed, and these are public key cryptosystems and secret key cryptosystems. A *public key cryptosystem* is an asymmetric cryptosystem where two different keys are employed: one for encryption and one for decryption. The fact that a person is able to encrypt a message does not mean that the person is able to decrypt a message.

In a *secret key cryptosystem* the same key is used for both encryption and decryption. Anyone who has knowledge on how to encrypt messages has sufficient knowledge to decrypt messages. The following notation is employed (Table 10.1).

The encryption and decryption algorithms satisfy the following equation:

$$Dd_k(C) = Dd_k(Ee_k(M)) = M$$

There are two different keys employed in a public key cryptosystem. These are the encryption key e_k and the decryption key d_k with $e_k \neq d_k$. It is called asymmetric since the encryption key differs from the decryption key.

There is just one key employed in a secret key cryptosystem, with the same key e_k. is used for both encryption and decryption. It is called *symmetric* since the encryption key is the same as the decryption key: i.e., $e_k. = d_k$.

Table 10.1 Notation in cryptography

Symbol	Description
M	Represents the message (plaintext)
C	Represents the encrypted message (cipher text)
e_k	Represents the encryption key
d_k	Represents the decryption key
E	Represents the encryption process
D	Represents the decryption process

10.4 Symmetric Key Systems

A symmetric key cryptosystem (Fig. 10.6) uses the same secret key for encryption and decryption. The sender and the receiver first need to agree a shared key prior to communication. This needs to be done over a secure channel to ensure that the shared key remains secret. Once this has been done they can begin to encrypt and decrypt messages using the secret key. Anyone who is able to encrypt a message has sufficient information to decrypt the message.

The encryption of a message is in effect a transformation from the space of messages m to the space of cryptosystems $\mathbb{C}$. That is, the encryption of a message with key k is an invertible transformation f such that:

$$f: m \overset{k}{\rightarrow} \mathbb{C}$$

The cipher text is given by $C = E_k(M)$ where $M \in m$ and $C \in \mathbb{C}$. The legitimate receiver of the message knows the secret key k (as it will have transmitted previously over a secure channel), and so the cipher text C can be decrypted by the inverse transformation f^{-1} defined by:

$$f^{-1}: \mathbb{C} \overset{k}{\rightarrow} m$$

Therefore, we have that $D_k(C) = D_k(E_k(M)) = M$ the original plaintext message.

There are advantages and disadvantages to symmetric key systems (Table 10.2), and these include

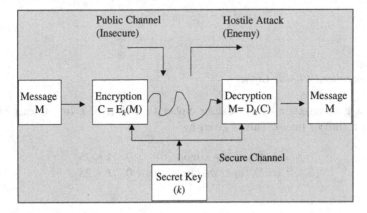

Fig. 10.6 Symmetric key cryptosystem

Table 10.2 Advantages and disadvantages of symmetric key systems

Advantages	Disadvantages
Encryption process is simple (as the same key is used for encryption and decryption)	A shared key must be agreed between two parties
It is faster than public key systems	Key exchange is difficult as there needs to be a secure channel between the two parties (to ensure that the key remains secret)
It uses less computer resources than public key systems	If a user has n trading partners then n secret keys must be maintained (one for each partner)
It uses a different key for communication with every different party	There are problems with the management and security of all of these keys (due to volume of keys that need to be maintained)
	Authenticity of origin or receipt cannot be proved (as key is shared)

Examples of Symmetric Key Systems

(i) *Caesar Cipher*

The Caesar cipher may be defined using modular arithmetic. It involves a shift of three places for each letter in the plaintext, and the alphabetic letters are represented by the numbers 0–25. The encryption is carried out by addition (modula 26). The encryption of a plaintext letter x to a cipher letter c is given by[2]:

$$c = x + 3 (\text{mod } 26)$$

Similarly, the decryption of a cipher letter c is given by:

$$x = c - 3 (\text{mod } 26)$$

(ii) *Generalized Caesar Cipher*

This is a generalization of the Caesar cipher to a shift of k (the Caesar cipher involves a shift of three). This is given by

$$f_k = E_k(x) \equiv x + k(\text{mod } 26) \qquad 0 \leq k \leq 25$$
$$f_k^{-1} = D_k(c) \equiv c - k(\text{mod } 26) \qquad 0 \leq k \leq 25$$

---

[2]Here x and c are variables rather than the alphabetic characters 'x' and 'c'.

(iii) *Affine Transformation*

This is a more general transformation and is defined by

$$f_{(a,b)} = E_{(a,b)}(x) \equiv ax + b \pmod{26} \qquad 0 \le a, b, x \le 25 \text{ and } \gcd(a, 26) = 1$$
$$f_{(a,b)}^{-1} = D_{(a,b)}(c) \equiv a^{-1}(c-b) \pmod{26} \qquad a^{-1} \text{ is the inverse of } a \bmod 26$$

(iv) *Block Ciphers*

Stream ciphers encrypt a single letter at a time and are easy to break. Block ciphers offer greater security, and the plaintext is split into groups of letters, and the encryption is performed on the block of letters rather than on a single letter.

The message is split into blocks of n-letters: $M_1, M_2, \ldots, M_k$, where each M_i ($1 \le i \le k$) is a block n-letters. The letters in the message are translated into their numerical equivalents, and the cipher text formed as follows:

$$C_i \equiv AM_i + B \pmod{N} \qquad i = 1, 2, \ldots k$$

$$\begin{pmatrix} a_{11} & a_{12} & a_{13} & \cdots & a_{1n} \\ a_{21} & a_{22} & a_{23} & \cdots & a_{2n} \\ a_{31} & a_{32} & a_{33} & \cdots & a_{3n} \\ \cdots & \cdots & \cdots & \cdots & \cdots \\ \cdots & \cdots & \cdots & \cdots & \cdots \\ a_{n1} & a_{n2} & a_{n3} & \cdots & a_{nn} \end{pmatrix} \begin{pmatrix} m_1 \\ m_2 \\ m_3 \\ \cdots \\ \cdots \\ m_n \end{pmatrix} + \begin{pmatrix} b_1 \\ b_2 \\ b_3 \\ \cdots \\ \cdots \\ b_n \end{pmatrix} = \begin{pmatrix} c_1 \\ c_2 \\ c_3 \\ \cdots \\ \cdots \\ c_n \end{pmatrix},$$

where (A, B) is the key, A is an invertible $n \times n$ matrix with $\gcd(\det(A), N) = 1$,[3] $M_i = (m_1, m_2, \ldots, m_n)^T$, $B = (b_1, b_2, \ldots, b_n)^T$, $C_i = (c_1, c_2, \ldots, c_n)^T$. The decryption is performed by

$$M_i \equiv A^{-1}(C_i - B) \pmod{N} \qquad i = 1, 2, \ldots, k$$

$$\begin{pmatrix} m_1 \\ m_2 \\ m_3 \\ \cdots \\ \cdots \\ m_n \end{pmatrix} = \begin{pmatrix} a_{11} & a_{12} & a_{13} & \cdots & a_{1n} \\ a_{21} & a_{22} & a_{23} & \cdots & a_{2n} \\ a_{31} & a_{32} & a_{33} & \cdots & a_{3n} \\ \cdots & \cdots & \cdots & \cdots & \cdots \\ \cdots & \cdots & \cdots & \cdots & \cdots \\ a_{n1} & a_{n2} & a_{n3} & \cdots & a_{nn} \end{pmatrix}^{-1} \begin{pmatrix} c_1 - b_1 \\ c_2 - b_2 \\ c_3 - b_3 \\ \cdots \\ \cdots \\ c_n - b_n \end{pmatrix}$$

---

[3]This requirement is to ensure that the matrix A is invertible.

(v) *Exponential Ciphers*

Pohlig and Hellman [1] invented the exponential cipher in 1976. This cipher is less vulnerable to frequency analysis than block ciphers.

Let p be a prime number and let M be the numerical representation of the plaintext, with each letter of the plaintext replaced with its two-digit representation (00–25). That is, A = 00, B = 01, ..., Z = 25.

M is divided into blocks M_i (these are equal size blocks of m letters where the block size is approximately the same number of digits as p). The number of letters m per block is chosen such that

$$\underbrace{2525\ldots25}_{m \text{ times}} < p < \underbrace{2525\ldots25}_{m+1 \text{ times}}$$

For example, for the prime 8191 a block size of $m = 2$ letters (4 digits) is chosen since:

$$2525 < 8191 < 252525$$

The secret encryption key is chosen to be an integer k such that $0 < k < p$ and $\gcd(k, p - 1) = 1$. Then the encryption of the block M_i is defined by

$$C_i = E_k(M_i) \equiv M_i^k \pmod{p}$$

The cipher text C_i is an integer such that $0 \leq C_i < p$.

The decryption of C_i involves first determining the inverse k^{-1} of the key k (mod $p - 1$), i.e., we determine k^{-1} such that $kk^{-1} \equiv 1 \pmod{p - 1}$. The secret key k was chosen so that $(k, p - 1) = 1$, and this means that there are integers d and n such that $kd = 1 + n(p - 1)$, and so k^{-1} is d and $kk^{-1} = 1 + n(p - 1)$. Therefore,

$$D_{k^{-1}}(C_i) \equiv C_i^{k^{-1}} \equiv (M_i^k)^{k^{-1}} \equiv M_i^{1+n(p-1)} \equiv M_i \pmod{p}$$

The fact that $M_i^{1+n(p-1)} \equiv M_i$ (mod p) follows from Euler's Theorem and Fermat's Little Theorem (Theorems 3.7 and 3.8), which were discussed in Chap. 3. Euler's Theorem states that for two positive integers a and n with $\gcd(a, n) = 1$ that $a^{\phi(n)} \equiv 1$ (mod n).

Clearly, for a prime p we have that $\phi(p) = p - 1$. This allows us to deduce that

$$M_i^{1+n(p-1)} \equiv M_i^1 M_i^{n(p-1)} \equiv M_i \left(M_i^{(p-1)} \right)^n \equiv M_i(1)^n \equiv M_i \pmod{p}$$

(vi) *Data Encryption Standard (DES)*

DES is a popular cryptographic system [2] used by governments and private companies around the world. It is based on a symmetric key algorithm and uses a shared secret key that is known only to the sender and receiver. It was designed by IBM and approved by the National Bureau of Standards (NBS[4]) in 1976. It is a block cipher and a message is split into 64-bit message blocks. The algorithm is employed in reverse to decrypt each cipher text block.

Today, DES is considered to be insecure for many applications as its key size (56 bits) is viewed as being too small, and the cipher has been broken in less than 24 h. This has led to it being withdrawn as a standard and replaced by the Advanced Encryption Standard (AES), which uses a larger key of 128 bits or 256 bits.

The DES algorithm uses the same secret 56-bit key for encryption and decryption. The key consists of 56 bits taken from a 64-bit key that includes 8 parity bits. The parity bits are at position 8, 16, ..., 64, and so every eighth bit of the 64-bit key is discarded leaving behind only the 56-bit key.

The algorithm is then applied to each 64-bit message block and the plaintext message block is converted into a 64-bit cipher text block. An initial permutation is first applied to M to create M', and M' is divided into a 32-bit left half L_0 and a 32-bit right half R_0. There are then 16 iterations, with the iterations having a left half and a right half:

$$L_i = R_{i-1}$$
$$R_i = L_{i-1} \oplus f(R_{i-1}, K_i)$$

The function f is a function that takes a 32-bit right half and a 48-bit round key K_i (each K_i contains a different subset of the 56-bit key) and produces a 32-bit output. Finally, the pre-cipher text (R_{16}, L_{16}) is permuted to yield the final cipher text C. The function f operates on half a message block and involves Table 10.3.

The decryption of the cipher text is similar to the encryption and it involves running the algorithm in reverse.

DES has been implemented on a microchip. However, it has been superseded in recent years by AES due to security concerns with its small 56-bit key size. The AES uses a key size of 128 bits or 256 bits.

10.5 Public Key Systems

A public key cryptosystem (Fig. 10.7) is an asymmetric key system where there is a separate key e_k for encryption and d_k decryption with $e_k \neq d_k$. Martin Hellman and Whitfield Diffie invented it in 1976. The fact that a person is able to encrypt a

---

[4]The NBS is now known as the National Institute of Standards and Technology (NIST).

Table 10.3 DES Encryption

Step	Description
1	Expansion of the 32-bit half block to 48 bits (by duplicating half of the bits)
2	The 48-bit result is combined with a 48-bit subkey of the secret key using an XOR operation
3	The 48-bit result is broken into 8 * 6 bits and passed through 8 substitution boxes to yield 8 * 4 = 32 bits (This is the core part of the encryption algorithm)
4	The 32-bit output is rearranged according to a fixed permutation

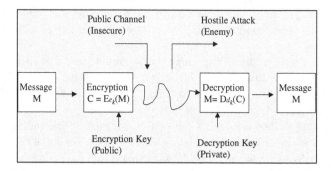

Fig. 10.7 Public key cryptosystem

Table 10.4 Public key encryption system

Item	Description
1	It uses the concept of a key pair (e_k, d_k)
2	One half of the pair can encrypt messages and the other half can decrypt messages
3	One key is private and one key is public
4	The private key is kept secret and the public key is published (but associated with trading partner)
5	The key pair is associated with exactly one trading partner

message does not mean that the person has sufficient information to decrypt messages.

The public key cryptosystem is based on the Table 10.4:

The advantages and disadvantages of public key cryptosystems Table 10.5:

The implementation of public key cryptosystems is based on *trapdoor one-way functions*. A function $f : X \rightarrow Y$ is a trapdoor one-way function if

- f is easy to computer
- f^{-1} is difficult to compute
- f^{-1} is easy to compute if a trapdoor (secret information associated with the function) becomes available.

Table 10.5 Advantages and disadvantages of public key cryptosystems

Advantages	Disadvantages
Only the private key needs to be kept secret	Public keys must be authenticated
The distribution of keys for encryption is convenient as everyone publishes their public key and the private key is kept private	It is slow and uses more computer resources
It provides message authentication as it allows the use of digital signatures (which enables the recipient to verify that the message is really from the particular sender)	Security Compromise is possible (if private key compromised)
The sender encodes with the private key that is known only to sender. The receiver decodes with the public key and therefore knows that the message is from the sender	Loss of private key may be irreparable (unable to decrypt messages)
Detection of tampering (digital signatures enable the receiver to detect whether message was altered in transit)	
Provides for nonrepudiation	

A function satisfying just the first two conditions above is termed a *one-way function*.

Examples of Trapdoor and One-way Functions

(i) The function $f : pq \rightarrow n$ (where p and q are primes) is a one-way function since it is easy to compute. However, the inverse function f^{-1} is difficult to compute problem for large n since there is no efficient algorithm to factorize a large integer into its prime factors (*integer factorization problem*).

(ii) The function $f_{g, N} : x \rightarrow g^x \pmod{N}$ is a one-way function since it is easy to compute. However, the inverse function f^{-1} is difficult to compute as there is no efficient method to determine x from the knowledge of $g^x \pmod{N}$ and g and N (*the discrete logarithm problem*).

(iii) The function $f_{k, N} : x \rightarrow x^k \pmod{N}$ (where $N = pq$ and p and q are primes) and $kk' \equiv 1 \pmod{\varphi(n)}$ is a trapdoor function. It is easy to compute but the inverse of f (the kth root modulo N) is difficult to compute. However, if the trapdoor k' is given then f can easily be inverted as $(x^k)^{k'} \equiv x \pmod{N}$

10.5.1 RSA Public Key Cryptosystem

Rivest, Shamir and Adleman proposed a practical public key cryptosystem (RSA) based on primality testing and integer factorization in the late 1970s. The RSA algorithm was filed as a patent (Patent No. 4,405, 829) at the U.S. Patent Office in December 1977. The RSA public key cryptosystem is based on the following assumptions:

- It is straightforward to find two large prime numbers.
- The integer factorization problem is infeasible for large numbers

The algorithm is based on mod n arithmetic, where n is a product of two large prime numbers.

The encryption of a plaintext message M to produce the cipher text C is given by

$$C \equiv M^e (\text{mod } n),$$

where e is the public encryption key, M is the plaintext, C is the cipher text, and n is the product of two large primes p and q. Both e and n are made public, and e is chosen such that $1 < e < \phi(n)$, where $\phi(n)$ is the number of positive integers that are relatively prime to n.

The cipher text C is decrypted by

$$M \equiv C^d (\text{mod } n),$$

where d is the private decryption key that is known only to the receiver, and $ed \equiv 1$ (mod $\phi(n)$) and d and $\phi(n)$ are kept private.

The calculation of $\phi(n)$ is easy if both p and q are known, as it is given by $\phi(n) = (p-1)(q-1)$. However, its calculation for large n is infeasible if p and q are unknown.

$$ed \equiv 1 \ (\text{mod } \phi(n))$$
$$\Rightarrow ed = 1 + k\phi(n) \text{for some } k \in \mathbb{Z}$$

We discussed Euler' Theorem in Chap. 3, and this result states that if a and n are positive integers with $\gcd(a, n) = 1$ then $a^{\phi(n)} \equiv 1 \ (\text{mod } n)$. Therefore, $M^{\phi(n)} \equiv 1$ (mod n) and $M^{k\phi(n)} \equiv 1 \ (\text{mod } n)$. The decryption of the cipher text is given by:

$$
\begin{aligned}
C^d (\text{mod } n) &\equiv M^{ed} (\text{mod } n) \\
&\equiv M^{1+k\phi(n)} (\text{mod } n) \\
&\equiv M^1 M^{k\phi(n)} (\text{mod } n) \\
&\equiv M.1 (\text{mod } n) \\
&\equiv M (\text{mod } n)
\end{aligned}
$$

10.5.2 Digital Signatures

The RSA public key cryptography may also be employed to obtain digital signatures. Suppose A wishes to send a secure message to B as well as a digital signature. This involves signature generation using the private key, and signature verification using the public key. The steps involved are: (Table 10.6):

Table 10.6 Steps for A to send secure message and signature to B

Step	Description
1	A uses B's public key to encrypt the message
2	A uses its private key to encrypt its signature
3	A sends the message and signature to B
4	B uses A's public key to decrypt A's signature
5	B uses its private key to decrypt A's message

The National Institute of Standards and Technology (NIST) proposed an algorithm for digital signatures in 1991. The algorithm is known as the Digital Signature Algorithm (DSA) and later became the Digital Signature Standard (DSS).

10.6 Review Questions

1. Discuss the early ciphers developed by Julius Caesar and Augustus. How effective were they at that period in history, and what are their weaknesses today?
2. Describe how the team at Bletchley Park cracked the German Enigma codes.
3. Explain the differences between a public key cryptosystem and a private key cryptosystem.
4. What are the advantages and disadvantages of private (symmetric) key cryptosystems?
5. Describe the various types of symmetric key systems.
6. What are the advantages and disadvantages of public key cryptosystems?
7. Describe public key cryptosystems including the RSA public key cryptosystem.
8. Describe how digital signatures may be generated.

10.7 Summary

This chapter provided a brief introduction to cryptography, which is the study of mathematical techniques that provide secrecy in the transmission of messages between computers. It was originally employed to protect communication between individuals, and today it is employed to solve security problems such as privacy and authentication over a communications channel.

It involves enciphering and deciphering messages, and theoretical results from number theory are employed to convert the original messages (or plaintext) into cipher text that is then transmitted over a secure channel to the intended recipient. The cipher text is meaningless to anyone other than the intended recipient, and the received cipher text is then decrypted to allow the recipient to read the message.

A public key cryptosystem is an asymmetric cryptosystem. It has two different encryption and decryption keys, and the fact that a person has knowledge on how to encrypt messages does not mean that the person has sufficient information to decrypt messages.

In a secret key cryptosystem the same key is used for both encryption and decryption. Anyone who has knowledge on how to encrypt messages has sufficient knowledge to decrypt messages, and it is essential that the key is kept secret between the two parties.

References

1. An Improved Algorithm for Computing Algorithms over GF(p) and its Cryptographic Significance. S. Pohlig and M. Hellman (1978). IEEE Transactions on Information Theory (24): 106–110.
2. Data Encryption Standard. FIPS-Pub 46. National Bureau of Standards. U.S. Department of Commerce. January 1977.

Coding Theory

11

Key Topics

Groups, Rings and Fields
Block Codes
Error Detection and Correction
Generation Matrix
Hamming Codes

11.1 Introduction

Coding theory is a practical branch of mathematics concerned with the reliable transmission of information over communication channels. It allows errors to be detected and corrected, which is essential when messages are transmitted through a noisy communication channel. The channel could be a telephone line, radio link or satellite link, and coding theory is applicable to mobile communications, and satellite communications. It is also applicable to storing information on storage systems such as the compact disc.

It includes theory and practical algorithms for error detection and correction, and it plays an important role in modern communication systems that require reliable and efficient transmission of information.

An error correcting code encodes the data by adding a certain amount of redundancy to the message. This enables the original message to be recovered if a small number of errors have occurred. The extra symbols added are also subject to errors, as accurate transmission cannot be guaranteed in a noisy channel.

© Springer International Publishing Switzerland 2016
G. O'Regan, *Guide to Discrete Mathematics*, Texts in Computer Science,
DOI 10.1007/978-3-319-44561-8_11

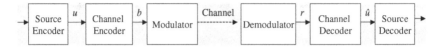

Fig. 11.1 Basic digital communication

The basic structure of a digital communication system is shown in Fig. 11.1. It includes transmission tasks such as source encoding, channel encoding and modulation; and receiving tasks such as demodulation, channel decoding and source decoding.

The modulator generates the signal that is used to transmit the sequence of symbols b across the channel. The transmitted signal may be altered due to the fact that there is noise in the channel, and the signal received is demodulated to yield the sequence of received symbols r.

The received symbol sequence r may differ from the transmitted symbol sequence b due to the noise in the channel, and therefore a channel code is employed to enable errors to be detected and corrected. The channel encoder introduces redundancy into the information sequence u, and the channel decoder uses the redundancy for error detection and correction. This enables the transmitted symbol sequence $\hat{u}$ to be estimated.

Shannon [1] showed that it is theoretically possible to produce an information transmission system with an error probability as small as required provided that the information rate is smaller than the channel capacity.

Coding theory uses several results from pure mathematics, and so first we briefly discuss the mathematical foundations of coding theory.

11.2 Mathematical Foundations

Coding theory is built from the results of modern algebra, and it uses abstract algebraic structures such as groups, rings, fields and vector spaces. These abstract structures provide a solid foundation for the discipline, and the main structures used include vector spaces and fields. A *group* is a non-empty set with a single binary operation, whereas *rings* and *fields* are algebraic structures with two binary operations satisfying various laws. A *vector space* consists of vectors over a field.

We discussed these abstract mathematical structures in Chap. 6, and presented examples of each structure. The representation of codewords is by n-dimensional vectors over the finite field F_q. A codeword vector v is represented as the n-tuple:

$$v = (a_0, a_1, \ldots a_{n-1})$$

where each $a_i \in F_q$. The set of all n-dimensional vectors is the n-dimensional vector space F_q^n with q^n elements. The addition of two vectors v and w, where $v = (a_0, a_1, \ldots a_{n-1})$ and $w = (b_0, b_1, \ldots b_{n-1})$ is given by:

$$v + w = (a_0 + b_0, a_1 + b_1, \ldots . a_{n-1} + b_{n-1})$$

The scalar multiplication of a vector $v = (a_0, a_1, \ldots . a_{n-1}) \in F_q^n$ by a scalar $\beta \in F_q$ is given by:

$$\beta v = (\beta a_0, \beta a_1, \ldots . \beta a_{n-1})$$

The set F_q^n is called the vector space over the finite field F_q. If the vector space properties above hold. A finite set of vectors $v_1, v_2, \ldots v_k$ is said to be *linearly independent* if

$$\beta_1 v_1 + \beta_2 v_2 + \cdots + \beta_k v_k = 0 \Rightarrow \beta_1 = \beta_2 = \ldots \beta_k = 0$$

Otherwise, the set of vectors $v_1, v_2, \ldots v_k$ is said to be *linearly dependent*.

The *dimension* (dim W) of a subspace $W \subseteq V$ is k if there are k linearly independent vectors in W but every $k + 1$ vectors are linearly dependent. A subset of a vector space is a *basis* for V if it consists of linearly independent vectors, and its linear span is V (i.e., the basis generates V). We shall employ the basis of the vector space of codewords to create the generator matrix to simplify the encoding of the information words. The linear span of a set of vectors $v_1, v_2, \ldots, v_k$ is defined as $\beta_1 v_1 + \beta_2 v_2 + \cdots + \beta_k v_k$.

11.3 Simple Channel Code

This section presents a simple example to illustrate the concept of an error correcting code. The example code presented is able to correct a single transmitted error only.

We consider the transmission of binary information over a noisy channel that leads to differences between the transmitted sequence and the received sequence. The differences between the transmitted and received sequence are illustrated by underlining the relevant digits in the example.

Sent   00101110

Received   00000110

Initially, it is assumed that the transmission is done without channel codes as follows:

Channel

00101110 ----------> 00000110

Next, the use of an encoder is considered and a triple repetition-encoding scheme is employed. That is, the binary symbol 0 is represented by the code word 000, and the binary symbol 1 is represented by the code word 111.

00101110→ Encoder →00000011100011111111000

Another words, if the symbol 0 is to be transmitted then the encoder emits the codeword 000, and similarly the encoder emits 111 if the symbol 1 is to be transmitted. Assuming that on average one symbol in four is incorrectly transmitted, then transmission with binary triple repetition may result in a received sequence such as:

00000011100011111111000→ Channel → 0̲1000001̲1010̲111010̲111010̲

The decoder tries to estimate the original sequence by using a *majority decision* on each 3-bit word. Any 3-bit word that contains more zeros than ones is decoded to 0, and similarly if it contains more ones than zero it is decoded to 1. The decoding algorithm yields:

0̲1000001̲1010̲111010̲111010̲ → Decoder → 00101̲010̲

In this example, the binary triple repetition code is able to correct a single error within a code word (as the majority decision is two to one). This helps to reduce the number of errors transmitted compared to unprotected transmission. In the first case where an encoder is not employed there are two errors, whereas there is just one error when the encoder is used.

However, there are disadvantages with this approach in that the transmission bandwidth has been significantly reduced. It now takes three times as long to transmit an information symbol with the triple replication code than with standard transmission. Therefore, it is desirable to find more efficient coding schemes.

11.4 Block Codes

There were two code words employed in the simple example above: namely 000 and 111. This is an example of a (n, k) code where the code words are of length $n = 3$, and the information words are of length $k = 1$ (as we were just encoding a single symbol 0 or 1). This is an example of a $(3, 1)$ block code, and the objective of this section is to generalize the simple coding scheme to more efficient and powerful channel codes.

The fundamentals of the q-nary (n, k) block codes (where q is the number of elements in the finite field F_q) involve converting an information block of length k to a codeword of length n. Consider an information sequence $u_0, u_1, u_2, \ldots$ of discrete information symbols where $u_i \in \{0, 1, \ldots q - 1\} = F_q$. The normal class of channel codes is when we are dealing with binary codes: i.e., $q = 2$. The information sequence is then grouped into blocks of length k as follows:

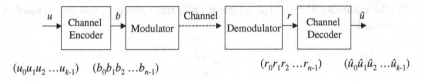

$$(u_0 u_1 u_2 \ldots u_{k\text{-}1}) \quad (b_0 b_1 b_2 \ldots b_{n\text{-}1}) \qquad\qquad (r_0 r_1 r_2 \ldots r_{n\text{-}1}) \quad (\hat{u}_0 \hat{u}_1 \hat{u}_2 \ldots \hat{u}_{k\text{-}1})$$

Fig. 11.2 Encoding and decoding of an (n, k) block

$$\underbrace{u_0 u_1 u_2 \ldots u_{k\text{-}1}}\ \underbrace{u_k u_{k+1} u_{k+2} \ldots u_{2k\text{-}1}}\ \underbrace{u_{2k} u_{2k+1} u_{2k+2} \ldots u_{3k\text{-}1}} \cdots \cdots$$

Each block is of length k (i.e., the information words are of length k), and it is then encoded separately into codewords of length n. For example, the block $u_k u_{k+1} u_{k+2} \ldots u_{2k-1}$ is encoded to the code word $b_n b_{n+1} b_{n+2} \ldots b_{2n-1}$ of length n where $b_i \in F_q$. Similarly, the information word $u_0 u_1 u_2 \ldots u_{k-1}$ is uniquely mapped to a code word $b_0 b_1 b_2 \ldots b_{n-1}$ of length n as follows:

$$(u_0 u_1 u_2 \ldots u_{k\text{-}1}) \rightarrow \boxed{\text{Encoder}} \rightarrow (b_0 b_1 b_2 \ldots b_{n\text{-}1})$$

These code words are then transmitted across the communication channel and the received words are then decoded. The received word $r = (r_0 r_1 r_2 \ldots r_{n-1})$ is decoded into the information word $\hat{u} = (\hat{u}_0 \hat{u}_1 \hat{u}_2 \ldots \hat{u}_{k-1})$.

$$(r_0 r_1 r_2 \ldots r_{n\text{-}1}) \rightarrow \boxed{\text{Decoder}} \rightarrow (\hat{u}_0 \hat{u}_1 \hat{u}_2 \ldots \hat{u}_{k\text{-}1})$$

Strictly speaking the decoding is done in two steps with the received n-block word r first decoded to the n-block codeword b^*. This is then decoded into the k-block information word $\hat{u}$. The encoding, transmission and decoding of an (n, k) block may be summarized as follows (Fig. 11.2):

A lookup table may be employed for the encoding to determine the code word b for each information word u. However, the size of the table grows exponentially with increasing information word length k, and so this is inefficient due to the large memory size required. We shall discuss later how a generator matrix provides an efficient encoding and decoding mechanism.

Notes

   (i) The codeword is of length n.
  (ii) The information word is of length k.
 (iii) The codeword length n is larger than the information word length k.
 (iv) A block (n, k) code is a code in which all codewords are of length n and all information words are of length k.
  (v) The number of possible information words is given by $M = q^k$ (where each information symbol can take one of q possible values and the length of the information word is k).

(vi) The code rate R in which information is transmitted across the channel is given by:

$$R = \frac{k}{n}$$

(vii) The weight of a codeword is $b = (b_0 b_1 b_2 \ldots b_{n-1})$ is given by the number of non-zero components of b. That is,

$$\text{wt}(b) = |\{i : b_i \neq 0,\ 0 \leq i < n\}|$$

(viii) The distance between two codewords $b = (b_0 b_1 b_2 \ldots b_{n-1})$ and $b' = (b_0' b_1' b_2' \ldots b_{n-1}')$ measures how close the codewords b and b' are to each other. It is given by the Hamming distance:

$$\text{dist}(b, b') = |\{i : b_i \neq b_i',\ 0 \leq i < n\}|$$

(ix) The minimum Hamming distance for a code B consisting of M codewords $b_1, \ldots, b_M$ is given by:

$$d = \min\{\text{dist}(b, b') : \text{where } b \neq b'\}$$

(x) The (n, k) block code $B = \{b_1, \ldots, b_M\}$ with $M\ (=q^k)$ codewords of length n and minimum Hamming distance d is denoted by $B(n, k, d)$.

11.4.1 Error Detection and Correction

The minimum Hamming distance offers a way to assess the error detection and correction capability of a channel code. Consider two codewords b and b' of an (n, k) block code $B(n, k, d)$.

Then, the distance between these two codewords is greater than or equal to the minimum Hamming distance d, and so errors can be detected as long as the erroneously received word is not equal to a codeword different from the transmitted code word.

That is, the *error detection capability* is guaranteed as long as the number of errors is less than the minimum Hamming distance d, and so the number of detectable errors is $d - 1$.

Any two codewords are of distance at least d and so if the number of errors is less than $d/2$ then the received word can be properly decoded to the codeword b. That is, the *error correction capability* is given by:

$$E_{\text{cor}} = \frac{d - 1}{2}$$

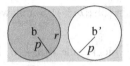

Fig. 11.3 Error correcting capability sphere

An error-correcting sphere (Fig. 11.3) may be employed to illustrate the error correction of a received word to the correct codeword b. This may be done when all received words are within the error-correcting sphere with radius p ($<d/2$).

If the received word r is different from b in less than $d/2$ positions, then it is decoded to b as it is more than $d/2$ positions from the next closest codeword. That is, b is the closest codeword to the received word r (provided that the error-correcting radius is less than $d/2$).

11.5   Linear Block Codes

Linear block codes have nice algebraic properties, and the codewords in a linear block code are considered to be vectors in the finite vector space F^n_q. The representation of codewords by vectors allows the nice algebraic properties of vector spaces to be used, and this simplifies the encoding of information words as a generator matrix may be employed to create the codewords.

An (n, k) block code $\boldsymbol{B}(n, k, d)$ with minimum Hamming distance d over the finite field F_q is called *linear* if $\boldsymbol{B}(n, k, d)$ is a subspace of the vector space F^n_q of dimension k. The number of codewords is then given by:

$$M = q^k$$

The rate of information (R) through the channel is given by:

$$R = \frac{k}{n}$$

Clearly, since $\boldsymbol{B}(n, k, d)$ is a subspace of F^n_q any linear combination of the codewords (vectors) will be a codeword. That is, for the codewords $b_1, b_2, ..., b_r$ we have that:

$$\alpha_1 b_1 + \alpha_2 b_2 + \cdots + \alpha_r b_r \in \boldsymbol{B}(n, k, d)$$

where $\alpha_1, \alpha_2, ..., \alpha_r \in F_q$ and $b_1, b_2, ..., b_r \in \boldsymbol{B}(n, k, d)$.

Clearly, the n-dimensional zero row vector $(0,0, ..., 0)$ is always a codeword, and so $(0, 0, ..., 0) \in \boldsymbol{B}(n, k, d)$. The minimum Hamming distance of a linear block code $\boldsymbol{B}(n, k, d)$ is equal to the minimum weight of the non-zero codewords: That is,

$$d = \min_{\forall b \neq b'} \{\text{dist}(b, b')\} = \min_{\forall b \neq 0} \text{wt}(b)$$

In summary, an (n, k) linear block code $\mathbf{B}(n, k, d)$ is:

1. A subspace of F^n_q.
2. The number of codewords is $M = q^k$.
3. The minimum Hamming distance d is the minimum weigh of the non-zero codewords.

The encoding of a specific k-dimensional information word $u = (u_0, u_1, \ldots u_{k-1})$ to a n-dimensional codeword $b = (b_0, b_1, \ldots, b_{n-1})$ may be done efficiently with a generator matrix. First, a basis $\{g_0, g_1, \ldots g_{k-1}\}$ of the k-dimensional subspace spanned by the linear block code is chosen, and this consists of k linearly independent n-dimensional vectors. Each basis element g_i (where $0 \leq i \leq k - 1$) is a n-dimensional vector:

$$g_i = (g_{i,0}, g_{i,1}, \ldots, g_{i,n-1})$$

The corresponding codeword $b = (b_0, b_1, \ldots, b_{n-1})$ is then a linear combination of the information word with the basis elements. That is,

$$b = u_0 g_0 + u_1 g_1 + \cdots + u_{k-1} g_{k-1}$$

where each information symbol $u_i \in F_q$. The generator matrix G is then constructed from the k linearly independent basis vectors as follows (Fig. 11.4):

The encoding of the k-dimensional information word u to the n-dimensional codeword b involves matrix multiplication (Fig. 11.5):

This may also be written as:

$$b = uG$$

$$G = \begin{pmatrix} g_0, \\ g_1, \\ g_2, \\ \cdots \\ \cdots \\ \cdots \\ g_{k-1} \end{pmatrix} = \begin{pmatrix} g_{0,0} & g_{0,1} & g_{0,2} & \cdots & g_{0,n-1} \\ g_{1,0} & g_{1,1} & g_{1,2} & \cdots & g_{1,n-1} \\ g_{2,0} & g_{2,1} & g_{2,2} & \cdots & g_{2,n-1} \\ \cdots & \cdots & \cdots & \cdots & \cdots \\ \cdots & \cdots & \cdots & \cdots & \cdots \\ \cdots & \cdots & \cdots & \cdots & \cdots \\ g_{k-1,0} & g_{k-1,1} & g_{k-1,2} & \cdots & g_{k-1,n-1} \end{pmatrix}$$

Fig. 11.4 Generator matrix

$$(u_0, u_1, ..., u_{k-1}) \begin{pmatrix} g_{0,0} & g_{0,1} & g_{0,2} & \cdots & g_{0,n-1} \\ g_{1,0} & g_{1,1} & g_{1,2} & \cdots & g_{1,n-1} \\ g_{2,0} & g_{2,1} & g_{2,2} & \cdots & g_{2,n-1} \\ \cdots & \cdots & \cdots & \cdots & \cdots \\ \cdots & \cdots & \cdots & \cdots & \cdots \\ \cdots & \cdots & \cdots & \cdots & \cdots \\ g_{k-1,0} & g_{k-1,1} & g_{k-1,2} & \cdots & g_{k-1,n-1} \end{pmatrix} = (b_0, b_1, ..., b_{n-1})$$

Fig. 11.5 Generation of codewords

$$I_k = \begin{pmatrix} 1 & 0 & 0 & \cdots & 0 \\ 0 & 1 & 0 & \cdots & 0 \\ 0 & 0 & 1 & \cdots & 0 \\ \cdots & \cdots & \cdots & \cdots & \cdots \\ \cdots & \cdots & \cdots & \cdots & \cdots \\ \cdots & \cdots & \cdots & \cdots & \cdots \\ 0 & 0 & 0 & \cdots & 1 \end{pmatrix}$$

Fig. 11.6 Identity matrix $(k \times k)$

Clearly, all $M = q^k$ codewords $b \in B(n, k, d)$ can be generated according to this rule, and so the matrix G is called the generator matrix. The generator matrix defines the linear block code $B(n, k, d)$.

There is an equivalent $k \times n$ generator matrix for $B(n, k, d)$ defined as:

$$G = I_k| \; A_{k,n-k}$$

where I_k is the $k \times k$ identity matrix (Fig. 11.6):

The encoding of the information word u yields the codeword b such that the first k symbols b_i of b are the same as the information symbols u_i $0 \leq i \leq k$.

$$b = uG = \left(u \,|\, u A_{k,n-k} \right)$$

The remaining $m = n - k$ symbols are generated from $u A_{k,n-k}$ and the last m symbols are the m parity check symbols. These are attached to the information vector u for the purpose of error detection and correction.

11.5.1 Parity Check Matrix

The linear block code $B(n, k, d)$ with generator matrix $G = (I_k | A_{k,n-k})$ may be defined equivalently by the $(n - k) \times n$ parity check matrix H, where this matrix is defined as:

$$H = \left(-A_{k,n-k}^T \mid I_{n-k} \right).$$

The generator matrix G and the parity check matrix H are orthogonal: i.e.,

$$HG^T = 0_{n-k,k}$$

The parity check orthogonality property holds if and only if the vector belongs to the linear block code. That is, for each code vector in $b \in B(n, k, d)$ we have

$$Hb^T = 0_{n-k,1}$$

and vice verse whenever the property holds for a vector r, then r is a valid codeword in $B(n, k, d)$. We present an example of a parity check matrix in Example 9.5 below.

11.5.2 Binary Hamming Code

The Hamming code is a linear code that has been employed in dynamic random access memory to detect and correct deteriorated data in memory. The generator matrix for the $B(7, 4, 3)$ binary Hamming code is given by (Fig. 11.7):

The information words are of length $k = 4$ and the codewords are of length $n = 7$. For example, it can be verified by matrix multiplication that the information word $(0, 0, 1, 1)$ is encoded into the codeword $(0, 0, 1, 1, 0, 0, 1)$.

That is, three parity bits 001 have been added to the information word $(0, 0, 1, 1)$ to yield the codeword $(0, 0, 1, 1, 0, 0, 1)$.

The minimum Hamming distance is $d = 3$, and the Hamming code can detect up to two errors, and it can correct one error.

Example 9.5 (**Parity Check Matrix—Hamming Code**) The objective of this example is to construct the Parity Check Matrix of the Binary Hamming Code (7, 4, 3), and to show an example of the parity check orthogonality property.

$$G = \begin{bmatrix} 1 & 0 & 0 & 0 & 0 & 1 & 1 \\ 0 & 1 & 0 & 0 & 1 & 0 & 1 \\ 0 & 0 & 1 & 0 & 1 & 1 & 0 \\ 0 & 0 & 0 & 1 & 1 & 1 & 1 \end{bmatrix}$$

Fig. 11.7 Hamming code $B(7, 4, 3)$ generator matrix

First, we construct the parity check matrix H which is given by $H = \left(-A^T_{k,n-k} \mid I_{n-k}\right)$ or another words $H = \left(-A^T_{4,3} \mid I_3\right)$. We first note that

$$A_{4,3} = \begin{pmatrix} 0 & 1 & 1 \\ 1 & 0 & 1 \\ 1 & 1 & 0 \\ 1 & 1 & 1 \end{pmatrix} \qquad A^T_{4,3} = \begin{pmatrix} 0 & 1 & 1 & 1 \\ 1 & 0 & 1 & 1 \\ 1 & 1 & 0 & 1 \end{pmatrix}$$

Therefore, H is given by:

$$H = \begin{pmatrix} 0 & -1 & -1 & -1 & 1 & 0 & 0 \\ -1 & 0 & -1 & -1 & 0 & 1 & 0 \\ -1 & -1 & 0 & -1 & 0 & 0 & 1 \end{pmatrix}$$

We noted that the encoding of the information word $u = (0011)$ yields the codeword $b = (0011001)$. Therefore, the calculation of Hb^T yields (recalling that addition is modulo two):

$$Hb^T = \begin{pmatrix} 0 & -1 & -1 & -1 & 1 & 0 & 0 \\ -1 & 0 & -1 & -1 & 0 & 1 & 0 \\ -1 & -1 & 0 & -1 & 0 & 0 & 1 \end{pmatrix} \begin{pmatrix} 0 \\ 0 \\ 1 \\ 1 \\ 0 \\ 0 \\ 1 \end{pmatrix} = \begin{pmatrix} 0 \\ 0 \\ 0 \end{pmatrix}$$

11.5.3 Binary Parity-Check Code

The binary parity-check code is a linear block code over the finite field F_2. The code takes a k-dimensional information word $u = (u_0, u_1,\ldots. u_{k-1})$ and generates the codeword $b = (b_0, b_1,\ldots, b_{k-1}, b_k)$ where $u_i = b_i$ $(0 \leq i \leq k - 1)$ and b_k is the parity bit chosen so that the resulting codeword is of even parity. That is,

$$b_k = u_0 + u_1 + \ldots + u_{k-1} = \sum_{i=0}^{k-1} u_i$$

11.6   Miscellaneous Codes in Use

There are many examples of codes in use such as repetition codes (such as the triple replication code considered earlier in Sect. 11.3); parity check codes where a parity symbol is attached such as the binary parity-check code; Hamming codes such as the (7, 4) code that was discussed in Sect. 11.5.2, and which has been applied for error correction of faulty memory.

The Reed-Muller codes form a class of error correcting codes that can correct more than one error. Cyclic codes are special linear block codes with efficient algebraic decoding algorithms. The BCH codes are an important class of cyclic codes, and the Reed Solomon codes are an example of a BCH code.

Convolution codes have been applied in the telecommunications field, for example, in GSM, UMTS and in satellite communications. They belong to the class of linear codes, but also employ a memory so that the output depends on the current input symbols and previous input. For more detailed information on coding theory see [2].

11.7   Review Questions

1. Describe the basic structure of a digital communication system.

2. Describe the mathematical structure known as the field. Give examples of fields.
3. Describe the mathematical structure known as the ring and give examples of rings. Give examples of zero divisors in rings.
4. Describe the mathematical structure known as the vector space and give examples.
5. Explain the terms linear independence and linear dependence and a basis.
6. Describe the encoding and decoding of an (n, k) block code where an information word of length k is converted to a codeword of length n.
7. Show how the minimum Hamming distance may be employed for error detection and error correction.
8. Describe linear block codes and show how a generator matrix may be employed to generate the codewords from the information words.

11.8   Summary

Coding theory is the branch of mathematics that is concerned with the reliable transmission of information over communication channels. It allows errors to be detected and corrected, and this is extremely useful when messages are transmitted through a noisy communication channel. This branch of mathematics includes theory and practical algorithms for error detection and correction.

The theoretical foundations of coding theory were considered, and its foundations lie in abstract algebra including group theory, ring theory, fields and vector spaces. The codewords are represented by n-dimensional vectors over a finite field F_q.

An error correcting code encodes the data by adding a certain amount of redundancy to the message so that the original message can be recovered if a small number of errors have occurred.

The fundamentals of block codes were discussed where an information word is of length k and a codeword is of length n. This led to the linear block codes $B(n, k, d)$ and a discussion on error detection and error correction capabilities of the codes.

The goal of this chapter was to give a flavour of coding theory, and the reader is referred to more specialised texts (e.g., [2]) for more detailed information.

References

1. A Mathematical Theory of Communication. Claude Shannon. Bell System Technical Journal, vol. 27, pp. 379–423. 1948.
2. Coding Theory. Algorithms, Architectures and Applications. André Neubauer, Jürgen Freunderberger and Volker Kühn. John Wiley & Sons. 2007.

Language Theory and Semantics

12

12.1 Introduction

There are two key parts to any programming language, and these are its syntax and semantics. The syntax is the grammar of the language and a program needs to be syntactically correct with respect to its grammar. The semantics of the language is deeper, and determines the meaning of what has been written by the programmer.

The difference between syntax and semantics may be illustrated by an example in a natural language. A sentence may be syntactically correct but semantically meaningless, or a sentence may have semantic meaning but be syntactically incorrect. For example, consider the sentence:

"I will go to Dublin yesterday"

© Springer International Publishing Switzerland 2016

185

G. O'Regan, *Guide to Discrete Mathematics*, Texts in Computer Science,
DOI 10.1007/978-3-319-44561-8_12

Then this sentence is syntactically valid but semantically meaningless. Similarly, if a speaker utters the sentence "Me Dublin yesterday" we would deduce that the speaker had visited Dublin the previous day even though the sentence is syntactically incorrect.

The semantics of a programming language determines what a syntactically valid program will compute. A programming language is therefore given by:

$$\text{Programming Language} = \text{Syntax} + \text{Semantics}$$

Many programming languages have been developed over the last 60 years including Plankalkül which was developed by Zuse in the 1940s; Fortran developed by IBM in the 1950s; Cobol was developed by a committee in the late 1950s; Algol 60 and Algol 68 were developed by an international committee in the 1960s; Pascal was developed by Wirth in the early 1970s; Ada was developed for the US military in the late 1970s; the C language was developed by Richie and Thompson at Bell Labs in the 1970s; C++ was developed by Stroustrup at Bell Labs in the early 1980s; and Java developed by Gosling at Sun Microsystems in the mid-1990s. A short description of a selection of programming languages in use is in [1].

A programming language needs to have a well-defined syntax and semantics, and the compiler preserves the semantics of the language. Compilers are programs that translate a program that is written in some programming language into another form. It involves syntax analysis and parsing to check the syntactic validity of the program; semantic analysis to determine what the program should do; optimization to improve the speed and performance; and code generation in some target language.

Alphabets are a fundamental building block in language theory, as words and language are generated from alphabets. They are discussed in the next section.

12.2 Alphabets and Words

An *alphabet* is a finite non-empty set A, and the elements of A are called letters. For example, consider the set A which consists of the letters a to z.

Words are finite strings of letters, and a set of words is generated from the alphabet. For example, the alphabet $A = \{a, b\}$ generates the following set of words:

$$\{\varepsilon, a, b, aa, ab, bb, ba, aaa, bbb, \ldots\}$$

Each word consists of an ordered list of one or more letters and the set of words of length two consists of all ordered lists of two letters[1]. It is given by

---

[1] ε denotes the empty word.

$$A^2 = \{aa, ab, bb, ba\}$$

Similarly, the set of words of length three is given by

$$A^3 = \{aaa, aab, abb, aba, baa, bab, bbb, bba\}$$

The set of all words over the alphabet A is given by the positive closure A^+, and it is defined by

Given any two words $w_1 = a_1a_2...a_k$ and $w_2 = b_1b_2...b_r$ then the concatenation of w_1 and w_2 is given by

$$w = w_1w_2 = a_1a_2...a_kb_1b_2...b_r$$

The empty word is a word of length zero and is denoted by ε. Clearly, $\varepsilon w = w\varepsilon = w$ for all w and so ε is the identity element under the concatenation operation. A^0 is used to denote the set containing the empty word $\{\varepsilon\}$, and the closure A^* ($=A^+ \cup \{\varepsilon\}$) denotes the infinite set of all words over A (including empty words). It is defined as:

$$A^* = \bigcup_{n=0}^{\infty} A^n$$

The mathematical structure (A^*, ^, ε) forms a monoid,[2] where ^ is the concatenation operator for words and the identity element is ε. The length of a word w is denoted by $|w|$ and the length of the empty word is zero: i.e., $|\varepsilon| = 0$.

A subset L of A^* is termed a formal language over A. Given two languages L_1, L_2 then the concatenation (or product) of L_1 and L_2 is defined by

$$L_1L_2 = \{w|w = w_1w_2 \quad \text{where } w_1 \in L_1 \text{ and } w_2 \in L_2\}$$

The positive closure of L and the closure of L may also be defined as

$$L^+ = \bigcup_{n=1}^{\infty} L^n \quad L^* = \bigcup_{n=0}^{\infty} L^n$$

12.3 Grammars

A formal grammar describes the syntax of a language, and we distinguish between *concrete* and *abstract syntax*. Concrete syntax describes the external appearance of programs as seen by the programmer, whereas abstract syntax aims to describe the

---

[2]Recall from chapter 6 that a monoid (M, *, *e*) is a structure that is closed and associative under the binary operation "*", and it has an identity element "*e*".

Fig. 12.1 Noah chomsky.
Courtesy of Duncan
rawlinson

essential structure of programs rather than the external form. In other words, abstract syntax aims to give the components of each language structure while leaving out the representation details (e.g., syntactic sugar). Backus Naur Form (BNF) notation is often used to specify the concrete syntax of a language. A grammar consists of

- A finite set of terminal symbols
- A finite set of nonterminal symbols
- A set of production rules
- A start symbol

A formal grammar generates a formal language, which is a set of finite length sequences of symbols created by applying the production rules of the grammar. The application of a production rule involves replacing symbols at the left-hand side of the rule with the symbols on the right-hand side of the rule. The formal language then consists of all words consisting of terminal symbols that are reached by a derivation (i.e., the application of production rules) starting from the start symbol of the grammar.

A construct that appears on the left-hand side of a production rule is termed a *nonterminal*, whereas a construct that only appears on the right-hand side of a production rule is termed a *terminal*. The set of nonterminals N is disjoint from the set of terminals A.

The theory of the syntax of programming languages is well established, and programming languages have a well-defined grammar that allows syntactically valid programs to be derived from the grammars.

Chomsky[3] (Fig. 12.1) was a famous linguist who classified a number of different types of grammar that occur.

---

[3]Chomsky made important contributions to linguistics and the theory of grammars. He is more widely known today as a critic of United States foreign policy.

Table 12.1 Chomsky hierarchy of grammars

Grammar type	Description
Type 0 grammar	Type 0-grammars include all formal grammars. They have production rules of the form $\alpha \rightarrow \beta$, where α and β are strings of terminals and nonterminals. They generate all languages that can be recognized by a Turing machine (discussed in Chap. 7)
Type 1 grammar (context sensitive)	These grammars generate the context-sensitive languages. They have production rules of the form $\alpha A \beta \rightarrow \alpha \gamma \beta$ where A is a nonterminal and α, β and γ are strings of terminals and nonterminals. They generate all languages that can be recognized by a linear bounded automaton[a]
Type 2 grammar (context free)	These grammars generate the context-free languages. These are defined by rules of the form $A \rightarrow \gamma$, where A is a nonterminal and γ is a string of terminals and nonterminals. These languages are recognized by a pushdown automaton[b] and are used to define the syntax of most programming languages
Type 3 grammar (regular grammars)	These grammars generate the regular languages (or regular expressions). These are defined by rules of the form $A \rightarrow a$ or $A \rightarrow aB$, where A and B are nonterminals and a is a single terminal. A finite state automaton recognizes these languages (discussed in Chap. 7), and regular expressions are used to define the lexical structure of programming languages

[a]A linear bounded automaton is a restricted form of a nondeterministic Turing machine in which a limited finite portion of the tape (a function of the length of the input) may be accessed
[b]A pushdown automaton is a finite automaton that can make use of a stack containing data, and it is discussed in Chap. 7

The Chomsky hierarchy (Table 12.1) consists of four levels including regular grammars; context free grammars; context sensitive grammars and unrestricted grammars. The grammars are distinguished by the production rules, which determine the type of language that is generated.

Regular grammars are used to generate the words that may appear in a programming language. This includes the identifiers (e.g., names for variables, functions and procedures); special symbols (e.g., addition, multiplication, etc.); and the reserved words of the language.

A rewriting system for context free grammars is a finite relation between N and $(A \cup N)^*$: i.e., a subset of $N \times (A \cup N)^*$: A production rule $<N> \rightarrow w$ is one element of this relation, and is an ordered pair $(<N>, w)$ where w is a word consisting of zero or more terminal and nonterminal letters. This production rule means that $<N>$ may be replaced by w.

12.3.1 Backus Naur Form

Backus Naur Form[4] (BNF) provides an elegant means of specifying the syntax of programming languages. It was originally employed to define the grammar for the Algol-60 programming language [2], and a variant was used by Wirth to specify the syntax of the Pascal programming language. BNF is widely used and accepted today as the way to specify the syntax of programming languages.

---

[4]Backus Naur Form is named after John Backus and Peter Naur. It was created as part of the design of the Algol 60 programming language, and is used to define the syntax rules of the language.

BNF specifications essentially describe the external appearance of programs as seen by the programmer. The grammar of a context-free grammar may then be input into a parser (e.g., Yacc), and the parser is used to determine if a program is syntactically correct or not.

A BNF specification consists of a set of production rules with each production rule describing the form of a class of language elements such expressions, statements and so on. A production rule is of the form

$$<\text{symbol}> ::= <\text{expression with symbols}>$$

where <symbol> is a *nonterminal*, and the expression consists of sequence of terminal and nonterminal symbols. A construct that has alternate forms appears more than once, and this is expressed by sequences separated by the vertical bar "|" (which indicates a choice). In other word, there is more than one possible substitution for the symbol on the left-hand side of the rule. Symbols that never appear on the left-hand side of a production rule are called *terminals*.

The following example defines the syntax of various statements in a sample programming language:

<loop statement> ::= <while loop> | <for loop>
<while loop> ::= while (<condition>) <statement>
<for loop> ::= for (<expression>) <statement>
<statement> ::= <assignment statement> | <loop statement>
<assignment statement> ::= <variable> := <expression>

This is a partial definition of the syntax of various statements in the language. It includes various nonterminals such as (<loop statement>, <while loop> and so on. The terminals include 'while', 'for', ':=', '("and")'. The production rules for <condition> and <expression> are not included.

The grammar of a context-free language (e.g. LL(1), LL(k), LR(1), LR(k)) grammar expressed in BNF notation) may be translated by a parser into a parse table. The parse table may then be employed to determine whether a particular program is valid with respect to its grammar.

Example 12.1 (Context-free grammar) The example considered is that of parenthesis matching in which there are two terminal symbols and one nonterminal symbol

$$S \rightarrow SS$$
$$S \rightarrow (S)$$
$$S \rightarrow ()$$

Then by starting with S and applying the rules we can construct

$$S \rightarrow SS \rightarrow (S)S \rightarrow (())S \rightarrow (())()$$

Example 12.2 (Context-free grammar) The example considered is that of expressions in a programming language. The definition is ambiguous as there is more than one derivation tree for some expressions (e.g., there are two parse trees for the expression $5 \times 3 + 1$ discussed below).

$$\begin{aligned}
\text{<expr>} &::= \text{<numeral>} \mid (\text{<expr>}) \\
&\mid (\text{<expr>} \text{<operator>} \text{<expr>}) \\
\text{<operator>} &::= + \mid - \mid \times \mid / \\
\text{<digit>} &::= 0 \mid 1 \mid \dots \mid 9 \\
\text{<numeral>} &::= \text{<digit>} \mid \text{<digit>} \text{<numeral>}
\end{aligned}$$

Example 12.3 (Regular Grammar) The definition of an identifier in most programming languages is similar to

$$\begin{aligned}
\text{<identifier>} &::= \text{<let>} \text{<letdig>} \\
\text{<letdig>} &::= \text{<let>} \mid \text{<dig>} \mid \varepsilon \\
\text{<letdig>} &::= \text{<let>} \text{<letdig>} \mid \text{<dig>} \text{<letdig>} \\
\text{<let>} &::= a \mid b \mid c \mid \dots \mid z \\
\text{<dig>} &::= 0 \mid 1 \mid \dots \mid 9
\end{aligned}$$

12.3.2 Parse Trees and Derivations

Let A and N be the terminal and nonterminal alphabet of a rewriting system and let $\text{<X>} \to w$ be a production. Let x be a word in $(A \cup N)^*$ with $x = u \text{<X>} v$ for some words $u, v \in (A \cup N)^*$. Then x is said to directly yield uwv and this is written as $x \Rightarrow uwv$.

This single substitution ($\Rightarrow$) can be extended by a finite number of productions ($\Rightarrow^*$), and this gives the set of words that can be obtained from a given word. This derivation is achieved by applying several production rules (one production rule is applied at a time) in the grammar.

That is, given $x, y \in (A \cup N)^*$ then x yields y (or y is a derivation of x) if $x = y$, or there exists a sequence of words $w_1, w_2, \dots, w_n \in (A \cup N)^*$ such that $x = w_1$, $y = w_n$ and $w_i \Rightarrow w_{i+1}$ for $1 \leq i \leq n - 1$. This is written as $x \Rightarrow^* y$.

The expression grammar presented in Example 12.2 is ambiguous, and this means that an expression such as $5 \times 3 + 1$ has more than one interpretation. (Figures 12.2 and 12.3). It is not clear from the grammar whether multiplication is performed first and then addition, or whether addition is performed first and then multiplication.

The first parse tree is given in Fig. 12.2, and the interpretation of the first parse tree is that multiplication is performed first and then addition (this is the normal interpretation of such expressions in programming languages as multiplication is a higher precedence operator than addition).

The interpretation of the second parse tree is that addition is performed first and then multiplication (Fig. 12.3). It may seem a little strange that one expression has

Fig. 12.2 Parse tree
5 × 3 + 1

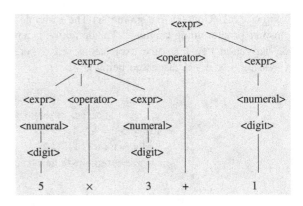

Fig. 12.3 Parse tree
5 × 3 + 1

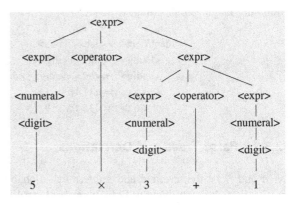

two parse trees and it shows that the grammar is ambiguous. This means that there is a choice for the compiler in evaluating the expression, and the compiler needs to assign the right meaning to the expression. For the expression grammar one solution would be for the language designer to alter the definition of the grammar to remove the ambiguity.

12.4   Programming Language Semantics

The formal semantics of a programming language is concerned with defining the actual meaning of a language. Language semantics is deeper than syntax, and the theory of the syntax of programming languages is well established. A programmer writes a program according to the rules of the language. The compiler first checks the program for syntactic correctness: i.e., it determines whether the program written is valid according to the rules of the grammar of the language. If the program is syntactically correct, then the compiler determines the meaning of what has been written and generates the corresponding machine code.[5]

---

[5]Of course, what the programmer has written may not be what the programmer had intended.

Table 12.2 Programming language semantics

Approach	Description
Axiomatic semantics	This involves giving meaning to phrases of the language using logical axioms It employs *pre-* and *post-condition assertion*s to specify what happens when the statement executes. The relationship between the initial assertion and the final assertion essentially gives the semantics of the code
Operational semantics	This approach describes how a valid program is interpreted as sequences of computational steps. These sequences then define the meaning of the program An abstract machine (SECD machine) may be defined to give meaning to phrases, and this is done by describing the transitions they induce on states of the machine
Denotational semantics	This approach provides meaning to programs in terms of mathematical objects such as integers, tuples and functions Each phrase in the language is translated into a mathematical object that is the *denotation* of the phrase

The compiler must preserve the semantics of the language: i.e., the semantics are not defined by the compiler, but rather the function of the compiler is to preserve the semantics of the language. Therefore, there is a need to have an unambiguous definition of the meaning of the language independently of the compiler, and the meaning is then preserved by the compiler.

A program's syntax[6] gives no information to the meaning of the program, and therefore there is a need to supplement the syntactic description of the language with a formal unambiguous definition of its semantics.

It is possible to utter syntactically correct but semantically meaningless sentences in a natural language. Similarly, it is possible to write syntactically correct programs that behave in quite a different way from the intention of the programmer.

The formal semantics of a language is given by a mathematical model that describes the possible computations described by the language. There are three main approaches to programming language semantics namely axiomatic semantics, operational semantics and denotational semantics (Table 12.2):

There are several applications of programming language semantics including language design, program verification, compiler writing and language standardization. The three main approaches to semantics are described in more detail below.

12.4.1 Axiomatic Semantics

Axiomatic semantics gives meaning to phrases of the language by describing the logical axioms that apply to them. It was developed by C.A.R. Hoare[7] in a famous paper "*An axiomatic basis for computer programming*" [3]. His axiomatic theory consists of *syntactic elements*, *axioms* and *rules of inference*.

---

[6]There are attribute (or affix) grammars that extend the syntactic description of the language with supplementary elements covering the semantics. The process of adding semantics to the syntactic description is termed decoration.

[7]Hoare was influenced by earlier work by Floyd on assigning meanings to programs using flowcharts [4].

The well-formed formulae that are of interest in axiomatic semantics are pre–post assertion formulae of the form $P\{a\}Q$, where a is an instruction in the language and P and Q are assertions: i.e., properties of the program objects that may be true or false.

An *assertion* is essentially a predicate that may be true in some states and false in other states. For example, the assertion $(x - y > 5)$ is true in the state in which the values of x and y are 7 and 1, respectively, and false in the state where x and y have values 4 and 2.

The pre- and post-condition assertions are employed to specify what happens when the statement executes. The relationship between the initial assertion and the final assertion gives the semantics of the code statement. The *pre-*and *post-condition* assertions are of the form

$$P\{a\}Q$$

The precondition P is a predicate (input assertion), and the postcondition Q is a predicate (output assertion). The braces separate the assertions from the program fragment. The well-formed formula $P\{a\}Q$ is itself a predicate that is either true or false.

This notation expresses the *partial correctness*[8] of a with respect to P and Q, and its meaning is that if statement a is executed in a state in which the predicate P is true and execution terminates, then it will result in a state in which assertion Q is satisfied.

The axiomatic semantics approach is described in more detail in [5], and the axiomatic semantics of a selection of statements is presented below.

• **Skip**

The skip statement does nothing and whatever condition is true on entry to the command is true on exit from the command. It's meaning is given by:

$$P\{skip\}P$$

• **Assignment**

The meaning of the assignment statement is given by the axiom

$$P_e^x\{x := e\}P$$

The meaning of the assignment statement is that P will be true after execution of the assignment statement if and only if the predicate P_e^x with the value of x replaced

---

[8]Total correctness is expressed using $\{P\}a\{Q\}$ amd program fragment a is totally correct for precondition P and postcondition Q if and only if whenever a is executed in any state in which P is satisfied then execution terminates, and the resulting state satisfies Q.

by e in P is true before execution (since x will contain the value of e after execution).

The notation P^x_e denotes the expression obtained by substituting e for all free occurrences of x in P.

- **Compound**

The meaning of the conditional command is:

$$\frac{P\{S_1\}Q, Q\{S_2\}R}{P\{S_1; S_2\}R}$$

The compound statement involves the execution of S_1 followed by the execution of S_2. The meaning of the compound statement is that R will be true after the execution of the compound statement S_1; S_2 provided that P is true, if it is established that Q will be true after the execution of S_1 provided that P is true, and that R is true after the execution of S_2 provided Q is true.

There needs to be at least one rule associated with every construct in the language in order to give its axiomatic semantics. The semantics of other programming language statements such as the 'while' statement and the 'if' statement are described in [5].

12.4.2 Operational Semantics

The operational semantics definition is similar to an interpreter, where the semantics of a language are expressed by a mechanism that makes it possible to determine the effect of any program in the language. The meaning of a program is given by the evaluation history that an interpreter produces when it interprets the program. The interpreter may be close to an executable programming language or it may be a mathematical language.

The operational semantics for a programming language describes how a valid program is interpreted as sequences of computational steps. The evaluation history then defines the meaning of the program, and this is a sequence of internal interpreter configurations.

One early use of operational semantics was the work done by John McCarthy in the late 1950s on the semantics of LISP in terms of the lambda calculus. The use of lambda calculus allows the meaning of a program to be expressed using a mathematical interpreter, and this offers precision through the use of mathematics.

The meaning of a program may be given in terms of a hypothetical or virtual machine that performs the set of actions that corresponds to the program. An abstract machine (SECD machine[9]) may be defined to give meaning to phrases in

---

[9]This virtual stack based machine was originally designed by Peter Landin to evaluate lambda calculus expressions, and it has since been used as a target for several compilers.

the language, and this is done by describing the transitions that they induce on states of the machine.

Operational semantics give an intuitive description of the programming language being studied, and its descriptions are close to real programs. It can play a useful role as a testing tool during the design of new languages, as it is relatively easy to design an interpreter to execute the description of example programs. This allows the effects of new languages or new language features to be simulated and studied through actual execution of the semantic descriptions prior to writing a compiler for the language. Another words, operational semantics can play a role in rapid prototyping during language design, and to get early feedback on the suitability of the language.

One disadvantage of the operational approach is that the meaning of the language is understood in terms of execution: i.e., in terms of interpreter configurations, rather than in an explicit *machine independent specification*. An operational description is just one way to execute programs. Another disadvantage is that the interpreters for non-trivial languages often tend to be large and complex.

A more detailed account of operational semantics is in [6, 7].

12.4.3 Denotational Semantics

Denotational semantics [7] expresses the semantics of a programming language by a translation schema that associates a meaning (denotation) with each program in the language. It maps a program directly to its meaning, and it was originally called mathematical semantics as it provides meaning to programs in terms of mathematical values such as integers, tuples and functions. That is, the meaning of a program is a mathematical object, and an interpreter is not employed. Instead, a valuation function is employed to map a program directly to its meaning, and the denotational description of a programming language is given by a set of *meaning functions* M associated with the constructs of the language (Fig. 12.4).

Each meaning function is of the form $M_T : T \rightarrow D_T$, where T is some construct in the language and D_T is some semantic domain. Many of the meaning functions will be "higher-order": i.e., functions that yield functions as results. The signature of the meaning function is from syntactic domains (i.e., T) to semantic domains (i.e., D_T). A valuation map $V_T : T \rightarrow \mathbf{B}$ may be employed to check the static semantics prior to giving a meaning of the language construct.[10]

A denotational definition is more abstract than an operational definition. It does not specify the computational steps and its exclusive focus is on the programs to the exclusion of the state and other data elements. The state is less visible in denotational specifications.

It was developed by Christopher Strachey and Dana Scott at the Programming Research Group at Oxford, England in the mid-1960s, and their approach to

---

[10]This is similar to what a compiler does in that if errors are found during the compilation phase, the compiler halts and displays the errors and does not continue with code generation.

Fig. 12.4 Denotational
semantics

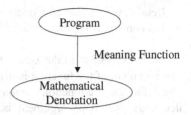

semantics is known as the Scott-Strachey approach [8]. It provided a mathematical
foundation for the semantics of programming languages.

Dana Scott's contributions included the formulation of domain theory, and this
allowed programs containing recursive functions and loops to be given a precise
semantics. Each phrase in the language is translated into a mathematical object that
is the *denotation* of the phrase. Denotational Semantics has been applied to language design and implementation.

12.5 Lambda Calculus

Functions (discussed in Chap. 2) are an essential part of mathematics, and they play
a key role in specifying the semantics of programming language constructs. We
discussed partial and total functions in Chap. 2, and a function was defined as a
special type of relation, and simple finite functions may be defined as an explicit set
of pairs: e.g.,

$$f \underline{\Delta} \{(a, 1), (b, 2), (c, 3)\}$$

However, for more complex functions there is a need to define the function more
abstractly, rather than listing all of its member pairs. This may be done in a similar
manner to set comprehension, where a set is defined in terms of a characteristic
property of its members.

Functions may be defined (by comprehension) through a powerful abstract
notation known as lambda calculus. This notation was introduced by Alonzo
Church in the 1930s to study computability, and lambda calculus provides an
abstract framework for describing mathematical functions and their evaluation. It
may be used to study function definition, function application, parameter passing
and recursion.

Any computable function can be expressed and evaluated using lambda calculus
or Turing machines, as these are equivalent formalisms. Lambda calculus uses a
small set of transformation rules, and these include

- Alpha-conversion rule (α-conversion)[11]

---

[11]This essentially expresses that the names of bound variables is unimportant.

- Beta-reduction rule (β-reduction)[12]
- Eta-conversion (η-conversion)[13]

Every expression in the λ-calculus stands for a function with a single argument. The argument of the function is itself a function with a single argument, and so on. The definition of a function is anonymous in the calculus. For example, the function that adds one to its argument is usually defined as $f(x) = x + 1$. However, in λ-calculus the function is defined as

$$succ \underline{\Delta} \lambda x \cdot x + 1$$

The name of the formal argument x is irrelevant and an equivalent definition of the function is $\lambda z \cdot z + 1$. The evaluation of a function f with respect to an argument (e.g. 3) is usually expressed by $f(3)$. In λ-calculus this would be written as $(\lambda x \cdot x + 1) 3$, and this evaluates to $3 + 1 = 4$. Function application is *left associative*: i.e., $f x y = (f x) y$. A function of two variables is expressed in lambda calculus as a function of one argument, which returns a function of one argument. This is known as *currying*: e.g., the function $f(x, y) = x + y$ is written as $\lambda x \cdot \lambda y \cdot x + y$. This is often abbreviated to $\lambda x y \cdot x + y$. λ-calculus is a simple mathematical system, and its syntax is defined as follows:

```
<exp> ::= <identifier>        |
          λ <identifier>.<exp> | --abstraction
          <exp> <exp>          | --application
          ( <exp> )
```

λ-Calculus's four lines of syntax plus *conversion* rules, are sufficient to define Booleans, integers, data structures and computations on them. It inspired Lisp and modern functional programming languages. The original calculus was untyped, but typed lambda calculi have been introduced in recent years. The typed lambda calculus allows the sets to which the function arguments apply to be specified. For example, the definition of the *plus* function is given as:

$$plus \underline{\Delta} \lambda a, b : \mathbb{N} \cdot a + b$$

The lambda calculus makes it possible to express properties of the function without reference to members of the base sets on which the function operates. It allows functional operations such as function composition to be applied, and one key benefit is that the calculus provides powerful support for higher order functions. This is important in the expression of the denotational semantics of the constructs of programming languages.

---

[12]This essentially expresses the idea of function application.

[13]This essentially expresses the idea that two functions are equal if and only if they give the same results for all arguments.

12.6 Lattices and Order

This section considers some of the mathematical structures used in the definition of the semantic domains used in denotational semantics. These mathematical structures may also be employed to give a secure foundation for recursion (discussed in Chap. 4), and it is essential that the conditions in which recursion may be used safely be understood.

It is natural to ask when presented with a recursive definition whether it means anything at all, and in some cases the answer is negative. Recursive definitions are a powerful and elegant way of giving the denotational semantics of language constructs. The mathematical structures considered in this section include partial orders, total orders, lattices, complete lattices and complete partial orders.

12.6.1 Partially Ordered Sets

A *partial order* $\leq$ on a set P is a binary relation such that for all x, y, $z \in P$ the following properties hold:

(i) $x \leq x$ *(Reflexivity)*
(ii) $x \leq y$ and $y \leq x \Rightarrow x = y$ *(Anti−symmetry)*
(iii) $x \leq y$ and $y \leq z \Rightarrow x \leq z$ *(Transitivity)*

A set P with an order relation $\leq$ is said to be a *partially ordered* set (Fig. 12.5).

Example 12.4 Consider the power set $\mathbb{P}X$, which consists of all the subsets of the set X with the ordering defined by set inclusion. That is, $A \leq B$ if and only if $A \subseteq B$ then $\subseteq$ is a partial order on $\mathbb{P}X$.

A partially ordered set is a *totally ordered* set *(also called chain)* if for all x, $y \in P$ then either $x \leq y$ or $y \leq x$. That is, any two elements of P are directly comparable.

A partially ordered set P is an *anti-chain* if for any x, y in P then $x \leq y$ only if $x = y$. That is, the only elements in P that are comparable to a particular element are the element itself.

Fig. 12.5 Pictorial representation of a partial order

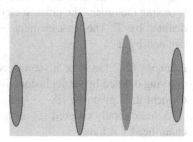

Maps between Ordered Sets

Let P and Q be partially ordered sets then a map ϕ from P to Q may preserve the order in P and Q. We distinguish between order preserving, order embedding and order isomorphism. These terms are defined as follows:

Order Preserving (or *Monotonic Increasing Function*)
A mapping $\phi: P \to Q$ is said to be order preserving if

$$x \leq y \Rightarrow \phi(x) \leq \phi(y)$$

Order Embedding
A mapping $\phi: P \to Q$ is said to be an order embedding if

$$x \leq y \text{ in } P \text{ if and only if } \phi(x) \leq \phi(y) \text{ in } Q.$$

Order Isomorphism
The mapping $\phi: P \to Q$ is an order isomorphism if and only if it is an order embedding mapping onto Q.

Dual of a Partially Ordered Set

The dual of a partially ordered set P (denoted P^{∂}) is a new partially ordered set formed from P where $x \leq y$ holds in P^{∂} if and only if $y \leq x$ holds in P (i.e., P^{∂} is obtained by reversing the order on P).

For each statement about P there is a corresponding statement about P^{∂}. Given any statement Φ about a partially ordered set, then the dual statement Φ^{∂} is obtained by replacing each occurrence of $\leq$ by $\geq$ and vice versa.

Duality Principle

Given that statement Φ is true of a partially ordered set P, then the statement Φ^{∂} is true of P^{∂}.

Maximal and Minimum Elements

Let P be a partially ordered set and let $Q \subseteq P$ then

(i) $a \in Q$ is a *maximal* element of Q if $a \leq x \in Q \Rightarrow a = x$.
(ii) $a \in Q$ is the *greatest* (or *maximum*) element of Q if $a \geq x$ for every $x \in Q$, and in that case we write $a = \max Q$

A *minimal* element of Q and the *least* (or *minimum*) are defined dually by reversing the order. The greatest element (if it exists) is called the top element and is denoted by $\top$. The least element (if it exists) is called the bottom element and is denoted by $\bot$.

Example 12.5 Let X be a set and consider $\mathbb{P}X$ the set of all subsets of X with the ordering defined by set inclusion. The top element $\top$ is given by X, and the bottom element $\bot$ is given by $\emptyset$.

A finite totally ordered set always has top and bottom elements, but an infinite chain need not have.

12.6.2 Lattices

Let P be a partially ordered set and let $S \subseteq P$. An element $x \in P$ is an upper bound of S if $s \leq x$ for all $s \in S$. A lower bound is defined similarly.

The set of all upper bounds for S is denoted by S^u, and the set of all lower bounds for S is denoted by S^l.

$$S^u = \{x \in P | (\forall s \in S) s \leq x\}$$
$$S^l = \{x \in P | (\forall s \in S) s \geq x\}$$

If S^u has a least element x then x is called the *least upper bound* of S. Similarly, if S^l has a greatest element x then x is called the *greatest lower bound* of S.

Another words, x is the least upper bound of S if

(i) x is an upper bound of S.
(ii) $x \leq y$ for all upper bounds y of S

The least upper bound of S is also called the *supremum* of S denoted (sup S), and the greatest lower bound is also called the infimum of S, and is denoted by inf S.

Join and Meet Operations
The *join* of x and y (denoted by $x \vee y$) is given by $\sup\{x, y\}$ when it exists. The *meet* of x and y (denoted by $x \wedge y$) is given by $\inf\{x, y\}$ when it exists.

The supremum of S is denoted by $\vee S$, and the infimum of S is denoted by $\in S$.

Definition
Let P be a non-empty partially ordered set then

(i) If $x \vee y$ and $x \wedge y$ exist for all $x, y \in P$ then P is called a *lattice*.
(ii) If $\vee S$ and $\wedge S$ exist for all $S \subseteq P$ then P is called a *complete lattice*

Every non-empty finite subset of a lattice has a meet and a join (inductive argument can be used), and every finite lattice is a complete lattice. Further, any complete lattice is bounded—i.e., it has top and bottom elements (Fig. 12.6).

Example 12.6 Let X be a set and consider $\mathbb{P}X$ the set of all subsets of X with the ordering defined by set inclusion. Then $\mathbb{P}X$ is a complete lattice in which

$$\vee \{A_i | i \in I\} = \cup A_i$$
$$\wedge \{A_i | i \in I\} = \cap A_i$$

Consider the set of natural numbers $\mathbb{N}$ and consider the usual ordering of $<$. Then $\mathbb{N}$ is a lattice with the join and meet operations defined as

Fig. 12.6 Pictorial
representation of a complete
lattice

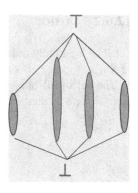

$$x \vee y = \max(x, y)$$
$$x \wedge y = \min(x, y)$$

Another possible definition of the meet and join operations are in terms of the greatest common multiple and least common divisor.

$$x \vee y = \text{lcm}(x, y)$$
$$x \wedge y = \text{gcd}(x, y)$$

12.6.3 Complete Partial Orders

Let S be a non-empty subset of a partially ordered set P. Then

(i) S is said to be a *directed set* if for every finite subset F of S there exists $z \in S$ such that $z \in F^u$.
(ii) S is said to be *consistent* if for every finite subset F of S there exists $z \in P$ such that $z \in F^u$

A partially ordered set P is a *complete partial order* (CPO) if:

(i) P has a bottom element $\bot$
(ii) $\bigvee D$ exists for each directed subset D of P

The simplest example of a directed set is a chain, and we note that any complete lattice is a complete partial order, and that any finite lattice is a complete lattice.

12.6.4 Recursion

Recursive definitions arise frequently in programs and offer an elegant way to define routines and data types. A recursive routine contains a direct or indirect call to itself, and a recursive data type contains a direct or indirect reference to specimens of the same type. Recursion needs to be used with care, as there is always a danger that the recursive definition may be circular (i.e., defines nothing). It is therefore important to investigate when a recursive definition may be used safely, and to give a mathematical definition of recursion.

The control flow in a recursive routine must contain at least one non-recursive branch since if all possible branches included a recursive form the routine could never terminate. The value of at least one argument in the recursive call is different from the initial value of the formal argument as otherwise the recursive call would result in the same sequence of events and therefore would never terminate.

The mathematical meaning of recursion is defined in terms of *fixed point theory*, which is concerned with determining solutions to equations of the form $x = \tau(x)$, where the function τ is of the form $\tau : X \rightarrow X$.

A recursive definition may be interpreted as a fixpoint equation of the form $f = \Phi(f)$; i.e., the fixpoint of a high-level functional Φ that takes a function as an argument. For example, consider the functional Φ defined as follows:

$$\Phi \triangleq \lambda f \lambda n \cdot \text{if } n = 0 \text{ then } 1 \text{ else } n^* f(n-1)$$

Then a fixpoint of Φ is a function f such that $f = \Phi(f)$ or another words

$$f = \lambda n \cdot \text{if } n = 0 \text{ then } 1 \text{ else } n^* f(n-1)$$

Clearly, the factorial function is a fixpoint of Φ, and it is the only total function that is a fixpoint. The solution of the equation $f = \Phi(f)$ (where Φ has a fixpoint) is determined as the limit f of the sequence of functions $f_0, f_1, f_2, \ldots$, where the f_i are defined inductively as

$$f_0 \triangleq \varnothing \quad \text{(the empty partial function)}$$
$$f_i \triangleq \Phi(f_{i-1})$$

Each f_i may be viewed as a successive approximation to the true solution f of the fixpoint equation, with each f_i bringing a little more information on the solution than its predecessor f_{i-1}.

The function f_i is defined for one more value than f_{i-1}, and gives the same result for any value for which they are both defined. The definition of the factorial function is thus built up as follows:

$$f_0 \underline{\Delta} \varnothing \qquad\qquad \text{(the empty partial function)}$$

$$f_1 \underline{\Delta} \{0 \rightarrow 1\}$$

$$f_2 \underline{\Delta} \{0 \rightarrow 1, 1 \rightarrow 1\}$$

$$f_3 \underline{\Delta} \{0 \rightarrow 1, 1 \rightarrow 1, 2 \rightarrow 2\}$$

$$f_4 \underline{\Delta} \{0 \rightarrow 1, 1 \rightarrow 1, 2 \rightarrow 2, 3 \rightarrow 6\}$$

$$\vdots \qquad\qquad \vdots$$

$$\vdots$$

For every i, the domain of f_i is the interval 1, 2, ... $i - 1$ and $f_i(n) = n!$ for any n in this interval. Another word f_i is the factorial function restricted to the interval 1, 2, ... $i - 1$. The sequence of f_i may be viewed as successive approximations of the true solution of the fixpoint equation (which is the factorial function), with each f_i bringing defined for one more value than its predecessor f_{i-1}, and defining the same result for any value for which they are both defined.

The candidate fixpoint f_∞ is the limit of the sequence of functions f_i, and is the union of all the elements in the sequence. It may be written as follows:

$$f_\infty \underline{\Delta} \varnothing \cup \Phi(\varnothing) \cup \Phi(\Phi(\varnothing)) \cup \ldots = \cup_{i:\mathbb{N}} f_i,$$

where the sequence f_i is defined inductively as

$$f_0 \underline{\Delta} \varnothing \qquad\qquad \text{(the empty partial function)}$$

$$f_{i+1} \underline{\Delta} f_i \cup \Phi(f_i)$$

This forms a subset chain where each element is a subset of the next, and it follows by induction that

$$f_{i+1} = \cup_{j:0\ldots i} \Phi(f_i)$$

A general technique for solving fixpoint equations of the form $h = \tau(h)$ for some functional τ is to start with the least-defined function $\varnothing$ and iterate with τ. The union of all the functions obtained as successive sequence elements is the fixpoint.

The conditions in which f_∞ is a fixpoint of Φ is the requirement for $\Phi(f_\infty) = f_\infty$. This is equivalent to:

$$\Phi(\cup_{i:\mathbb{N}} f_i) = \cup_{i:\mathbb{N}} f_i$$

$$\Phi(\cup_{i:\mathbb{N}} f_i) = \cup_{i:\mathbb{N}} \Phi(f_i)$$

A sufficient point for Φ to have a fixpoint is that the property $\Phi(\cup_{i:\mathbb{N}} f_i) = \cup_{i:\mathbb{N}} \Phi(f_i)$ holds for any subset chain f_i.

We discussed recursion earlier in Chap. 4, and a more detailed account on the mathematics of recursion is in Chap. 8 of [7].

12.7 Review Questions

1. Explain the difference between syntax and semantics.
2. Describe the Chomsky hierarchy of grammars and give examples of each type.
3. Show that a grammar may be ambiguous leading to two difference parse trees. What problems does this create and how should it be dealt with?
4. Describe axiomatic semantics, operation semantics and denotational semantics and explain the differences between them.
5. Explain partial orders, lattices and complete partial orders. Give examples of each.
6. Show how the meaning of recursion is defined with fixpoint theory.

12.8 Summary

This chapter considered two key parts to any programming language namely syntax and semantics. The syntax of the language is concerned with the production of grammatically correct programs in the language, whereas the semantics of the language is deeper and is concerned with the meaning of what has been written by the programmer.

There are several approaches to defining the semantics of programming languages, and these include axiomatic, operational and denotational semantics. Axiomatic semantics is concerned with defining properties of the language in terms of axioms; operational semantics is concerned with defining the meaning of the language in terms of an interpreter; and denotational semantics is concerned with defining the meaning of the phrases in a language by the denotation or mathematical meaning of the phrase.

Compilers are programs that translate a program that is written in some programming language into another form. It involves syntax analysis and parsing to check the syntactic validity of the program; semantic analysis to determine what the program should do; optimization to improve the speed and performance of the compiler; and code generation in some target language.

Various mathematical structures including partial orders, total orders, lattices and complete partial orders were considered. These are useful in the definition of the denotational semantics of a language, and in giving a mathematical interpretation of recursion.

References

1. Introduction to the History of Computing. Gerard O'Regan. Springer Verlag. 2016.
2. Report on the Algorithmic Language, ALGOL 60. Edited by P. Naur. Communications of the ACM, 3(5):299–314, 1960.
3. An Axiomatic Basis for Computer Programming. C.A.R. Hoare. Communications of the ACM. 12(10):576–585. 1969.
4. Assigning Meanings to Programs. Robert Floyd. Proceedings of Symposia in Applied Mathematics, (19), 19-32. 1967.
5. Mathematical Approaches to Software Quality. Gerard O' Regan. Springer. 2006.
6. A Structural Approach to Operational Semantics. Gordon Plotkin. Technical Report DAIM FN-19. Computer Science Department. AarhusUniversity, Denmark. 1981.
7. Introduction to the Theory of Programming Languages. Bertrand Meyer. Prentice Hall. 1990.
8. Denotational Semantics. The Scott-Strachey Approach to Programming Language Theory. Joseph Stoy. MIT Press. 1977.

Computability and Decidability

13

13.1 Introduction

It is impossible for a human or machine to write out all of the members of an infinite countable set, such as the set of natural numbers $\mathbb{N}$. However, humans can do something quite useful in the case of certain enumerable infinite sets: they can give explicit instructions (that may be followed by a machine or another human) to produce the nth member of the set for an arbitrary finite n. The problem remains that for all but a finite number of values of n it will be physically impossible for any human or machine to actually carry out the computation, due to the limitations on the time available for computation, the speed at which the individual steps in the computation may be carried out, and due to finite materials.

The intuitive meaning of computability is in terms of an algorithm (or effective procedure) that specifies a set of instructions to be followed to complete the task. Another words, a function f is *computable* if there exists an algorithm that produces the value of f correctly for each possible argument of f. The computation of f for a particular argument x just involves following the instructions in the algorithm, and

© Springer International Publishing Switzerland 2016   207
G. O'Regan, *Guide to Discrete Mathematics*, Texts in Computer Science,
DOI 10.1007/978-3-319-44561-8_13

it produces the result $f(x)$ in a finite number of steps if x is in the domain of f. If x is not in the domain of f then the algorithm may produce an answer saying so or it might run forever never halting. A computer program implements an algorithm.

The concept of computability may be made precise in several equivalent ways such as Church's *lambda calculus*, *recursive function theory* or by the theoretical *Turing machines.*[1] These are all equivalent and perhaps the most well known is the Turing machine (discussed in Chap. 7). This is a mathematical machine with a potentially infinite tape divided into frames (or cells) in which very basic operations can be carried out. The set of functions that are computable are those that are computable by a Turing machine.

Decidability is an important topic in contemporary mathematics. Church and Turing independently showed in 1936 that mathematics is not decidable. In other words, there is no mechanical procedure (i.e., algorithm) to determine whether an arbitrary mathematical proposition is true or false, and so the only way is to determine the truth or falsity of a statement is try to solve the problem. The fact that there is no a general method to solve all instances of a specific problem, as well as the impossibility of proving or disproving certain statements within a formal system may suggest limitations to human and machine knowledge.

13.2 Logicism and Formalism

Gottlob Frege (Fig. 14.8) was a nineteenth century German mathematician and logician who invented a formal system which is the basis of modern predicate logic. It included axioms, definitions, universal and existential quantification and formalization of proof. His objective was to show that mathematics was reducible to logic (logicism) but his project failed as one of the axioms that he had added to his system proved to be inconsistent.

This inconsistency was pointed out by Bertrand Russell, and it is known as *Russell's paradox.*[2] Russell later introduced the theory of types to deal with the paradox, and he jointly published *Principia Mathematica* with Whitehead as an attempt to derive the truths of arithmetic from a set of logical axioms and rules of inference.

The sentences of Frege's logical system denote the truth values of true or false. The sentences may include expressions such as equality ($x = y$), and this returns true if x is the same as y, and false otherwise. Similarly, a more complex expression such as $f(x, y, z) = w$ is true if $f(x, y, z)$ is identical with w, and false otherwise. Frege represented statements such as "5 is a prime" by "$P(5)$" where $P()$ is termed a concept. The statement $P(x)$ returns true if x is prime and false otherwise. His

---

[1]The Church-Turing Thesis states that anything that is computable is computable by a Turing Machine.

[2]Russell's paradox (discussed in Chap. 2) considers the question as to whether the set of all sets that contain themselves as members is a set. In either case there is a contradiction.

Fig. 13.1 David Hilbert

approach was to represent a predicate as a function of one variable which returns a Boolean value of true or false.

Formalism was proposed by Hilbert (Fig. 13.1) as a foundation for mathematics in the early twentieth century. The motivation for the programme was to provide secure foundations for mathematics, and to resolve the contradictions in the formalization of set theory identified by Russell's paradox. The presence of a contradiction in a theory means the collapse of the whole theory, and so it was seen as essential that there be a proof of the consistency of the formal system. The methods of proof in mathematics are formalized with axioms and rules of inference.

Formalism is a formal system that contains meaningless symbols together with rules for manipulating them. The individual formulas are certain finite sequences of symbols obeying the syntactic rules of the formal language. A formal system consists of

- A formal language
- A set of axioms
- Rules of inference.

The expressions in a formal system are terms, and a term may be simple or complex. A simple term may be an object such as a number, and a complex term may be an arithmetic expression such as $4^3 + 1$. A complex term is formed via functions, and the expression above uses two functions namely the cube function with argument 4 and the plus function with two arguments.

A formal system is generally intended to represent some aspect of the real world. A rule of inference relates a set of formulas $(P_1, P_2, ..., P_k)$ called the premises to the consequence formula P called the conclusion. For each rule of inference there is a finite procedure for determining whether a given formula Q is an immediate consequence of the rule from the given formulas $(P_1, P_2, ..., P_k)$. A *proof* in a formal system consists of a finite sequence of formulae, where each formula is either an axiom or derived from one or more preceding formulae in the sequence by one of the rules of inference.

Hilbert's programme was concerned with the formalization of mathematics (i.e. the axiomatization of mathematics) together with a proof that the axiomatization was consistent (i.e., there is no formula A such that both A and $\neg A$ are deducible in the calculus). The specific objectives of Hilbert's programme were to

- Provide a formalism of mathematics.
- Show that the formalization of mathematics was *complete*: i.e. all mathematical truths can be proved in the formal system.
- Provide a proof that the formal system is *consistent* (i.e. that no contradictions may be derived).
- Show that mathematics is *decidable*: i.e. there is an algorithm to determine the truth of falsity of any mathematical statement.

The formalist movement in mathematics led to the formalization of large parts of mathematics, where theorems could be proved using just a few mechanical rules. The two most comprehensive formal systems developed were *Principia Mathematica* by Russell and Whitehead, and the axiomatization of set theory by Zermelo–Fraenkel (subsequently developed further by von Neumann).

Principia Mathematica is a comprehensive three volume work on the logical foundations of mathematics written by Bertrand Russell and Alfred Whitehead between 1910 and 1913. Its goal was to show that all of the concepts of mathematics can be expressed in logic, and that all of the theorems of mathematics can be proved using only the logical axioms and rules of inference of logic. It covered set theory, ordinal numbers and real numbers, and it showed that in principle that large parts of mathematics could be developed using *logicism*.

It avoided the problems with contradictions that arose with Frege's system by introducing the theory of types in the system. The theory of types meant that one could no longer speak of the set of all sets, as a set of elements is of a different type from that of each of its elements, and so Russell's paradox was avoided. It remained an open question at the time as to whether the *Principia* were consistent and complete. That is, is it possible to derive all the truths of arithmetic in the system and is it possible to derive a contradiction from the Principia's axioms? However, it was clear from the three volume work that the development of mathematics using the approach of the Principia was extremely lengthy and time consuming.

13.3 Decidability

The question remained whether these axioms and rules of inference are sufficient to decide any mathematical question that can be expressed in these systems. Hilbert believed that every mathematical problem could be solved, and that the truth or falsity of any mathematical proposition could be determined in a finite number of steps. He outlined twenty-three key problems in 1900 that needed to be solved by mathematicians in the twentieth century.

He believed that the formalism of mathematics would allow a mechanical procedure (or algorithm) to determine whether a particular statement was true or false. The problem of the decidability of mathematics is known as the decision problem (*Entscheidungsproblem*).

The question of the decidability of mathematics had been considered by Leibnitz in the seventeenth century. He had constructed a mechanical calculating machine, and wondered if a machine could be built that could determine whether particular mathematical statements are true or false.

Definition 13.1 (*Decidability*) Mathematics is decidable if the truth or falsity of any mathematical proposition may be determined by an algorithm.

Church and Turing independently showed this to be impossible in 1936. Church developed the lambda calculus in the 1930s as a tool to study computability,[3] and he showed that anything that is computable by the lambda calculus. Turing showed that decidability was related to the halting problem for Turing machines, and that therefore if first-order logic was decidable then the halting problem for Turing machines could be solved. However, he had already proved that there was no general algorithm to determine whether a given Turing machine halts. Therefore, first-order logic is undecidable.

The question as to whether a given Turing machine halts or not can be formulated as a first-order statement. If a general decision procedure exists for first-order logic, then the statement of whether a given Turing machine halts or not is within the scope of the decision algorithm. However, Turing had already proved that the halting problem for Turing machines is not computable: i.e. it is not possible algorithmically to decide whether or not any given Turing machine will halt or not. Therefore, since there is no general algorithm that can decide whether any given Turing machine halts, there is no general decision procedure for first-order logic. The only way to determine whether a statement is true or false is to try to solve it. However, if one tries but does not succeed this does not prove that an answer does not exist

There are first-order theories that are decidable. However, first-order logic that includes Peano's axioms of arithmetic (or any formal system that includes addition and multiplication) cannot be decided by an algorithm. That is, there is no algorithm to determine whether an arbitrary mathematical proposition is true or false. Propositional logic is decidable as there is a procedure (e.g. using a truth table) to determine whether an arbitrary formula is valid[4] in the calculus.

Gödel (Fig. 13.2) proved that first-order predicate calculus is *complete*. i.e. all truths in the predicate calculus can be proved in the language of the calculus.

---

[3]The Church Turing Thesis states that anytime that is computable is computable by Lambda Calculus or equivalently by a Turing Machine.

[4]A well-formed formula is valid if it follows from the axioms of first-order logic. A formula is valid if and only if it is true in every interpretation of the formula in the model.

Fig. 13.2 Kurt Gödel

Definition 13.2 (*Completeness*) A formal system is complete if all the truths in the system can be derived from the axioms and rules of inference.

Gödel's *first incompleteness theorem* showed that first-order arithmetic is incomplete; i.e. there are truths in first-order arithmetic that cannot be proved in the language of the axiomatization of first-order arithmetic. Gödel's *second incompleteness theorem* showed that any formal system extending basic arithmetic cannot prove its own consistency within the formal system.

Definition 13.3 (*Consistency*) A formal system is consistent if there is no formula A such that A and $\neg A$ are provable in the system (i.e. there are no contradictions in the system).

13.4 Computability

Alonzo Church (Fig. 13.3) developed the lambda calculus in the mid 1930s, as part of his work into the foundations of mathematics. Turing published a key paper on computability in 1936, which introduced the theoretical machine known as the Turing machine. This machine is computationally equivalent to the lambda

Fig. 13.3 Alonzo Church

calculus, and is capable of performing any conceivable mathematical problem that has an algorithm.

Definition 13.4 (*Algorithm*)An algorithm(or effective procedure) is a finite set of unambiguous instructions to perform a specific task.

A function is *computable* if there is an effective procedure or algorithm to compute f for each value of its domain. The algorithm is finite in length and sufficiently detailed so that a person can execute the instructions in the algorithm. The execution of the algorithm will halt in a finite number of steps to produce the value of $f(x)$ for all x in the domain of f. However, if x is not in the domain of f then the algorithm may produce an answer saying so, or it may get stuck, or it may run forever never halting.

The *Church–Turing Thesis* that states that *any computable function may be computed by a Turing machine*. There is overwhelming evidence in support of this thesis, including the fact that alternative formalizations of computability in terms of lambda calculus, recursive function theory, and post-systems have all been shown to be equivalent to Turing machines.

A Turing machine (discussed previously in Chap. 7) consists of a head and a potentially infinite tape that is divided into cells. Each cell on the tape may be either blank or printed with a symbol from a finite alphabet of symbols. The input tape may initially be blank or have a finite number of cells containing symbols.

At any step, the head can read the contents of a frame. The head may erase a symbol on the tape, leave it unchanged, or replace it with another symbol. It may then move one position to the right, one position to the left, or not at all. If the frame is blank, the head can either leave the frame blank or print one of the symbols.

Turing believed that a human with finite equipment and with an unlimited supply of paper could do every calculation. The unlimited supply of paper is formalized in the Turing machine by a tape marked off in cells.

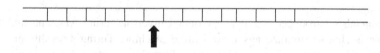

We gave a formal definition of a Turing machine as a 7-tuple $M = (Q, \Gamma, b, \Sigma, \delta, q_0, F)$ in Chap. 7. We noted that the Turing machine is a simple theoretical machine, but it is equivalent to an actual physical computer in the sense that they both compute exactly the same set of functions. A Turing machine is easier to analyse and prove things about than a real computer.

A Turing machine is essentially a finite state machine (FSM) with an unbounded tape. The machine may read from and write to the tape and the tape provides memory and acts as the store. The finite state machine is essentially the control unit of the machine, whereas the tape is a potentially infinite and unbounded store. A real computer has a large but finite store whereas the store in a Turing machine is potentially infinite. However, the store in a real computer may be extended with backing tapes and disks, and in a sense may be regarded as unbounded. The maximum amount of tape that may be read or written within n steps is n.

A Turing machine has an associated set of rules that defines its behaviour. These rules are defined by the transition function that specifies the actions that a machine will perform with respect to a particular input. The behaviour will depend on the current state of the machine and the contents of the tape.

A Turing machine may be programmed to solve any problem for which there is an algorithm. However, if the problem is unsolvable then the machine will either stop in a non-accepting state or compute forever. The solvability of a problem may not be determined beforehand, but, there is, of course, some answer (i.e. either the machine either halts or it computes forever).

Turing showed that there was no solution to the decision problem (*Entscheidungsproblem*) posed by Hilbert. Hilbert believed that the truth or falsity of a mathematical problem may always be determined by a mechanical procedure, and he believed that first-order logic is decidable: i.e. there is a decision procedure to determine if an arbitrary formula is a theorem of the logical system.

Turing was skeptical on the decidability of first-order logic, and the Turing machine played a key role in refuting Hilbert's claim of the decidability of first-order logic.

The question as to whether a given Turing machine halts or not can be formulated as a first-order statement. If a general decision procedure exists for first-order logic, then the statement of whether a given Turing machine halts or not is within the scope of the decision algorithm. However, Turing had already proved that the halting problem for Turing machines is not computable: i.e. it is not possible algorithmically to decide whether a give Turing machine will halt or not. Therefore, there is no general algorithm that can decide whether a given Turing machine halts. In other words, there is no general decision procedure for first-order logic. The only way to determine whether a statement is true or false is to try to solve it.

Turing also introduced the concept of a Universal Turing Machine and this machine is able to simulate any other Turing machine. Turing's results on computability were proved independently of Church's lambda calculus equivalent results in computability. Turing's studied at Princeton University in 1937 and 1938 and was awarded a PhD from the university in 1938. His research supervisor was Alonzo Church.[5]

Question 13.1 (Halting Problem)
Given an arbitrary program is there an algorithm to decide whether the program will finish running or will continue running forever? Another words, given a

---

[5]Alonzo Church was a famous American mathematician and logician who developed the lambda calculus. He also showed that Peano arithmetic and first order logic were undecidable. Lambda calculus is equivalent to Turing machines and whatever may be computed is computable by Lambda calculus or a Turing machine.

program and an input will the program eventually halt and produce an output or will it run forever?

Note (Halting Problem)

The halting problem was one of the first problems that was shown to be undecidable: i.e. there is no general decision procedure or algorithm that may be applied to an arbitrary program and input to decide whether the program halts or not when run with that input.

Proof We assume that there is an algorithm (i.e. a computable function (i, j)) that takes any program i (program i refers to the ith program in the enumeration of all the programs) and arbitrary input j to the program such that

$$H(i,j) = \begin{cases} 1 & \text{If program } i \text{ halts on input } j. \\ 0 & \text{otherwise} \end{cases}$$

We then employ a diagonalization argument[6] to show that every computable total function f with two arguments differs from the desired function H. First, we construct a partial function g from any computable function f with two arguments such that g is computable by some program e.

$$g(i) = \begin{cases} 0 & \text{if } f(i, i) = 0 \\ \text{undefined} & \text{otherwise} \end{cases}$$

There is a program e that computes g and this program is one of the programs in which the halting problem is defined. One of the following two cases must hold

$$g(e) = f(e, e) = 0 \tag{13.1}$$

In this case $H(e, e) = 1$ because e halts on input e.

$$g(e) \text{ is undefined and } f(e, e) \neq 0. \tag{13.2}$$

In this case $H(e, e) = 0$ because the program e does not halt on input e.

In either case, f is not the same function as H. Further, since f was an arbitrary total computable function all such functions must differ from H. Hence, the function H is not computable and there is no such algorithm to determine whether an arbitrary Turing machine halts for an input x. Therefore, the halting problem is not decidable.

---

[6]This is similar to Cantor's diagonalization argument that shows that the Real numbers are uncountable. This argument assumes that it is possible to enumerate all real numbers between 0 and 1, and it then constructs a number whose nth decimal differs from the nth decimal position in the nth number in the enumeration. If this holds for all n then the newly defined number is not among the enumerated numbers.

13.5 Computational Complexity

An algorithm is of little practical use if it takes millions of years to compute particular instances. There is a need to consider the efficiency of the algorithm due to practical considerations. Chapter 10 discussed cryptography and the RSA algorithm, and the security of the RSA encryption algorithm is due to the fact that there is no known efficient algorithm to determine the prime factors of a large number.

There are often slow and fast algorithms for the same problem, and a measure of complexity is the number of steps in a computation. An algorithm is of *time complexity* $f(n)$ if for all n and all inputs of length n the execution of the algorithm takes at most $f(n)$ steps.

An algorithm is said to be *polynomially bounded* if there is a polynomial $p(n)$ such that for all n and all inputs of length n the execution of the algorithm takes at most $p(n)$ steps. The notation P is used for all problems that can be solved in polynomial time.

A problem is said to be *computationally intractable* if it may not be solved in polynomial time—there is no known algorithm to solve the problem in polynomial time.

A problem L is said to be in the set NP (non-deterministic polynomial time problems) if any given solution to L can be verified quickly in polynomial time. A non-deterministic Turing machine may have several possibilities for its behaviour, and an input may give rise to several computations.

A problem is *NP complete* if it is in the set NP of non-deterministic polynomial time problems and it is also in the class of *NP hard* problems. A key characteristic to NP complete problems is that there is no known fast solution to them, and the time required to solve the problem using known algorithms increases quickly as the size of the problem grows. Often, the time required to solve the problem is in billions or trillions of years. Although any given solution can be verified quickly there is no known efficient way to find a solution.

13.6 Review Questions

1. Explain computability and decidability.
2. What were the goals of logicism and formalism and how successful were these movement in mathematics?
3. What is a formal system?
4. Explain the difference between consistency, completeness and decidability.

5. Describe a Turing machine and explain its significance in computability.
6. Describe the halting problem and show that it is undecidable.
7. Discuss the complexity of an algorithm and explain terms such as 'polynomial bounded', 'computationally intractable' and 'NP complete'.

13.7 Summary

This chapter provided an introduction to computability and decidability. The intuitive meaning of computability is that in terms of an algorithm (or effective procedure) that specifies a set of instructions to be followed to solve the problem. Another words, a function f is computable if there exists an algorithm that produces the value of f correctly for each possible argument of f. The computation of f for a particular argument x just involves following the instructions in the algorithm, and it produces the result $f(x)$ in a finite number of steps if x is in the domain of f.

The concept of computability may be made precise in several equivalent ways such as Church's lambda calculus, recursive function theory or by the theoretical Turing machines. The Turing machine is a mathematical machine with a potentially infinite tape divided into frames (or cells) in which very basic operations can be carried out. The set of functions that are computable are those that are computable by a Turing machine.

A formal system contains meaningless symbols together with rules for manipulating them, and is generally intended to represent some aspect of the real world. The individual formulas are certain finite sequences of symbols obeying the syntactic rules of the formal language. A formal system consists of a formal language, a set of axioms and rules of inference

Church and Turing independently showed in 1936 that mathematics is not decidable. In other words it is not possible to determine the truth or falsity of any mathematical proposition by an algorithm.

Turing had already proved that the halting problem for Turing machines is not computable: i.e. it is not possible algorithmically to decide whether a given Turing machine will halt or not. He then applied this result to first-order logic to show that it is undecidable. That is, the only way to determine whether a statement is true or false is to try to solve it.

The complexity of an algorithm was discussed, and it was noted that an algorithm is of little practical use if it takes millions of years to compute the solution. There is a need to consider the efficiency of the algorithm due to practical considerations. The class of polynomial time bound problems and non-deterministic polynomial time problems were considered, and it was noted that the security of various cryptographic algorithms is due to the fact that there are no time efficient algorithms to determine the prime factors of large integers.

The reader is referred to [1] for a more detailed account of decidability and computability.

Reference

1. Cornerstones of Undecideability. Grzegorz Rozenberg and Arto Salomaa. Prentice Hall. 1994.

A Short History of Logic

14

14.1 Introduction

Logic is concerned with reasoning and with establishing the validity of arguments. It allows conclusions to be deduced from premises according to logical rules, and the logical argument establishes the truth of the conclusion provided that the premises are true.

The origins of logic are with the Greeks who were interested in the nature of truth. The sophists (e.g., Protagoras and, Gorgias) were teachers of rhetoric, who taught their pupils techniques in winning an argument and convincing an audience. Plato explores the nature of truth in some of his dialogues, and he is critical of the position of the sophists who argue that there is no absolute truth, and that truth instead is always relative to some frame of reference. The classic sophist position is stated by Protagoras "Man is the measure of all things: of things which are, that they are, and of things which are not, that they are not." Another word: what is true for you is true for you, and what is true for me is true for me.

Socrates had a reputation for demolishing an opponent's position, and the Socratean enquiry consisted of questions and answers in which the opponent would be led to a conclusion incompatible with his original position. The approach was similar to a *reductio ad absurdum* argument, although Socrates was a moral philosopher who did no theoretical work on logic.

Aristotle did important work on logic, and he developed a system of logic, *syllogistic logic*, that remained in use up to the nineteenth century. Syllogistic logic is a 'term-logic', with letters used to stand for the individual terms. A syllogism consists of two premises and a conclusion, where the conclusion is a valid deduction from the two premises. Aristotle also did some early work on modal logic, and was the founder of the field.

The Stoics developed an early form of propositional logic, where the assertibles (propositions) have a truth value such that at any time they are either true or false. The assertibles may be simple or non-simple, and various connectives such as conjunctions, disjunctions and implication are used in forming more complex assertibles.

George Boole developed his symbolic logic in the mid-1800s, and it later formed the foundation for digital computing. Boole argued that logic should be considered as a separate branch of mathematics, rather than a part of philosophy. He argued that there are mathematical laws to express the operation of reasoning in the human mind, and he showed how Aristotle's syllogistic logic could be reduced to a set of algebraic equations.

Logic plays a key role in reasoning and deduction in mathematics, but it is considered a separate discipline to mathematics. There were attempts in the early twentieth century to show that all mathematics can be derived from formal logic, and that the formal system of mathematics would be complete, with all the truths of mathematics provable in the system (see Chap. 13). However, this program failed when the Austrian logician, Kurt Goedel, showed that there are truths in the formal system of arithmetic that cannot be proved within the system (i.e. first-order arithmetic is incomplete).

14.2   Syllogistic Logic

Early work on logic was done by Aristotle in the fourth century B.C. in the *Organon* [1]. Aristotle regarded logic as a useful tool of enquiry into any subject, and he developed *syllogistic logic*. This is a form of reasoning in which a conclusion is drawn from two premises, where each premise is in a subject–predicate form. A common or middle term is present in each of the two premises but not in the conclusion. For example:

Table 14.1 Types of
syllogistic premises

Type	Symbol	Example
Universal affirmative	G A M	All greeks are mortal
Universal negative	G E M	No greek is mortal
Particular affirmative	G I M	Some greek is mortal
Particular negative	G O M	Some greek is not mortal

Table 14.2 Forms of
syllogistic premises

	Form (i)	Form (ii)	Form (iii)	Form (iv)
Premise 1	M P	P M	P M	M P
Premise 2	M S	S M	M S	S M
Conclusion	S P	S P	S P	S P

All Greeks are mortal.
Socrates is a Greek

———————————

Therefore Socrates is mortal

The common (or middle) term in this example is 'Greek'. It occurs in both premises but not in the conclusion. The above argument is valid, and Aristotle studied and classified the various types of syllogistic arguments to determine those that were valid or invalid. Each premise contains a subject and a predicate, and the middle term may act as subject or a predicate. Each premise is a positive or negative affirmation, and an affirmation may be universal or particular. The universal and particular affirmations and negatives are described in the table below (Table 14.1):

This leads to four basic forms of syllogistic arguments (Table 14.2) where the middle is the subject of both premises; the predicate of both premises; and the subject of one premise and the predicate of the other premise.

There are four types of premises (A, E, I, O) and therefore 16 sets of premise pairs for each of the forms above. However, only some of these premise pairs will yield a valid conclusion. Aristotle went through every possible premise pair to determine if a valid argument may be derived. The syllogistic argument above is of form (iv) and is valid

G A M
S I G

———————————

S I M

Syllogistic logic is a 'term-logic' with letters used to stand for the individual terms. Syllogistic logic was the first attempt at a science of logic and it remained in use up to the nineteenth century. There are many limitations to what it may express, and on its suitability as a representation of how the mind works.

14.3   Paradoxes and Fallacies

A paradox is a statement that apparently contradicts itself, and it presents a situation that appears to defy logic. Some logical paradoxes have a solution, whereas others are contradictions or invalid arguments. There are many examples of paradoxes, and they often arise due to self-reference in which one or more statements refer to each other. We discuss several paradoxes such as the *liar paradox* and the *sorites paradox*, which were invented by Eubulides of Miletus, and the *barber paradox*, which was introduced by Russell to explain the contradictions in naïve set theory.

An example of the *liar paradox* is the statement "Everything that I say is false", which is made by the liar. This looks like a normal sentence but it is also saying something about itself as a sentence. If the statement is true, then the statement must be false, since the meaning of the sentence is that every statement (including the current statement) made by the liar is false. If the current statement is false, then the statement that everything that I say is false is false, and so this must be a true statement.

The *Epimenides paradox* is a variant of the liar paradox. Epimenides was a Cretan who allegedly stated "All Cretans are liars". If the statement is true, then since Epimenides is Cretan, he must be a liar, and so the statement is false and we have a contradiction. However, if we assume that the statement is false and that Epimenides is lying about all Cretan being liars, then we may deduce (without contradiction) that there is at least one Cretan who is truthful. So in this case the paradox can be avoided.

The *sorites paradox* (paradox of the heap) involves a heap of sand in which grains are individually removed. It is assumed that removing a single grain of sand does not turn a heap into a non-heap, and the paradox is to consider what happens after when the process is repeated often enough. Is a single remaining grain a heap? When does it change from being a heap to a non-heap? This paradox may be avoided by specifying a fixed boundary of the number of grains of sand required to form a heap, or to define a heap as a collection of multiple grains (≥ 2 grains). Then any collection of grains of sand less than this boundary is not a heap.

The *barber paradox* is a variant of Russell's paradox (a contradiction in naïve set theory), which was discussed in chapter two. In a village there is a barber who shaves everyone who does not shave himself, and no one else. Who shaves the barber? The answer to this question results in a contradiction, as the barber cannot shave himself, since he shaves only those who do not shave themselves. Further, as the barber does not shave himself then he falls into the group of people who would be shaved by the barber (himself). Therefore, we conclude that there is no such barber.

The purpose of a debate is to convince an audience of the correctness of your position, and to challenge and undermine your opponent's position. Often, the arguments made are factual, but occasionally individuals skilled in rhetoric and persuasion introduce bad arguments as a way to persuade the audience. Aristotle studied and classified bad arguments (known as *fallacies*), and these include fallacies

Table 14.3 Fallacies in arguments

Fallacy	Description/example
Hasty/accident generalization	This is a bad argument that involves a generalization that disregards exceptions
Slippery slope	This argument outlines a chain reaction leading to a highly undesirable situation that will occur if a certain situation is allowed. The claim is that even if one step is taken onto the slippery slope then we will fall all the way down to the bottom
Against the person Ad Hominem	The focus of this argument is to attack the person rather than the argument that the person has made
Appeal to people Ad Populum	This argument involves an appeal to popular belief to support an argument, with a claim that the majority of the population supports this argument. However, popular opinion is not always correct
Appeal to authority (Ad Verecundiam)	This argument is when an appeal is made to an authoritative figure to support an argument, and where the authority is not an expert in this area
Appeal to pity (Ad Misericordiam)	This is where the arguer tries to get people to accept a conclusion by making them fell sorry for someone
Appeal to ignorance	The arguer makes the case that there is no conclusive evidence on the issue at hand and that therefore his conclusion should be accepted
Straw man argument	The arguer sets up a version of an opponent's position of his argument and defeats this watered down version of his opponent's position
Begging the question	This is a circular argument where the arguer relies on a premise that says the same thing as the conclusion and without providing any real evidence for the conclusion
Red herring	The arguer goes off on a tangent that has nothing to do with the argument in question
False dichotomy	The arguer presents the case that there are only two possible outcomes (often there are more). One of the possible outcomes is then eliminated leading to the desired outcome. The argument suggests that there is only one outcome

such as the *ad hominem* argument; the *appeal to authority* argument; and the *straw man* argument. The fallacies are described in more detail in Table 14.3 below.

14.4 Stoic Logic

The Stoic school[1] was founded in the Hellenistic period by Zeno of Citium (in Cyprus) in the late fourth/early third century B.C (Fig. 14.1). The school presented its philosophy as a way of life, and it emphasized ethics as the main focus of human knowledge. The Stoics stressed the importance of living a good life in harmony with nature.

---

[1]The origin of the word Stoic is from the *Stoa Poikile* (Στοα Ποιλικη), which was a covered walkway in the Agora of Athens. Zeno taught his philosophy in a public space at this location, and his followers became known as Stoics.

Fig. 14.1 Zeno of Citium

The Stoics recognized the importance of reason and logic, and Chrysippus, the head of the Stoics in the third century B.C., developed an early version of propositional logic. This was a system of deduction in which the smallest unanalyzed expressions are assertibles (Stoic equivalent of propositions). The assertibles have a truth value such that at any moment of time they are either true or false. True assertibles are viewed as facts in the Stoic system of logic, and false assertibles are defined as the contradictories of true ones.

Truth is temporal and assertions may change their truth value over time. The assertibles may be simple or non-simple (more than one assertible), and there may be present tense, past tense and future tense assertibles. Chrysippus distinguished between simple and compound propositions, and he introduced a set of logical connectives for conjunction, disjunction and implication that are used to form non-simple assertibles from existing assertibles.

The conjunction connective is of the form '*both… and…*', and it has two conjuncts. The disjunction connective is of the form '*either… or… or…*', and it consists of two or more disjuncts. Conditionals are formed from the connective '*if…,*' and they consist of an antecedent and a consequence.

His deductive system included various logical argument forms such as *modus ponens* and *modus tollens*. His propositional logic differed from syllogistic logic, in that the Stoic logic was based on propositions (or statements) as distinct from Aristotle's term-logic. However, he could express the universal affirmation in syllogistic logic (e.g., All A's are B) by rephrasing it as a conditional statement that if something is A then it is B.

Chrysippus's propositional logic did not replace Aristotle's syllogistic logic, and syllogistic logic remained in use up to the mid-nineteenth century. George Boole developed his symbolic logic in the mid-1800s, and his logic later formed the foundation for digital computing. Boole's symbolic logic is discussed in the next section.

Fig. 14.2 George Boole

14.5 Boole's Symbolic Logic

George Boole (Fig. 14.2) was born in Lincoln, England in 1815. His father (a cobbler who was interested in mathematics and optical instruments) taught him mathematics, and showed him how to make optical instruments. Boole inherited his father's interest in knowledge, and he was self-taught in mathematics and Greek. He taught at various schools near Lincoln, and he developed his mathematical knowledge by working his way through Newton's Principia, as well as applying himself to the work of mathematicians such as Laplace and Lagrange.

He published regular papers from his early twenties, and these included contributions to probability theory, differential equations, and finite differences. He developed his symbolic algebra, which is the foundation for modern computing, and he is considered (along with Babbage) to be one of the grandfathers of computing. His work was theoretical, and he never actually built a computer or calculating machine. *However, Boole's symbolic logic was the perfect mathematical model for switching theory, and for the design of digital circuits.*

Boole became interested in formulating a calculus of reasoning, and he published a pamphlet titled "Mathematical Analysis of Logic" in 1847 [2]. This short book developed novel ideas on a logical method, and he argued that logic should be considered as a separate branch of mathematics, rather than a part of philosophy. He argued that there are mathematical laws to express the operation of reasoning in the human mind, and he showed how Aristotle's syllogistic logic could be reduced to a set of algebraic equations. He corresponded regularly on logic with Augustus De Morgan.[2]

He introduced two quantities '0' and '1' with the quantity 1 used to represent the universe of thinkable objects (i.e. the universal set), and the quantity 0 represents the absence of any objects (i.e. the empty set). He then employed symbols such as

---

[2]De Morgan was a nineteenth century British mathematician based at University College London. De Morgan's laws in Set Theory and Logic state that: $(A \cup B)^c = A^c \cap B^c$ and $\neg (A \vee B) \equiv \neg A \wedge \neg B$.

x, y, z, etc., to represent collections or classes of objects given by the meaning attached to adjectives and nouns. Next, he introduced three operators ($+$, $-$, and $\times$) that combined classes of objects.

The expression xy (i.e. x multiplied by y or $x \times y$) combines the two classes x, y to form the new class xy (i.e. the class whose objects satisfy the two meanings represented by the classes x *and* y). Similarly, the expression $x + y$ combines the two classes x, y to form the new class $x + y$ (that satisfies either the meaning represented by class x *or* class y). The expression $x - y$ combines the two classes x, y to form the new class $x - y$. This represents the class (that satisfies the meaning represented by class x but not class y. The expression $(1 - x)$ represents objects that do not have the attribute that represents class x.

Thus, if $x =$ black and $y =$ sheep, then xy represents the class of black sheep. Similarly, $(1 - x)$ would represents the class obtained by the operation of selecting all things in the world except black things; $x(1 - y)$ represents the class of all things that are black but not sheep; and $(1 - x)(1 - y)$ would give us all things that are neither sheep nor black.

He showed that these symbols obeyed a rich collection of algebraic laws and could be added, multiplied, etc., in a manner that is similar to real numbers. These symbols may be used to reduce propositions to equations, and algebraic rules may be employed to solve the equations. The rules include the following:

1.	$x + 0 = x$	(Additive Identity)
2.	$x + (y + z) = (x + y) + z$	(Associative)
3.	$x + y = y + x$	(Commutative)
4.	$x + (1 - x) = 1$	
5.	$x \cdot 1 = x$	(Multiplicative Identity)
6.	$x \cdot 0 = 0$	
7.	$x + 1 = 1$	
8.	$xy = yx$	(Commutative)
9.	$x(yz) = (xy)z$	(Associative)
10.	$x(y + z) = xy + xz$	(Distributive)
11.	$x(y - z) = xy - xz$	(Distributive)
12.	$x^2 = x$	(Idempotent)

These operations are similar to the modern laws of set theory with the set union operation represented by '$+$', and the set intersection operation is represented by multiplication. The universal set is represented by '1' and the empty by '0'. The associative and distributive laws hold. Finally, the set complement operation is given by $(1 - x)$.

Boole applied the symbols to encode Aristotle's Syllogistic Logic, and he showed how the syllogisms could be reduced to equations. This allowed conclusions to be derived from premises by eliminating the middle term in the syllogism. He refined his ideas on logic further in his book "An Investigation of the Laws of

Thought" [3]. This book aimed to identify the fundamental laws underlying reasoning in the human mind, and to give expression to these laws in the symbolic language of a calculus.

He considered the equation $x^2 = x$ to be a fundamental laws of thought. It allows the principle of contradiction to be expressed (i.e. for an entity to possess an attribute and at the same time not to possess it):

$$x^2 = x$$
$$\Rightarrow x - x^2 = 0$$
$$\Rightarrow x(1-x) = 0$$

For example, if x represents the class of horses then $(1 - x)$ represents the class of 'not-horses'. The product of two classes represents a class whose members are common to both classes. Hence, $x(1 - x)$ represents the class whose members are at once both horses and 'not-horses', and the equation $x(1 - x) = 0$ expresses that fact that there is no such class. That is, it is the empty set.

Boole contributed to other areas in mathematics including differential equations, finite differences,[3] and to the development of probability theory. Des McHale has written an interesting biography of Boole [4]. Boole's logic appeared to have no practical use, but this changed with Claude Shannon's 1937 Master's Thesis, which showed its applicability to switching theory and to the design of digital circuits.

14.5.1 Switching Circuits and Boolean Algebra

Claude Shannon showed in his famous Master's Thesis that Boole's symbolic algebra provided the perfect mathematical model for switching theory and for the design of digital circuits. It may be employed to optimize the design of systems of electromechanical relays, and circuits with relays solve Boolean algebra problems. The use of the properties of electrical switches to process logic is the basic concept that underlies all modern electronic digital computers. Digital computers use the binary digits 0 and 1, and Boolean logical operations may be implemented by electronic AND, OR and NOT gates. More complex circuits (e.g., arithmetic) may be designed from these fundamental building blocks.

Modern electronic computers use millions (billions) of transistors that act as switches and can change state rapidly. The use of switches to represent binary values is the foundation of modern computing. A high voltage represents the binary value 1 with low voltage representing the binary value 0.

A silicon chip may contain billions of tiny electronic switches arranged into logical gates. The basic logic gates are AND, OR and NOT. These gates may be combined in various ways to allow the computer to perform more complex tasks such as binary arithmetic. Each gate has binary value inputs and outputs.

---

[3]Finite Differences are a numerical method used in solving differential equations.

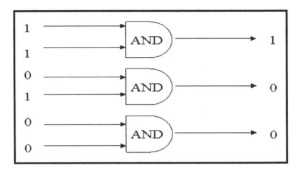

Fig. 14.3 Binary AND operation

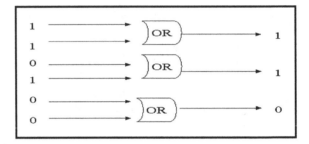

Fig. 14.4 Binary OR operation

Fig. 14.5 NOT operation

The example in Fig. 14.3 is that of an 'AND' gate which produces the binary value 1 as output only, if both inputs are 1. Otherwise, the result will be the binary value 0. Figure 14.4 is an 'OR' gate which produces the binary value 1 as output if any of its inputs is 1. Otherwise, it will produce the binary value 0.

Finally, a NOT gate (Fig. 14.5) accepts only a single input which it reverses. That is, if the input is '1' the value '0' is produced and vice versa.

The logic gates may be combined to form more complex circuits. The example in Fig. 14.6 is that of a half-adder of 1 + 0. The inputs to the top OR gate are 1 and 0 which yields the result of 1. The inputs to the bottom AND gate are 1 and 0 which yields the result 0, which is then inverted through the NOT gate to yield binary 1. Finally, the last AND gate receives two 1's as input and the binary value 1 is the result of the addition.

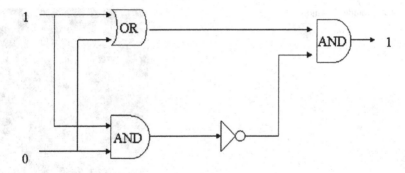

Fig. 14.6 Half-adder

The half-adder computes the addition of two arbitrary binary digits, but it does not calculate the carry. It may be extended to a full adder that provides a carry for addition.

14.6 Application of Symbolic Logic to Digital Computing

Claude Shannon (Fig. 14.7) was an American mathematician and engineer who made fundamental contributions to computing. He was the first person[4] to see the applicability of Boolean algebra to simplify the design of circuits and telephone routing switches. He showed that Boole's symbolic logic developed in the nineteenth century provided the perfect mathematical model for switching theory and for the subsequent design of digital circuits and computers.

His influential Master's Thesis is a key milestone in computing, and it shows how to lay out circuits according to Boolean principles. It provides the theoretical foundation of switching circuits, and his insight of using the properties of electrical switches to do Boolean logic is the basic concept that underlies all electronic digital computers.

Shannon realized that you could combine switches in circuits in such a manner as to carry out symbolic logic operations. This allowed binary arithmetic and more complex mathematical operations to be performed by relay circuits. He designed a circuit, which could add binary numbers, and he later designed circuits that could make comparisons and thus is capable of performing a conditional statement. This was the birth of digital logic and the digital computing age.

Vannevar Bush [5] was Shannon's supervisor at MIT, and Shannon's initial work was to improve Bush's mechanical computing device known as the Differential Analyser. This machine had a complicated control circuit that was composed

---

[4]Victor Shestakov at Moscow State University also proposed a theory of electric switches based on Boolean algebra around the same time as Shannon. However, his results were published in Russian in 1941 whereas Shannon's were published in 1937.

Fig. 14.7 Claude Shannon

of one hundred switches that could be automatically opened and closed by an electromagnet. Shannon's insight was his realization that an electronic circuit is similar to Boole's symbolic algebra, and he showed how Boolean algebra could be employed to optimize the design of systems of electromechanical relays used in the analogue computer. He also realized that circuits with relays could solve Boolean algebra problems.

He showed in his Master's thesis "A Symbolic Analysis of Relay and Switching Circuits" [6] that the binary digits (i.e. 0 and 1) can be represented by electrical switches. The implications of true and false being denoted by the binary digits one and zero were enormous, since it allowed binary arithmetic and more complex mathematical operations to be performed by relay circuits. This provided electronics engineers with the mathematical tool they needed to design digital electronic circuits, and provided the foundation of digital electronic design.

The design of circuits and telephone routing switches could be simplified with Boolean algebra. Shannon showed how to lay out circuitry according to Boolean principles, and his Master's thesis became the foundation for the practical design of digital circuits. These circuits are fundamental to the operation of modern computers and telecommunication systems, and his insight of using the properties of electrical switches to do Boolean logic is the basic concept that underlies all electronic digital computers.

14.7   Frege

Gottlob Frege (Fig. 14.8) was a German mathematician and logician who is considered (along with Boole) to be one of the founders of modern logic. He also made important contributions to the foundations of mathematics, and he attempted to show that all of the basic truths of mathematics (or at least of arithmetic) could be derived from a limited set of logical axioms (this approach is known as logicism).

Fig. 14.8 Gottlob Frege

He invented predicate logic and the universal and existential quantifiers, and predicate logic was a significant advance on Aristotle's syllogistic logic. Predicate logic is described in more detail in Chap. 15.

Frege's first logical system, the 1879 Begriffsschrift, contained nine axioms and one rule of inference. It was the first axiomization of logic, and it was complete in its treatment of propositional logic and first-order predicate logic. He published several important books on logic, including Begriffsschrift, in 1879; Die Grundlagen der Arithmetik (The Foundations of Arithmetic) in 1884; and the two-volume work Grundgesetze der Arithmetik (Basic Laws of Arithmetic), which were published in 1893 and 1903. These books described his invention of axiomatic predicate logic; the use of quantified variables; and the application of his logic to the foundations of arithmetic.

Frege presented his predicate logic in his books, and he began to use it to define the natural numbers and their properties. He had intended producing three volumes of the Basic Laws of Arithmetic, with the later volumes dealing with the real numbers and their properties. However, Bertrand Russell discovered a contradiction in Frege's system (see Russell's paradox in Chap. 2), which he communicated to Frege shortly before the publication of the second volume. Frege was astounded by the contradiction and he struggled to find a satisfactory solution, and Russell introduced the theory of types in the Principia Mathematica as a solution.

14.8 Review Questions

1. What is logic?
2. What is a fallacy?

3. Give examples of fallacies in arguments in natural language (e.g., in politics, marketing, debates).
4. Investigate some of the early paradoxes (for example the Tortoise and Achilles paradox or the arrow in flight paradox) and give your interpretation of the paradox.
5. What is syllogistic logic and explain its relevance.
6. What is stoic logic and explain its relevance.
7. Explain the significance of the equation $x^2 = x$ in Boole's symbolic logic.
8. Describe how Boole's symbolic logic provided the foundation for digital computing.
9. Describe Frege's contributions to logic.

14.9   Summary

This chapter gave a short introduction to logic, and logic is concerned with reasoning and with establishing the validity of arguments. It allows conclusions to be deduced from premises according to logical rules, and the logical argument establishes the truth of the conclusion provided that the premises are true.

The origins of logic are with the Greeks who were interested in the nature of truth. Socrates had a reputation for demolishing an opponents position (it meant that he did not win any friends with in debate), and the Socratean enquiry consisted of questions and answers in which the opponent would be led to a conclusion incompatible with his original position. His approach was similar to a reductio ad absurdum argument, and as its effect was to show that his opponent's position was incoherent,

Aristotle did important work on logic, and he developed a system of logic, syllogistic logic, that remained in use up to the nineteenth century. Syllogistic logic is a 'term-logic', with letters used to stand for the individual terms. A syllogism consists of two premises and a conclusion, where the conclusion is a valid deduction from the two premises. He also did some early work on modal logic.

The Stoics developed an early form of propositional logic, where the assertibles (propositions) have a truth value such that at any time they are either true or false.

George Boole developed his symbolic logic in the mid-1800s, and it later formed the foundation for digital computing. Boole argued that logic should be considered as a separate branch of mathematics, rather than a part of philosophy. He argued that there are mathematical laws to express the operation of reasoning in the human mind, and he showed how Aristotle's syllogistic logic could be reduced to a set of algebraic equations.

Gottlob Frege made important contributions to logic and to the foundations of mathematics. He attempted to show that all of the basic truths of mathematics (or at least of arithmetic) could be derived from a limited set of logical axioms (this approach is known as logicism). He invented predicate logic and the universal and existential quantifiers, and predicate logic was a significant advance on Aristotle's syllogistic logic

References

1. Aristotle the Philosopher. J.L. Ackrill. Clarendon Press Oxford. 1994.
2. The Calculus of Logic. George Boole. Cambridge and Dublin Mathematical Journal. Vol. III (1848), pp. 183–98.
3. An Investigation into the Laws of Thought. George Boole. Dover Publications. 1958. (First published in 1854).
4. Boole. Des McHale. Cork University Press. 1985.
5. Giants of Computing. Gerard O' Regan. Springer Verlag. 2013.
6. A Symbolic Analysis of Relay and Switching Circuits. Claude Shannon. Masters Thesis. Massachusetts Institute of Technology. 1937.

Propositional and Predicate Logic

15

Key Topics

Propositions
Truth Tables
Semantic Tableaux
Natural Deduction
Proof
Predicates
Universal Quantifiers
Existential Quantifiers

15.1 Introduction

Logic is the study of reasoning and the validity of arguments, and it is concerned with the truth of statements (propositions) and the nature of truth. Formal logic is concerned with the form of arguments and the principles of valid inference. Valid arguments are truth preserving, and for a valid deductive argument the conclusion will always be true if the premises are true.

Propositional logic is the study of propositions, where a proposition is a statement that is either true or false. Propositions may be combined with other propositions (with a logical connective) to form compound propositions. Truth tables are used to give operational definitions of the most important logical connectives, and they provide a mechanism to determine the truth values of more complicated logical expressions.

© Springer International Publishing Switzerland 2016
G. O'Regan, *Guide to Discrete Mathematics*, Texts in Computer Science,
DOI 10.1007/978-3-319-44561-8_15

Propositional logic may be used to encode simple arguments that are expressed in natural language, and to determine their validity. The validity of an argument may be determined from truth tables, or using the inference rules such as modus ponens to establish the conclusion via deductive steps.

Predicate logic allows complex facts about the world to be represented, and new facts may be determined via deductive reasoning. Predicate calculus includes predicates, variables and quantifiers, and a *predicate* is a characteristic or property that the subject of a statement can have. A predicate may include variables, and statements with variables become propositions once the variables are assigned values.

The universal quantifier is used to express a statement such as that all members of the domain of discourse have property P. This is written as $(\forall x) P(x)$, and it expresses the statement that the property $P(x)$ is true for all x. The existential quantifier states that there is at least one member of the domain of discourse that has property P. This is written as $(\exists x)P(x)$.

15.2 Propositional Logic

Propositional logic is the study of propositions where a proposition is a statement that is either true or false. There are many examples of propositions such as '1 + 1 = 2' which is a true proposition, and the statement that 'Today is Wednesday' which is true if today is Wednesday and false otherwise. The statement $x > 0$ is not a proposition as it contains a variable x, and it is only meaningful to consider its truth or falsity only when a value is assigned to x. Once the variable x is assigned a value it becomes a proposition. The statement 'This sentence is false' is not a proposition as it contains a self-reference that contradicts itself. Clearly, if it the statement is true it is false, and if is false it is true.

A propositional variable may be used to stand for a proposition (e.g. let the variable P stand for the proposition '2 + 2 = 4' which is a true proposition). A propositional variable takes the value or false. The negation of a proposition P (denoted $\neg P$) is the proposition that is true if and only if P is false, and is false if and only if P is true.

A well-formed formula (*wff*) in propositional logic is a syntactically correct formula created according to the syntactic rules of the underlying calculus. A well-formed formula is built up from variables, constants, terms and logical connectives such as conjunction (and), disjunction (or), implication (if... then...), equivalence (if and only if) and negation. A distinguished subset of these well formed formulae is the *axioms* of the calculus, and there are *rules of inference* that allow the truth of new formulae to be derived from the axioms and from formulae that have already demonstrated to be true in the calculus.

Table 15.1 Truth table for formula W

A	B	W (A, B)
T	T	T
T	F	F
F	T	F
F	F	T

A formula in propositional calculus may contain several propositional variables, and the truth or falsity of the individual variables needs to be known prior to determining the truth or falsity of the logical formula.

Each propositional variable has two possible values, and a formula with n-propositional variables has 2^n values associated with the n-propositional variables. The set of values associated with the n variables may be used derive a truth table with 2^n rows and $n + 1$ columns. Each row gives each of the 2^n truth values that the n variables may take, and column $n + 1$ gives the result of the logical expression for that set of values of the propositional variables. For example, the propositional formula W defined in the truth table above (Table 15.1) has two propositional variables A and B, with $2^2 = 4$ rows for each of the values that the two propositional variables may take. There are $2 + 1 = 3$ columns with W defined in the third column.

A rich set of connectives is employed in the calculus to combine propositions and to build up the well-formed formulae. This includes the conjunction of two propositions $(A \wedge B)$, the disjunction of two propositions $(A \vee B)$ and the implication of two propositions $(A \rightarrow B)$. These connectives allow compound propositions to be formed, and the truth of the compound propositions is determined from the truth values of its constituent propositions and the rules associated with the logical connective. The meaning of the logical connectives is given by truth tables.[1]

Mathematical Logic is concerned with inference, and it involves proceeding in a methodical way from the axioms and using the rules of inference to derive further truths. The rules of inference allow new propositions to be deduced from a set of existing propositions. A valid argument (or deduction) is truth preserving: i.e. for a valid logical argument if the set of premises is true then the conclusion (i.e. the deduced proposition) will also be true. The rules of inference include rules such as *modus ponens*, and this rule states that given the truths of the proposition A, and the proposition $A \rightarrow B$, then the truth of proposition B may be deduced.

The propositional calculus is employed in reasoning about propositions, and it may be applied to formalize arguments in natural language. *Boolean algebra* is used in computer science, and it is named after George Boole, who was the first professor of mathematics at Queens College, Cork.[2] His symbolic logic (discussed in Chap. 14) is the foundation for modern computing.

---

[1] Basic truth tables were first used by Frege, and developed further by Post and Wittgenstein.

[2] This institution is now known as University College Cork and has approximately 18,000 students.

Table 15.2 Conjunction

A	B	$A \wedge B$
T	T	T
T	F	F
F	T	F
F	F	F

Table 15.3 Disjunction

A	B	$A \vee B$
T	T	T
T	F	T
F	T	T
F	F	F

15.2.1 Truth Tables

Truth tables give operational definitions of the most important logical connectives, and they provide a mechanism to determine the truth values of more complicated compound expressions. Compound expressions are formed from propositions and connectives, and the truth values of a compound expression containing several propositional variables is determined from the underlying propositional variables and the logical connectives.

The conjunction of A and B (denoted $A \wedge B$) is true if and only if both A and B are true, and is false in all other cases (Table 15.2). The disjunction of two propositions A and B (denoted $A \vee B$) is true if at least one of A and B are true, and false in all other cases (Table 15.3). The disjunction operator is known as the *'inclusive or'* operator as it is also true when both A and B are true; there is also an *exclusive or* operator that is true exactly when one of A or B is true, and is false otherwise.

Example 15.1 Consider proposition A given by "An orange is a fruit" and the proposition B given by "2 + 2 = 5" then A is true and B is false. Therefore

(i) $A \wedge B$ (i.e. An orange is a fruit and 2 + 2 = 5) is false
(ii) $A \vee B$ (i.e. An orange is a fruit or 2 + 2 = 5) is true

The implication operation $(A \rightarrow B)$ is true if whenever A is true means that B is also true; and also whenever A is false (Table 15.4). It is equivalent (as shown by a

Table 15.4 Implication

A	B	$A \rightarrow B$
T	T	T
T	F	F
F	T	T
F	F	T

Table 15.5 Equivalence

A	B	$A \leftrightarrow B$
T	T	T
T	F	F
F	T	F
F	F	T

Table 15.6 Not operation

A	$\neg A$
T	F
F	T

truth table) to $\neg A \vee B$. The equivalence operation ($A \leftrightarrow B$) is true whenever both A and B are true, or whenever both A and B are false (Table 15.5).

The not operator ($\neg$) is a unary operator (i.e. it has one argument) and is such that $\neg A$ is true when A is false, and is false when A is true (Table 15.6).

Example 15.2 Consider proposition A given by 'Jaffa cakes are biscuits' and the proposition B given by '2 + 2 = 5' then A is true and B is false. Therefore

(i) $A \rightarrow B$ (i.e. Jaffa cakes are biscuits implies 2 + 2 = 5) is false
(ii) $A \leftrightarrow B$ (i.e. Jaffa cakes are biscuits is equivalent to 2 + 2 = 5) is false
(iii) $\neg B$ (i.e. 2 + 2 ≠ 5) is true.

Creating a Truth Table

The truth table for a well-formed formula $W(P_1, P_2, ..., P_n)$ is a table with 2^n rows and $n + 1$ columns. Each row lists a different combination of truth values of the propositions $P_1, P_2, ..., P_n$ followed by the corresponding truth value of W.

The example above (Table 15.7) gives the truth table for a formula W with three propositional variables (meaning that there are $2^3 = 8$ rows in the truth table).

Table 15.7 Truth table for $W(P, Q, R)$

P	Q	R	W(P, Q, R)
T	T	T	F
T	T	F	F
T	F	T	F
T	F	F	T
F	T	T	T
F	T	F	F
F	F	T	F
F	F	F	F

15.2.2   Properties of Propositional Calculus

There are many well-known properties of the propositional calculus such as the commutative, associative and distributive properties. These ease the evaluation of complex expressions, and allow logical expressions to be simplified.

The *commutative property* holds for the conjunction and disjunction operators, and it states that the order of evaluation of the two propositions may be reversed without affecting the resulting truth value: i.e.

$$A \wedge B = B \wedge A$$
$$A \vee B = B \vee A$$

The *associative property* holds for the conjunction and disjunction operators. This means that order of evaluation of a sub-expression does not affect the resulting truth value, i.e.

$$(A \wedge B) \wedge C = A \wedge (B \wedge C)$$
$$(A \vee B) \vee C = A \vee (B \vee C)$$

The conjunction operator *distributes* over the disjunction operator and vice versa.

$$A \wedge (B \vee C) = (A \wedge B) \vee (A \wedge C)$$
$$A \vee (B \wedge C) = (A \vee B) \wedge (A \vee C)$$

The result of the logical conjunction of two propositions is false if one of the propositions is false (irrespective of the value of the other proposition).

$$A \wedge F = F \wedge A = F$$

The result of the logical disjunction of two propositions is true if one of the propositions is true (irrespective of the value of the other proposition).

$$A \vee T = T \vee A = T$$

The result of the logical disjunction of two propositions, where one of the propositions is known to be false is given by the truth value of the other proposition. That is, the Boolean value 'F' acts as the identity for the disjunction operation.

$$A \vee F = A = F \vee A$$

The result of the logical conjunction of two propositions, where one of the propositions is known to be true, is given by the truth value of the other proposition. That is, the Boolean value 'T' acts as the identity for the conjunction operation.

$$A \wedge T = A = T \wedge A$$

The $\wedge$ and $\vee$ operators are *idempotent*. That is, when the arguments of the conjunction or disjunction operator are the same proposition A the result is A. The idempotent property allows expressions to be simplified.

$$A \wedge A = A$$

$$A \vee A = A$$

The *law of the excluded middle* is a fundamental property of the propositional calculus. It states that a proposition A is either true or false: i.e. there is no third logical value.

$$A \vee \neg A$$

We mentioned earlier that $A \rightarrow B$ is logically equivalent to $\neg A \vee B$ (same truth table), and clearly $\neg A \vee B$ is the same as $\neg A \vee \neg\neg B = \neg\neg B \vee \neg A$ which is logically equivalent to $\neg B \rightarrow \neg A$. Another words, $A \rightarrow B$ is logically equivalent to $\neg B \rightarrow \neg A$, and this is known as the *contrapositive*.

De Morgan was a contemporary of Boole in the nineteenth century, and the following law is known as De Morgan's law.

$$\neg(A \wedge B) \equiv \neg A \vee \neg B$$
$$\neg(A \vee B) \equiv \neg A \wedge \neg B$$

Certain well-formed formulae are true for all values of their constituent variables. This can be seen from the truth table when the last column of the truth table consists entirely of true values. A proposition that is true for all values of its constituent propositional variables is known as a *tautology*. An example of a tautology is the proposition $A \vee \neg A$ (Table 15.8)

A proposition that is false for all values of its constituent propositional variables is known as a *contradiction*. An example of a contradiction is the proposition $A \wedge \neg A$.

Table 15.8 Tautology $B \vee \neg B$

B	$\neg B$	$B \vee \neg B$
T	F	T
F	T	T

15.2.3 Proof in Propositional Calculus

Logic enables further truths to be derived from existing truths by rules of inference that are truth preserving. Propositional calculus is both *complete* and *consistent*. The completeness property means that all true propositions are deducible in the calculus, and the consistency property means that there is no formula A such that both A and $\neg A$ are deducible in the calculus.

An argument in propositional logic consists of a sequence of formulae that are the premises of the argument and a further formula that is the conclusion of the argument. One elementary way to see if the argument is valid is to produce a truth table to determine if the conclusion is true whenever all of the premises are true.

Consider a set of premises $P_1, P_2, \ldots P_n$ and conclusion Q. Then to determine if the argument is valid using a truth table involves adding a column in the truth table for each premise $P_1, P_2, \ldots P_n$, and then to identify the rows in the truth table for which these premises are all true. The truth value of the conclusion Q is examined in each of these rows, and if Q is true for each case for which $P_1, P_2, \ldots P_n$ are all true then the argument is valid. This is equivalent to $P_1 \wedge P_2 \wedge \ldots \wedge P_n \rightarrow Q$ is a tautology.

An alternate approach to proof with truth tables is to assume the negation of the desired conclusion (i.e. $\neg Q$) and to show that the premises and the negation of the conclusion result in a contradiction (i.e. $P_1 \wedge P_2 \wedge \ldots \wedge P_n \wedge \neg Q$) is a contradiction.

The use of truth tables becomes cumbersome when there are a large number of variables involved, as there are 2^n truth table entries for n propositional variables.

Procedure for Proof by Truth Table

   (i)   Consider argument $P_1, P_2, \ldots, P_n$ with conclusion Q
  (ii)   Draw truth table with column in truth table for each premise $P_1, P_2, \ldots, P_n$
 (iii)   Identify rows in truth table for when these premises are all true.
 (iv)   Examine truth value of Q for these rows.
  (v)   If Q is true for each case that $P_1, P_2, \ldots P_n$ are true then the argument is valid.
 (vi)   That is $P_1 \wedge P_2 \wedge \ldots \wedge P_n \rightarrow Q$ is a tautology

Example 15.3 (**Truth Tables**) Consider the argument adapted from [1] and determine if it is valid.

If the pianist plays the concerto then crowds will come if the prices are not too high.

If the pianist plays the concerto then the prices will not be too high

Therefore, if the pianist plays the concerto then crowds will come.

Solution
We will adopt a common proof technique that involves showing that the negation of the conclusion is incompatible (inconsistent) with the premises, and from this we deduce the conclusion must be true. First, we encode the argument in propositional logic:

Table 15.9 Proof of argument with a truth table

P	C	H	$\neg H$	$\neg H \to C$	$P \to (\neg H \to C)$	$P \to \neg H$	$P \to C$	$\neg(P \to C)$	*
T	T	T	F	T	T	F	T	F	F
T	T	F	T	T	T	T	T	F	F
T	F	T	F	T	T	F	F	T	F
T	F	F	T	F	F	T	F	T	F
F	T	T	F	T	T	T	T	F	F
F	T	F	T	T	T	T	T	F	F
F	F	T	F	T	T	T	T	F	F
F	F	F	T	F	T	T	T	F	F

Let P stand for 'The pianist plays the concerto'; C stands for 'Crowds will come'; and H stands for 'Prices are too high'. Then the argument may be expressed in propositional logic as

$$P \to (\neg H \to C)$$
$$P \to \neg H$$
$$P \to C$$

Then we negate the conclusion $P \to C$ and check the consistency of $P \to (\neg H \to C) \wedge (P \to \neg H) \wedge \neg (P \to C)$* using a truth table (Table 15.9).

It can be seen from the last column in the truth table that the negation of the conclusion is incompatible with the premises, and therefore it cannot be the case that the premises are true and the conclusion false. Therefore, the conclusion must be true whenever the premises are true, and we conclude that the argument is valid.

Logical Equivalence and Logical Implication
The laws of mathematical reasoning are truth preserving, and are concerned with deriving further truths from existing truths. Logical reasoning is concerned with moving from one line in mathematical argument to another, and involves deducing the truth of another statement Q from the truth of P.

The statement Q maybe in some sense be logically equivalent to P and this allows the truth of Q to be immediately deduced. In other cases the truth of P is sufficiently strong to deduce the truth of Q; in other words P logically implies Q. This leads naturally to a discussion of the concepts of logical equivalence ($W_1 \equiv W_2$) and logical implication ($W_1 \vdash W_2$).

Logical Equivalence
Two well-formed formulae W_1 and W_2 with the same propositional variables (P, Q, R ...) are logically equivalent ($W_1 \equiv W_2$) if they are always simultaneously true or false for any given truth values of the propositional variables.

If two well-formed formulae are logically equivalent then it does not matter which of W_1 and W_2 is used, and $W_1 \leftrightarrow W_2$ is a tautology. In Table 15.10 above we see that $P \wedge Q$ is logically equivalent to $\neg(\neg P \vee \neg Q)$.

Table 15.10 Logical
equivalence of two WFFs

P	Q	P∧Q	¬P	¬Q	¬P∨¬Q	¬P∨¬Q
T	T	T	F	F	F	T
T	F	F	F	T	T	F
F	T	F	T	F	T	F
F	F	F	T	T	T	F

Table 15.11 Logical
implication of two WFFs

PQR	(P∧Q)∨(Q∧¬R)	(Q∨R)
TTT	T	T
TTF	T	T
TFT	F	T
TFF	F	F
FTT	F	T
FTF	T	T
FFT	F	T
FFF	F	F

Logical Implication

For two well-formed formulae W_1 and W_2 with the same propositional variables (P, Q, R ...) W_1 logically implies W_2 ($W_1 \vdash W_2$) if any assignment to the propositional variables which makes W_1 true also makes W_2 true (Table 15.11). That is, $W_1 \rightarrow W_2$ is a tautology.

Example 15.4 Show by truth tables that $(P \wedge Q) \vee (Q \wedge \neg R) \vdash (Q \vee R)$.

The formula $(P \wedge Q) \vee (Q \wedge \neg R)$ is true on rows 1, 2 and 6 and formula $(Q \vee R)$ is also true on these rows. Therefore $(P \wedge Q) \vee (Q \wedge \neg R) \vdash (Q \vee R)$.

15.2.4 Semantic Tableaux in Propositional Logic

We showed in example 15.3 how truth tables may be used to demonstrate the validity of a logical argument. However, the problem with truth tables is that they can get extremely large very quickly (as the size of the table is 2^n where n is the number of propositional variables), and so in this section we will consider an alternate approach known as semantic tableaux.

The basic idea of semantic tableaux is to determine if it is possible for a conclusion to be false when all of the premises are true. If this is not possible, then the conclusion must be true when the premises are true, and so the conclusion is *semantically entailed* by the premises. The method of semantic tableaux is a technique to expose inconsistencies in a set of logical formulae, by identifying conflicting logical expressions.

Table 15.12 Rules of semantic tableaux

Rule No.	Definition	Description
1.	$A \wedge B$ A B	If $A \wedge B$ is true then both A and B are true, and may be added to the branch containing $A \wedge B$
2.	$A \vee B$ $A \qquad B$	If $A \vee B$ is true then either A or B is true, and we add two new branches to the tableaux, one containing A and one containing B
3.	$A \to B$ $\neg A \qquad B$	If $A \to B$ is true then either $\neg A$ or B is true, and we add two new branches to the tableaux, one containing $\neg A$ and one containing B
4.	$A \leftrightarrow B$ $A \wedge B \qquad \neg A \wedge \neg B$	If $A \leftrightarrow B$ is true then either $A \wedge B$ or $\neg A \wedge \neg B$ is true, and we add two new branches, one containing $A \wedge B$ and one containing $\neg A \wedge \neg B$
5.	$\neg\neg A$ A	If $\neg\neg A$ is true then A may be added to the branch containing $\neg\neg A$
6.	$\neg(A \wedge B)$ $\neg A \qquad \neg B$	If $\neg(A \wedge B)$ is true then either $\neg A$ or $\neg B$ is true, and we add two new branches to the tableaux, one containing $\neg A$ and one containing $\neg B$
7.	$\neg(A \vee B)$ $\neg A$ $\neg B$	If $\neg(A \vee B)$ is true then both $\neg A$ and $\neg B$ are true, and may be added to the branch containing $\neg(A \vee B)$
8.	$\neg(A \to B)$ A $\neg B$	If $\neg(A \to B)$ is true then both A and $\neg B$ are true, and may be added to the branch containing $\neg(A \to B)$

We present a short summary of the rules of semantic tableaux in Table 15.12, and we then proceed to provide a proof for Example 15.3 using semantic tableaux instead of a truth table.

Whenever a logical expression A and its negation $\neg A$ appear in a branch of the tableau, then an inconsistency has been identified in that branch, and the branch is said to be *closed*. If all of the branches of the semantic tableaux are closed, then the logical propositions from which the tableau was formed are mutually inconsistent, and cannot be true together.

The method of proof is to negate the conclusion, and to show that all branches in the semantic tableau are closed, and that therefore it is not possible for the premises of the argument to be true and for the conclusion to be false. Therefore, the argument is valid and the conclusion follows from the premises.

Example 15.5 (**Semantic Tableaux**) Perform the proof for Example 15.3 using semantic tableaux.

Solution

We formalized the argument previously as

$$\begin{array}{ll}
\text{(Premise 1)} & P \rightarrow (\neg H \rightarrow C) \\
\text{(Premise 2)} & P \rightarrow \neg H \\
\text{(Conclusion)} & P \rightarrow C
\end{array}$$

We negate the conclusion to get $\neg(P \rightarrow C)$ and we show that all branches in the semantic tableau are closed, and that therefore it is not possible for the premises of the argument to be true and for the conclusion false. Therefore, the argument is valid, and the truth of the conclusion follows from the truth of the premises.

We have showed that all branches in the semantic tableau are closed, and that therefore it is not possible for the premises of the argument to be true and for the conclusion false. Therefore, the argument is valid as required.

15.2.5 Natural Deduction

The German mathematician, Gerhard Gentzen (Fig. 15.1), developed a method for logical deduction known as *'Natural Deduction'*, and his formal approach to natural deduction aimed to be as close as possible to natural reasoning. Gentzen worked as an assistant to David Hilbert at the University of Göttingen, and he died of malnutrition in Prague at the end of the Second World War.

Natural deduction includes rules for $\wedge$, $\vee$, $\rightarrow$ introduction and elimination and also for *reductio ab absurdum*. There are ten inference rules in the Natural Deduction system, and they include two inference rules for each of the five logical

Fig. 15.1 Gerhard gentzen

Table 15.13 Natural deduction rules

Rule	Definition	Description
$\wedge$ I	$$\frac{P_1, P_2, \ldots P_n}{P_1 \wedge P_2 \wedge \ldots \wedge P_n}$$	Given the truth of propositions $P_1, P_2, \ldots P_n$ then the truth of the conjunction $P_1 \wedge P_2 \wedge \ldots \wedge P_n$ follows. This rule shows how conjunction can be introduced
$\wedge$ E	$$\frac{P_1 \wedge P_2 \wedge \ldots \wedge P_n}{P_i}$$ where $i \in \{1, \ldots, n\}$	Given the truth the conjunction $P_1 \wedge P_2 \wedge \ldots \wedge P_n$ then the truth of proposition P_i follows. This rule shows how a conjunction can be eliminated
$\vee$ I	$$\frac{P_i}{P_1 \vee P_2 \vee \ldots \vee P_n}$$	Given the truth of propositions P_i then the truth of the disjunction $P_1 \vee P_2 \vee \ldots \vee P_n$ follows. This rule shows how a disjunction can be introduced
$\vee$ E	$$\frac{P_1 \vee \ldots \vee P_n, P_1 \rightarrow E, \ldots P_n \rightarrow E}{E}$$	Given the truth of the disjunction $P_1 \vee P_2 \vee \ldots \vee P_n$, and that each disjunct implies E, then the truth of E follows. This rule shows how a disjunction can be eliminated
$\rightarrow$ I	From $P_1, P_2, \ldots P_n$ infer P $$\frac{}{P_1 \wedge P_2 \wedge \ldots \wedge P_n \rightarrow P}$$	This rule states that if we have a theorem that allows P to be inferred from the truth of premises $P_1, P_2, \ldots P_n$ (or previously proved) then we can deduce $(P_1 \wedge P_2 \wedge \ldots \wedge P_n) \rightarrow P$. This is known as the *Deduction Theorem*
$\rightarrow$ E	$$\frac{P_i \rightarrow P_j, P_i}{P_j}$$	This rule is known as *modus ponens*. The consequence of an implication follows if the antecedent is true (or has been previously proved)
$\equiv$ I	$$\frac{P_i \rightarrow P_j, P_j \rightarrow P_i}{P_i \leftrightarrow P_j}$$	If proposition P_i implies proposition P_j and vice versa then they are equivalent (i.e. $P_i \leftrightarrow P_j$)
$\equiv$ E	$$\frac{P_i \leftrightarrow P_j}{P_i \rightarrow P_j, P_j \rightarrow P_i}$$	If proposition P_i is equivalent to proposition P_j then proposition P_i implies proposition P_j and vice versa
$\neg$ I	From P infer $P_1 \wedge \neg P_1$ $$\frac{}{\neg P}$$	If the proposition P allows a contradiction to be derived, then $\neg P$ is deduced. This is an example of a *proof by contradiction*
$\neg$ E	From $\neg P$ infer $P_1 \wedge \neg P_1$ $$\frac{}{\neg P}$$	If the proposition $\neg P$ allows a contradiction to be derived, then P is deduced. This is an example of a proof by contradiction

operators—$\wedge$, $\vee$, $\neg$, $\rightarrow$ and $\leftrightarrow$. There are two inference rules per operator (an introduction rule and an elimination rule), and the rules are defined in Table 15.13:

Natural deduction may be employed in logical reasoning and is described in detail in [1, 2].

15.2.6 Sketch of Formalization of Propositional Calculus

Truth tables provide an informal approach to proof and the proof is provided in terms of the meanings of the propositions and logical connectives. The formalization of propositional logic includes the definition of an alphabet of symbols and well-formed formulae of the calculus, the axioms of the calculus and rules of inference for logical deduction.

The deduction of a new formulae Q is via a sequence of well-formed formulae $P_1, P_2, \ldots P_n$ (where $P_n = Q$) such that each P_i is either an axiom, a hypothesis or deducible from an earlier pair of formula P_j, P_k, (where P_k is of the form $P_j \Rightarrow P_i$) and modus ponens. *Modus ponens* is a rule of inference that states that given propositions A, and $A \Rightarrow B$ then proposition B may be deduced. The deduction of a formula Q from a set of hypothesis H is denoted by $H \vdash Q$, and where Q is deducible from the axioms alone this is denoted by $\vdash Q$.

The *deduction theorem* of propositional logic states that if $H \cup \{P\} \vdash Q$, then $H \vdash P \rightarrow Q$, and the converse of the theorem is also true: i.e. if $H \vdash P \rightarrow Q$ then $H \cup \{P\} \vdash Q$. Formalism (this approach was developed by the German mathematician, David Hilbert) allows reasoning about symbols according to rules, and to derive theorems from formulae irrespective of the meanings of the symbols and formulae.

Propositional calculus is *sound*; i.e. any theorem derived using the Hilbert approach is true. Further, the calculus is also *complete*, and every tautology has a proof (i.e. is a theorem in the formal system). The propositional calculus is *consistent*: (i.e. it is not possible that both the well-formed formula A and $\neg A$ are deducible in the calculus).

Propositional calculus is *decidable*: i.e. there is an algorithm (truth table) to determine for any well-formed formula A whether A is a theorem of the formal system. The Hilbert style system is slightly cumbersome in conducting proof and is quite different from the normal use of logic in mathematical deduction.

15.2.7 Applications of Propositional Calculus

Propositional calculus may be employed in reasoning with arguments in natural language. First, the premises and conclusion of the argument are identified and formalized into propositions. Propositional logic is then employed to determine if the conclusion is a valid deduction from the premises.

Consider, for example, the following argument that aims to prove that Superman does not exist.

If Superman were able and willing to prevent evil, he would do so. If Superman were unable to prevent evil he would be impotent; if he were unwilling to prevent evil he would be malevolent; Superman does not prevent evil. If superman exists he is neither malevolent nor impotent; therefore Superman does not exist.

First, letters are employed to represent the propositions as follows:

a: Superman is able to prevent evil
w: Superman is willing to prevent evil
i: Superman is impotent
m: Superman is malevolent
p: Superman prevents evil
e: Superman exists

Then, the argument above is formalized in propositional logic as follows:

Premises	
P_1	$(a \wedge w) \to p$
P_2	$(\neg a \to i) \wedge (\neg w \to m)$
P_3	$\neg p$
P_4	$e \to \neg i \wedge \neg m$
Conclusion	$P_1 \wedge P_2 \wedge P_3 \wedge P_4 \Rightarrow \neg e$

Proof that Superman does not exist

1.	$a \wedge w \to p$	Premise 1
2.	$(\neg a \to i) \wedge (\neg w \to m)$	Premise 2
3.	$\neg p$	Premise 3
4.	$e \to (\neg i \wedge \neg m)$	Premise 4
5.	$\neg p \to \neg(a \wedge w)$	1, Contrapositive
6.	$\neg(a \wedge w)$	3, 5 Modus Ponens
7.	$\neg a \vee \neg w$	6, De Morgan's Law
8.	$\neg (\neg i \wedge \neg m) \to \neg e$	4, Contrapositive
9.	$i \vee m \to \neg e$	8, De Morgan's Law
10.	$(\neg a \to i)$	2, $\wedge$ Elimination
11.	$(\neg w \to m)$	2, $\wedge$ Elimination
12.	$\neg \neg a \vee i$	10, A$\to$ B equivalent to $\neg$A$\vee$ B
13.	$\neg \neg a \vee i \vee m$	11, $\vee$ Introduction
14.	$\neg \neg a \vee (i \vee m)$	
15.	$\neg a \to (i \vee m)$	14, A $\to$ B equivalent to $\neg$A$\vee$ B
16.	$\neg \neg w \vee m$	11, A $\to$ B equivalent to $\neg$A$\vee$ B
17.	$\neg \neg w \vee (i \vee m)$	
18.	$\neg w \to (i \vee m)$	17, A $\to$ B equivalent to $\neg$A$\vee$ B
19.	$(i \vee m)$	7, 15, 18 $\vee$Elimination
20.	$\neg e$	9, 19 Modus Ponens

Second Proof

1.	$\neg p$	P_3
2.	$\neg(a \wedge w) \vee p$	P_1 ($A \rightarrow B \equiv \neg A \vee B$)
3.	$\neg(a \wedge w)$	1, 2 $A \vee B$, $\neg B \vdash A$
4.	$\neg a \vee \neg w$	3, De Morgan's Law
5.	$(\neg a \rightarrow i)$	P_2 ($\wedge$-Elimination)
6.	$\neg a \rightarrow i \vee m$	5, $x \rightarrow y \vdash x \rightarrow y \vee z$
7.	$(\neg w \rightarrow m)$	P_2 ($\wedge$-Elimination)
8.	$\neg w \rightarrow i \vee m$	7, $x \rightarrow y \vdash x \rightarrow y \vee z$
9.	$(\neg a \vee \neg w) \rightarrow (i \vee m)$	8, $x \rightarrow z$, $y \rightarrow z \vdash x \vee y \rightarrow z$
10.	$(i \vee m)$	4, 9 Modus Ponens
11.	$e \rightarrow \neg(i \vee m)$	P_4 (De Morgan's Law)
12.	$\neg e \vee \neg (i \vee m)$	11, ($A \rightarrow B \equiv \neg A \vee B$)
13.	$\neg e$	10, 12 $A \vee B$, $\neg B \vdash A$

Therefore, the conclusion that Superman does not exist is a valid deduction from the given premises.

15.2.8 Limitations of Propositional Calculus

The propositional calculus deals with propositions only. It is incapable of dealing with the syllogism 'All Greeks are mortal; Socrates is a Greek; therefore Socrates is mortal'. This would be expressed in propositional calculus as three propositions *A*, *B* therefore *C*, where *A* stands for 'All Greeks are mortal', *B* stands for 'Socrates is a Greek' and *C* stands for 'Socrates is mortal'. Propositional logic does not allow the conclusion that all Greeks are mortal to be derived from the two premises.

Predicate calculus deals with these limitations by employing variables and terms, and using universal and existential quantification to express that a particular property is true of all (or at least one) values of a variable. Predicate calculus is discussed in the next section.

15.3 Predicate Calculus

Predicate logic is a richer system than propositional logic, and it allows complex facts about the world to be represented. It allows new facts about the world to be derived in a way that guarantees that if the initial facts are true then the conclusions are true. Predicate calculus includes predicates, variables, constants and quantifiers.

A *predicate* is a characteristic or property that an object can have, and we are predicating some property of the object. For example, *"Socrates is a Greek"* could be expressed as $G(s)$, with capital letters standing for predicates and small letters standing for objects. A predicate may include variables, and a statement with a variable becomes a proposition once the variables are assigned values. For example, $G(x)$ states that the variable x is a Greek, whereas $G(s)$ is an assignment of values to x. The set of values that the variables may take is termed the universe of discourse, and the variables take values from this set.

Predicate calculus employs quantifiers to express properties such as all members of the domain have a particular property: e.g., $(\forall x)P(x)$, or that there is at least one member that has a particular property: e.g. $(\exists x)P(x)$. These are referred to as the *universal and existential quantifiers.*

The syllogism 'All Greeks are mortal; Socrates is a Greek; therefore Socrates is mortal' may be easily expressed in predicate calculus by

$$(\forall x)(G(x) \rightarrow M(x))$$
$$G(s)$$

$$M(s)$$

In this example, the predicate $G(x)$ stands for x is a Greek and the predicate $M(x)$ stands for x is mortal. The formula $G(x) \rightarrow M(x)$ states that if x is a Greek then x is mortal, and the formula $(\forall x)(G(x) \rightarrow M(x))$ states for any x that if x is a Greek then x is mortal. The formula $G(s)$ states that Socrates is a Greek and the formula $M(s)$ states that Socrates is mortal.

Example 15.6 (**Predicates**) A predicate may have one or more variables. A predicate that has only one variable (i.e. a unary or 1-place predicate) is often related to sets; a predicate with two variables (a 2-place predicate) is a relation; and a predicate with n variables (a n-place predicate) is a n-ary relation. Propositions do not contain variables and so they are 0-place predicates. The following are examples of predicates:

  (i) The predicate *Prime*(x) states that x is a prime number (with the natural numbers being the universe of discourse).
 (ii) *Lawyer*(a) may stand for a is a lawyer.
(iii) Mean(m, x, y) states that m is the mean
 (iv) of x and y: i.e. $m = \frac{1}{2}(x + y)$.
 (iv) LT$(x, 6)$ states that x is less than 6.
  (v) $G(x, \pi)$ states that x is greater than π (where is the constant 3.14159)
 (vi) $G(x, y)$ states that x is greater than y.
(vii) EQ(x, y) states that x is equal to y.
(viii) LE(x, y) states that x is less than or equal to y.
 (ix) Real(x) states that x is a real number.

(x)   Father(x, y) states that x is the father of y.

(xi) $\neg(\exists x)(Prime(x) \wedge B(x, 32, 36))$ states that there is no prime number between 32 and 36.

Universal and Existential Quantification

The universal quantifier is used to express a statement such as that all members of the domain have property P. This is written as $(\forall x)P(x)$ and expresses the statement that the property $P(x)$ is true for all x. Similarly, $(\forall x_1, x_2, \ldots, x_n)\, P(x_1, x_2, \ldots, x_n)$ states that property $P(x_1, x_2, \ldots, x_n)$ is true for all $x_1, x_2, \ldots, x_n$. Clearly, the predicate $(\forall x)\, P(a, b)$ is identical to $P(a, b)$ since it contains no variables, and the predicate $(\forall y \in \mathbb{N})\,(x \leq y)$ is true if $x = 1$ and false otherwise.

The existential quantifier states that there is at least one member in the domain of discourse that has property P. This is written as $(\exists x)P(x)$ and the predicate $(\exists x_1, x_2, \ldots, x_n)\, P(x_1, x_2, \ldots, x_n)$ states that there is at least one value of $(x_1, x_2, \ldots, x_n)$ such that $P(x_1, x_2, \ldots, x_n)$ is true.

Example 15.7 (**Quantifiers**)

(i) $(\exists p)\, (Prime(p) \wedge p > 1,000,000)$ is true
      It expresses the fact that there is at least one prime number greater than a million, which is true as there are an infinite number of primes.

(ii) $(\forall x)\, (\exists y)\, x < y$ is true
      This predicate expresses the fact that given any number x we can always find a larger number: e.g. take $y = x + 1$.

(iii) $(\exists y)\, (\forall x)\, x < y$ is false
      This predicate expresses the statement that there is a natural number y such that all natural numbers are less than y. Clearly, this statement is false since there is no largest natural number, and so the predicate $(\exists y)\, (\forall x)\, x < y$ is false.

Comment 15.1

It is important to be careful with the order in which quantifiers are written, as the meaning of a statement may be completely changed by the simple transposition of two quantifiers.

The well-formed formulae in the predicate calculus are built from terms and predicates, and the rules for building the formulae are described briefly in Sect. 15.3.1. Examples of well-formed formulae include

$$(\forall x)(x > 2)$$
$$(\exists x)x^2 = 2$$
$$(\forall x)\,(x > 2 \wedge x < 10)$$
$$(\exists y)x^2 = y$$
$$(\forall x)\,(\forall y)\,Love(y, x) \qquad \text{(everyone is loved by someone)}$$
$$(\exists y)\,(\forall x)Love(y, x) \qquad \text{(someone loves everyone)}$$

The formula $(\forall x)(x > 2)$ states that every x is greater than the constant 2; $(\exists x)\, x^2 = 2$ states that there is an x that is the square root of 2; $(\forall x)\,(\exists y)\, x^2 = y$ states that for every x there is a y such that the square of x is y.

15.3.1 Sketch of Formalization of Predicate Calculus

The formalization of predicate calculus includes the definition of an alphabet of symbols (including constants and variables), the definition of function and predicate letters, logical connectives and quantifiers. This leads to the definitions of the terms and well-formed formulae of the calculus.

The predicate calculus is built from an alphabet of constants, variables, function letters, predicate letters and logical connectives (including the logical connectives discussed in propositional logic, and universal and existential quantifiers).

The definition of terms and well-formed formulae specifies the syntax of the predicate calculus, and the set of well-formed formulae gives the language of the predicate calculus. The terms and well-formed formulae are built from the symbols, and these symbols are not given meaning in the formal definition of the syntax.

The language defined by the calculus needs to be given an *interpretation* in order to give a meaning to the terms and formulae of the calculus. The interpretation needs to define the domain of values of the constants and variables, provide meaning to the function letters, the predicate letters and the logical connectives.

Terms are built from constants, variables and function letters. A constant or variable is a term, and if $t_1, t_2, \ldots, t_k$ are terms, then $f_i^k(t_1, t_2, \ldots, t_k)$ is a term (where f_i^k is a k-ary function letter). Examples of terms include

x^2 where x is a variable and square is a $1 - $ ary function letter

$x^2 + y^2$ where $x^2 + y^2$ is shorthand for the function add(square(x), square(y))

         where add is a $2 - $ ary function letter and square is a $1 - $ ary function

The well-formed formulae are built from terms as follows. If P_i^k is a k-ary predicate letter, $t_1, t_2, \ldots, t_k$ are terms, then $P_i^k(t_1, t_2, \ldots, t_k)$ is a well-formed formula. If A and B are well-formed formulae then so are $\neg A, A \wedge B, A \vee B, A \rightarrow B, A \leftrightarrow B, (\forall x)A$ and $(\exists x)A$.

There is a set of axioms for predicate calculus and two rules of inference used for the deduction of new formulae from the existing axioms and previously deduced formulae. The deduction of a new formula Q is via a sequence of well-formed formulae P_1, P_2, ... P_n (where $P_n = Q$) such that each P_i is either an axiom, a hypothesis or deducible from one or more of the earlier formulae in the sequence.

The two rules of inference are *modus ponens* and *generalization*. Modus ponens is a rule of inference that states that given predicate formulae A, and $A \Rightarrow B$ then the predicate formula B may be deduced. Generalization is a rule of inference that states that given predicate formula A, then the formula $(\forall x)A$ may be deduced where x is any variable.

The deduction of a formula Q from a set of hypothesis H is denoted by $H \vdash Q$, and where Q is deducible from the axioms alone this is denoted by $\vdash Q$. The *deduction theorem* states that if $H \cup \{P\} \vdash Q$ then $H \vdash P \to Q$[3] and the converse of the theorem is also true: i.e. if $H \vdash P \to Q$ then $H \cup \{P\} \vdash Q$.

The approach allows reasoning about symbols according to rules, and to derive theorems from formulae irrespective of the meanings of the symbols and formulae. Predicate calculus is *sound*: i.e. any theorem derived using the approach is true, and the calculus is also *complete*.

Scope of Quantifiers

The scope of the quantifier $(\forall x)$ in the well-formed formula $(\forall x)A$ is A. Similarly, the scope of the quantifier $(\exists x)$ in the well-formed formula $(\exists x)B$ is B. The variable x that occurs within the scope of the quantifier is said to be a *bound variable*. If a variable is not within the scope of a quantifier it is *free*.

Example 15.8 (**Scope of Quantifiers**)

(i) x is free in the well-formed formula $\forall y \, (x^2 + y > 5)$
(ii) x is bound in the well-formed formula $\forall x \, (x^2 > 2)$

A well-formed formula is *closed* if it has no free variables. The substitution of a term t for x in A can only take place only when no free variable in t will become bound by a quantifier in A through the substitution. Otherwise, the interpretation of A would be altered by the substitution.

A term t is free for x in A if no free occurrence of x occurs within the scope of a quantifier $(\forall y)$ or $(\exists y)$ where y is free in t. This means that the term t may be substituted for x without altering the interpretation of the well-formed formula A.

For example, suppose A is $\forall y \, (x^2 + y^2 > 2)$ and the term t is y, then t is not free for x in A as the substitution of t for x in A will cause the free variable y in t to become bound by the quantifier $\forall y$ in A, thereby altering the meaning of the formula to $\forall y \, (y^2 + y^2 > 2)$.

---

[3]This is stated more formally that if $H \cup \{P\} \vdash Q$ by a deduction containing no application of generalization to a variable that occurs free in P then $H \vdash P \to Q$.

15.3.2 Interpretation and Valuation Functions

An *interpretation* gives meaning to a formula and it consists of a *domain of discourse* and a *valuation function*. If the formula is a sentence (i.e. does not contain any free variables) then the given interpretation of the formula is either true or false. If a formula has free variables, then the truth or falsity of the formula depends on the values given to the free variables. A formula with free variables essentially describes a relation say, $R(x_1, x_2, \ldots x_n)$ such that $R(x_1, x_2, \ldots x_n)$ is true if $(x_1, x_2, \ldots x_n)$ is in relation R. If the formula is true irrespective of the values given to the free variables, then the formula is true in the interpretation.

A *valuation* (meaning) *function gives meaning to the logical symbols and connectives*. Thus, associated with each constant c is a constant c_Σ in some universe of values Σ; with each function symbol f of arity k, we have a function symbol f_Σ in Σ and $f_\Sigma: \Sigma^k \to \Sigma$; and for each predicate symbol P of arity k a relation $P_\Sigma \subseteq \Sigma^k$. *The valuation function, in effect, gives the semantics of the language of the predicate calculus L.*

The truth of a predicate P is then defined in terms of the meanings of the terms, the meanings of the functions, predicate symbols, and the normal meanings of the connectives.

Mendelson [3] provides a technical definition of truth in terms of *satisfaction* (with respect to an interpretation M). Intuitively a formula F is *satisfiable* if it is *true* (in the intuitive sense) for some assignment of the free variables in the formula F. If a formula F is satisfied for every possible assignment to the free variables in F, then it is *true* (in the technical sense) for the interpretation M. An analogous definition is provided for *false* in the interpretation M.

A formula is *valid* if it is true in every interpretation; however, as there may be an uncountable number of interpretations, it may not be possible to check this requirement in practice. M is said to be a model for a set of formulae if and only if every formula is true in M.

There is a distinction between proof theoretic and model theoretic approaches in predicate calculus. *Proof theoretic* is essentially syntactic, and there is a list of axioms with rules of inference. The theorems of the calculus are logically derived (i.e. $\vdash A$) and the logical truths are as a result of the syntax or form of the formulae, rather than the *meaning* of the formulae. *Model theoretical*, in contrast is essentially semantic. The truth derives from the meaning of the symbols and connectives, rather than the logical structure of the formulae. This is written as $\vdash_M A$.

A calculus is *sound* if all of the logically valid theorems are true in the interpretation, i.e. proof theoretic $\Rightarrow$ model theoretic. A calculus is *complete* if all the truths in an interpretation are provable in the calculus, i.e. model theoretic $\Rightarrow$ proof theoretic. A calculus is *consistent* if there is no formula A such that $\vdash A$ and $\vdash \neg A$.

The predicate calculus is sound, complete and consistent. *Predicate calculus is not decidable*: i.e. there is no algorithm to determine for any well-formed formula A whether A is a theorem of the formal system. The undecidability of the predicate

calculus may be demonstrated by showing that if the predicate calculus is decidable
then the halting problem (of Turing machines) is solvable. We discussed the halting
problem in Chap. 13.

15.3.3 Properties of Predicate Calculus

The following are properties of the predicate calculus.

(i) $(\forall x)P(x) \equiv (\forall y)P(y)$

(ii) $(\forall x)P(x) \equiv \neg(\exists x)\neg P(x)$

(iii) $(\exists x)P(x) \equiv \neg(\forall x)\neg P(x)$

(iv) $(\exists x)P(x) \equiv (\exists y)P(y)$

(v) $(\forall x)(\forall y)P(x,y) \equiv (\forall y)(\forall x)P(x,y)$

(vi) $(\exists x)(P(x) \vee Q(x)) \equiv (\exists x)P(x) \vee (\exists y)Q(y)$

(vii) $(\forall x)P(x) \wedge Q(x)) \equiv (\forall x)P(x) \wedge (\forall y)Q(y)$

15.3.4 Applications of Predicate Calculus

The predicate calculus is may be employed to formally state the system requirements of a proposed system. It may be used to conduct formal proof to verify the
presence or absence of certain properties in a specification. It may also be employed
to define piecewise defined functions such as $f(x, y)$ where $f(x, y)$ is defined by

$$f(x,y) = x^2 - y^2 \quad \text{where } x \leq 0 \wedge y < 0;$$
$$f(x,y) = x^2 + y^2 \quad \text{where } x > 0 \wedge y < 0;$$
$$f(x,y) = x + y \quad \text{where } x \geq 0 \wedge y = 0;$$
$$f(x,y) = x - y \quad \text{where } x < 0 \wedge y = 0;$$
$$f(x,y) = x + y \quad \text{where } x \leq 0 \wedge y > 0;$$
$$f(x,y) = x^2 + y^2 \quad \text{where } x > 0 \wedge y > 0$$

The predicate calculus may be employed for program verification, and to show
that a code fragment satisfies its specification. The statement that a program F is
correct with respect to its precondition P and postcondition Q is written as $P\{F\}$
Q. The objective of program verification is to show that if the precondition is true
before execution of the code fragment, then this implies that the postcondition is
true after execution of the code fragment.

A program fragment a is *partially correct* for precondition P and postcondition Q if and only if whenever a is executed in any state in which P is satisfied and execution terminates, then the resulting state satisfies Q. Partial correctness is denoted by $P\{F\}Q$, and Hoare's Axiomatic Semantics is based on partial correctness. It requires proof that the postcondition is satisfied if the program terminates.

A program fragment a is *totally correct* for precondition P and postcondition Q, if and only if whenever a is executed in any state in which P is satisfied then the execution terminates and the resulting state satisfies Q. It is denoted by $\{P\}F\{Q\}$, and Dijkstra's calculus of weakest preconditions is based on total correctness [2, 4]. It is required to prove that if the precondition is satisfied then the program terminates and the postcondition is satisfied

15.3.5 Semantic Tableaux in Predicate Calculus

We discussed the use of semantic tableaux for determining the validity of arguments in propositional logic earlier in this chapter, and its approach is to negate the conclusion of an argument and to show that this results in inconsistency with the premises of the argument.

The use of semantic tableaux is similar with predicate logic, except that there are some additional rules to consider. As before, if all branches of a semantic tableau are closed, then the premises and the negation of the conclusion are mutually inconsistent, and all branches in the tableau are closed. From this, we deduce that the conclusion must be true.

The rules of semantic tableaux for propositional logic were presented in Table 15.12, and the additional rules specific to predicate logic are detailed in Table 15.14.

Example 15.9 (**Semantic Tableaux**) Show that the syllogism 'All Greeks are mortal; Socrates is a Greek; therefore Socrates is mortal' is a valid argument in predicate calculus.

Table 15.14 Extra rules of semantic tableaux (for predicate calculus)

Rule No.	Definition	Description
1.	$(\forall x)\, A(x)$ $A(t)$ where t is a term	Universal instantiation
2.	$(\exists x)\, A(x)$ $A(t)$ where t is a term that has not been used in the derivation so far	Rule of Existential instantiation. The term "t" is often a constant "a"
3.	$\neg(\forall x)\, A(x)$ $(\exists x)\, \neg A(x)$	
4.	$\neg(\exists x)\, A(x)$ $(\forall x)\neg A(x)$	

Solution

We expressed this argument previously as $(\forall x)(G(x) \rightarrow M(x))$; $G(s)$; $M(s)$. Therefore, we negate the conclusion (i.e. $\neg M(s)$), and try to construct a closed tableau.

$$(\forall x)(G(x) \rightarrow M(x))$$
$$G(s)$$
$$\neg M(s).$$
$$G(s) \rightarrow M(s) \qquad\qquad \text{Universal Instantiation}$$
$$\wedge$$
$$\neg G(s) \quad M(s)$$
$$\text{-----} \qquad \text{--------}$$
$$closed \qquad closed$$

Therefore, as the tableau is closed we deduce that the negation of the conclusion is inconsistent with the premises, and that therefore the conclusion follows from the premises.

Example 15.10 (**Semantic Tableaux**) Determine whether the following argument is valid.

All lecturers are motivated
Anyone who is motivated and clever will teach well
Joanne is a clever lecturer
Therefore, Joanne will teach well.

Solution

We encode the argument as follows

$L(x)$ stands for 'x is a lecturer'
$M(x)$ stands for 'x is motivated'
$C(x)$ stands for 'x is clever'
$W(x)$ stands for 'x will teach well'

We therefore wish to show that

$$(\forall x)(L(x) \rightarrow M(x)) \wedge (\forall x)((M(x) \wedge C(x))$$
$$\rightarrow W(x)) \wedge L(joanne) \wedge C(joanne) \models W(joanne)$$

Therefore, we negate the conclusion (i.e. $\neg W(joanne)$) and try to construct a closed tableau.

1. $(\forall x)(L(x) \rightarrow M(x))$
2. $(\forall x)((M(x) \wedge C(x)) \rightarrow W(x))$
3. $L(joanne)$
4. $C(joanne)$
5. $\neg W(joanne)$
6. $L(joanne) \rightarrow M(joanne)$ Universal Instantiation (line 1)
7. $(M(joanne) \wedge C(joanne)) \rightarrow W(joanne)$ Universal Instantiation (line 2)

/ \
8. $\neg L(joanne)$ $M(joanne)$ From line 6

     Closed

/ \
9. $\neg (M(joanne) \wedge C(joanne))$ $W(joanne)$ From line 7

         Closed

/ \
10. $\neg M(joanne)$ $\neg C(joanne)$

-----------------   -------------

    Closed   Closed

Therefore, since the tableau is closed we deduce that the argument is valid.

15.4 Review Questions

1. Draw a truth table to show that $\neg (P \rightarrow Q) \equiv P \wedge \neg Q$

2. Translate the sentence 'Execution of program P begun with $x < 0$ will not terminate' into propositional form.

3. Prove the following theorems using the inference rules of natural deduction

   a. From b infer $b \vee \neg c$
   b. From $b \Rightarrow (c \wedge d)$, b infer d

4. Explain the difference between the universal and the existential quantifier.

5. Express the following statements in the predicate calculus

   a. All natural numbers are greater than 10
   b. There is at least one natural number between 5 and 10
   c. There is a prime number between 100 and 200.

6. Which of the following predicates are true?

    a. $\forall i \in \{10, \ldots, 50\}.i^2 < 2000 \land i < 100$
    b. $\exists i \in \mathbb{N}.i > 5 \land i^2 = 25$
    c. $\exists i \in \mathbb{N}.i^2 = 25$

7. Use semantic tableaux to show that $(A \rightarrow A) \lor (B \land \neg B)$ is true
8. Determine if the following argument is valid.
   If Pilar lives in Cork, she lives in Ireland. Pilar lives in Cork. Therefore, Pilar lives in Ireland.

15.5 Summary

This chapter considered propositional and predicate calculus. Propositional logic is the study of propositions, and a proposition is a statement that is either true or false. A formula in propositional calculus may contain several variables, and the truth or falsity of the individual variables, and the meanings of the logical connectives determines the truth or falsity of the logical formula.

A rich set of connectives is employed in propositional calculus to combine propositions and to build up the well-formed formulae of the calculus. This includes the conjunction of two propositions $(A \land B)$, the disjunction of two propositions $(A \lor B)$, and the implication of two propositions $(A \Rightarrow B)$. These connectives allow compound propositions to be formed, and the truth of the compound propositions is determined from the truth values of the constituent propositions and the rules associated with the logical connectives. The meaning of the logical connectives is given by truth tables.

Propositional calculus is both complete and consistent with all true propositions deducible in the calculus, and there is no formula A such that both A and $\neg A$ are deducible in the calculus.

An argument in propositional logic consists of a sequence of formulae that are the premises of the argument and a further formula that is the conclusion of the argument. One elementary way to see if the argument is valid is to produce a truth table to determine if the conclusion is true whenever all of the premises are true. Other ways are to use semantic tableaux or natural deduction.

Predicates are statements involving variables and these statements become propositions once the variables are assigned values. Predicate calculus allows expressions such as all members of the domain have a particular property to be expressed formally: e.g., $(\forall x)Px$, or that there is at least one member that has a particular property: e.g., $(\exists x)Px$.

Predicate calculus may be employed to specify the requirements for a proposed system and to give the definition of a piecewise defined function. Semantic tableaux may be used for determining the validity of arguments in propositional or predicate logic, and its approach is to negate the conclusion of an argument and to show that this results in inconsistency with the premises of the argument.

References

1. The Essence of Logic. John Kelly. Prentice Hall. 1997.
2. The Science of Programming. David Gries. Springer Verlag. Berlin. 1981.
3. Introduction to Mathematical Logic. Elliot Mendelson. Wadsworth and Cole/Brook, Advanced Books & Software. 1987.
4. A Disciple of Programming. E.W. Dijkstra. Prentice Hall. 1976.

Advanced Topics in Logic

16

Key Topics

Fuzzy Logic
Intuitionist Logic
Temporal Logic
Undefined values
Theorem Provers
Logic of partial functions
Logic and AI

16.1 Introduction

In this chapter, we consider some advanced topics in logic including fuzzy logic, temporal logic, intuitionist logic, undefined values, logic and AI and theorem provers. Fuzzy logic is an extension of classical logic that acts as a mathematical model for vagueness, and it handles the concept of partial truth where truth values lie between completely true and completely false. Temporal logic is concerned with the expression of properties that have time dependencies, and it allows temporal properties about the past, present and future to be expressed.

Brouwer and others developed intuitionist logic as the logical foundation for intuitionism, which was a controversial theory of the foundations of mathematics based on a rejection of the law of the excluded middle and an insistence on constructive existence. Martin Löf successfully applied it to type theory in the 1970s.

© Springer International Publishing Switzerland 2016
G. O'Regan, *Guide to Discrete Mathematics*, Texts in Computer Science,
DOI 10.1007/978-3-319-44561-8_16

Partial functions arise naturally in computer science, and such functions may fail to be defined for one or more values in their domain. One approach to dealing with partial functions is to employ a precondition, which restricts the application of the function to values where it is defined. We consider three approaches to deal with undefined values, including the logic of partial functions; Dijkstra's approach with his *cand* and *cor* operators; and Parnas's approach which preserves a classical two-valued logic.

We examine the contribution of logic to the AI field, and the work done by theorem provers to supporting proof.

16.2   Fuzzy Logic

Fuzzy logic is a branch of *many-valued logic* that allows inferences to be made when dealing with vagueness, and it can handle problems with imprecise or incomplete data. It differs from classical two-valued propositional logic, in that it is based on degrees of truth, rather than on the standard binary truth values of "true or false" (1 or 0) of propositional logic. That is, while statements made in propositional logic are either true or false (1 or 0), the truth value of a statement made in fuzzy logic is a value between 0 and 1. Its value expresses the extent to which the statement is true, with a value of 1 expressing absolute truth, and a value of 0 expressing absolute falsity.

Fuzzy logic uses *degrees of truth* as a mathematical model for vagueness, and this is useful since statements made in natural language are often vague and have a certain (rather than an absolute) degree of truth. It is an extension of classical logic to handle the concept of partial truth, where the truth value lies between completely true and completely false. Lofti Zadeh developed fuzzy logic at Berkley in the 1960s, and it has been successfully applied to Expert Systems and other areas of Artificial Intelligence.

For example, consider the statement "John is tall". If John were 6 foot, 4 in. then we would say that this is a true statement (with a truth value of 1) since John is well above average height. However, if John is 5 ft, 9 in. tall (around average height) then this statement has a degree of truth, and this could be indicated by a fuzzy truth valued of 0.6. Similarly, the statement that today is sunny may be assigned a truth value of 1 if there are no clouds, 0.8 if there are a small number of clouds, and 0 if it is raining all day.

Propositions in fuzzy logic may be combined together to form compound propositions. Suppose X and Y are propositions in fuzzy logic, then compound propositions may be formed from the conjunction, disjunction and implication operators. The usual definition in fuzzy logic of the truth values of the compound propositions formed from X and Y is given by

$$\text{Truth } (\neg X) = 1 - \text{Truth}(X)$$
$$\text{Truth } (X \text{ and } Y) = \min(\text{Truth}(X), \text{ Truth}(Y))$$
$$\text{Truth } (X \text{ or } Y) = \max(\text{Truth}(X), \text{ Truth}(Y))$$
$$\text{Truth } (X \rightarrow Y) = \text{Truth}(\neg X \text{ or } Y))$$

There is another way in which the operators may be defined in terms of multiplication

$$\text{Truth}(X \text{ and } Y) = \text{Truth}(X) * \text{Truth}(Y)$$
$$\text{Truth}(X \text{ or } Y) = 1 - (1 - \text{Truth}(X)) * (1 - \text{Truth}(Y))$$
$$\text{Truth}(X \rightarrow Y) = \max\{z \mid \text{Truth}(X) * z \leq \text{Truth}(Y)\} \text{ where } 0 \leq z \leq 1$$

Under these definitions, fuzzy logic is an extension of classical two-valued logic, which preserves the usual meaning of the logical connectives of propositional logic when the fuzzy values are just $\{0, 1\}$.

Fuzzy logic has been very useful in expert system and artificial intelligence applications. The first fuzzy logic controller was developed in England in the mid-1970s. It has been applied to the aerospace and automotive sectors, and also to the medical, robotics and transport sectors.

16.3 Temporal Logic

Temporal logic is concerned with the expression of properties that have time dependencies, and the various temporal logics can express facts about the past, present and future. Temporal logic has been applied to specify temporal properties of natural language, artificial intelligence as well as the specification and verification of program and system behaviour. It provides a language to encode temporal knowledge in artificial intelligence applications, and it plays a useful role in the formal specification and verification of temporal properties (e.g. liveness and fairness) in safety critical systems.

The statements made in temporal logic can have a truth value that varies over time. In other words, sometimes the statement is true and sometimes it is false, but it is never true or false at the same time. The two main types of temporal logics are *linear time logics* (reason about a single time line), and *branching time logics* (reason about multiple timelines).

The roots of temporal logic lie in work done by Aristotle in the fourth century B. C., when he considered whether a truth value should be given to a statement about a future event that may or may not occur. For example, what truth value (if any) should be given to the statement that *'There will be a sea battle tomorrow'*. Aristotle argued against assigning a truth value to such statements in the present time.

Newtonian mechanics assumes an absolute concept of time independent of space, and this viewpoint remained dominant until the development of the theory of relativity in the early twentieth century (when space-time became the dominant paradigm).

Arthur Prior began analyzing and formalizing the truth values of statements concerning future events in the 1950s, and he introduced Tense Logic (a temporal logic) in the early 1960s. Tense logic contains four modal operators (strong and weak) that express events in the future or in the past

> -P (It has at some time been the case that)
>
> -F (It will be at some time be the case that)
>
> -H (It has always been the case that)
>
> -G (It will always be the case that)

The P and F operators are known as weak tense operators, while the H and G operators known as strong tense operators. The two pairs of operators are interdefinable via the equivalences

$$P\phi \cong \neg H\neg\phi$$
$$H\phi, \cong \neg P\neg\phi$$
$$F\phi \cong \neg G\neg\phi$$
$$G\phi, \cong \neg F\neg\phi$$

The set of formulae in Prior's temporal logic may be defined recursively, and they include the connectives used in classical logic (e.g. $\neg$, $\wedge$, $\vee$, $\rightarrow$, $\leftrightarrow$). We can express a property ϕ that is always true as $A\phi \cong H\phi \wedge \phi \wedge G\phi$ and a property that is sometimes true as $E\phi \cong P\phi \vee \phi \vee F\phi$. Various extensions of Prior's tense logic have been proposed to enhance its expressiveness. These include the binary *since* temporal operator 'S', and the binary *until* temporal operator 'U'. For example, the meaning of $\phi S\psi$ is that ϕ has been true since a time when ψ was true.

Temporal logics are applicable in the specification of computer systems, and a specification may require *safety*, *fairness* and *liveness properties* to be expressed. For example, a fairness property may state that it will always be the case that a certain property will hold sometime in the future. The specification of temporal properties often involves the use of special temporal operators.

We discuss common temporal operators that are used, including an operator to express properties that will always be true; properties that will eventually be true; and a property that will be true in the next time instance. For example

> $\Box P$ $-P$ is always true
>
> $\Diamond$ $-P$ will be true sometime in the future
>
> $\bigcirc$ $-P$ is true in the next time instant(*discrete time*)

Linear temporal logic (LTL) was introduced by Pnueli in the late 1970s. This linear time logic is useful in expressing safety and liveness properties. Branching time logics assume a non-deterministic branching future for time (with a deterministic, linear past). Computation tree logic (CTL and CTL*) were introduced in the early 1980s by Emerson and others.

It is also possible to express temporal operations directly in classical mathematics, and the well-known computer scientist, Parnas, prefers this approach. He is critical of computer scientists for introducing unnecessary formalisms when classical mathematics already possesses the ability to do this. For example, the value of a function f at a time instance prior to the current time t is defined as

$$\text{Prior}(f, t) = \lim_{\varepsilon \to 0} f(t - \varepsilon)$$

For more detailed information on temporal logic the reader is referred to the excellent article on temporal logic in [1].

16.4 Intuitionist Logic

The controversial school of intuitionist mathematics was founded by the Dutch mathematician, L. E. J. Brouwer, who was a famous topologist, and well known for his fixpoint theorem in topology. This constructive approach to mathematics proved to be highly controversial, as its acceptance as a foundation of mathematics would have led to the rejection of many accepted theorems in classical mathematics (including his own fixed point theorem).

Brouwer was deeply interested in the foundations of mathematics, and the problems arising from the paradoxes of set theory. He was determined to provide a secure foundation for mathematics, and his view was that an existence theorem in mathematics that demonstrates the proof of a mathematical object has no validity, unless the proof is constructive and accompanied by a procedure to construct the object. He therefore rejected indirect proof and the law of the excluded middle ($P \lor \neg P$) or equivalently ($\neg\neg P \to P$), and he insisted on an explicit construction of the mathematical object.

The problem with the law of the excluded middle (LEM) arises in dealing with properties of infinite sets. For finite sets, one can decide if all elements of the set possess a certain property P by testing each one. However, this procedure is no longer possible for infinite sets. We may know that a certain element of the infinite set does not possess the property, or it may be the actual method of construction of the set allows us to prove that every element has the property. However, the application of the law of the excluded middle is invalid for infinite sets, as we cannot conclude from the situation where not all elements of an infinite set possesses a property P that there exists at least one element which does not have the property P. In other words, the law of the excluded middle may only be applied in cases where the conclusion can be reached in a finite number of steps.

Consequently, if the Brouwer view of the world was accepted then many of the classical theorems of mathematics (including his own well-known results in topology) could no longer be said to be true. His approach to the foundations of mathematics hardly made him popular with other mathematicians (the differences were so fundamental that it was more like a war), and intuitionism never became mainstream in mathematics. It led to deep and bitter divisions between Hilbert and Brouwer, with Hilbert accusing Brouwer (and Weyl) of trying to overthrow everything that did not suit them in mathematics, and that intuitionism was treason to science. Hilbert argued that a suitable foundation for mathematics should aim to preserve most of mathematics. Brouwer described Hilbert's formalist program as a false theory that would produce nothing of mathematical value. For Brouwer, 'to exist' is synonymous with 'constructive existence', and constructive mathematics is relevant to computer science, as a program may be viewed as the result obtained from a constructive proof of its specification.

Brouwer developed one of the more unusual logics that have been invented (intuitionist logic), in which many of the results of classical mathematics were no longer true. Intuitionist logic may be considered the logical basis of constructive mathematics, and formal systems for intuitionist propositional and predicate logic were developed by Heyting and others [2].

Consider a hypothetical mathematical property $P(x)$ of which there is no known proof (i.e. it is unknown whether $P(x)$ is true or false for arbitrary x where x ranges over the natural numbers). Therefore, the statement $\forall x \, (P(x) \vee \neg \, P(x))$ cannot be asserted with the present state of knowledge, as neither $P(x)$ or $\neg P(x)$ has been proved. That is, unproved statements in intuitionist logic are not given an intermediate truth value, and they remain of an unknown truth value until they have been either proved or disproved.

The intuitionist interpretation of the logical connectives is different from classical propositional logic. A sentence of the form $A \vee B$ asserts that either a proof of A or a proof of B has been constructed, and $A \vee B$ is not equivalent to $\neg \, (\neg A \wedge \neg B)$. Similarly, a proof of $A \wedge B$ is a pair whose first component is a proof of A, and whose second component is a proof of B. The statement $\forall x \, \neg P(x)$ is not equivalent to $\exists x \, P(x)$ in intuitionist logic.

Intuitionist logic was applied to Type Theory by Martin Löf in the 1970s [3]. Intuitionist type theory is based on an analogy between propositions and types, where $A \wedge B$ is identified with $A \times B$, the Cartesian product of A and B. The elements in the set $A \times B$ are of the form (a, b) where $a \in A$ and $b \in B$. The expression $A \vee B$ is identified with $A + B$, the disjoint union of A and B. The elements in the set $A + B$ are got from tagging elements from A and B, and they are of the form inl(a) for $a \in A$, and inr(b) for $b \in B$. The left and right injections are denoted by inl and inr.

16.5   Undefined Values

Total functions $f: X \rightarrow Y$ are functions that are defined for every element in their domain, and total functions are widely used in mathematics. However, there are functions that are undefined for one or more elements in their domain, and one example is the function $y = 1/x$. This function is undefined at $x = 0$.

Partial functions arise naturally in computer science, and such functions may fail to be defined for one or more values in their domain. One approach to dealing with partial functions is to employ a precondition, which restricts the application of the function to where it is defined. This makes it possible to define a new set (a proper subset of the domain of the function) for which the function is total over the new set.

Undefined terms often arise[1] and need to be dealt with. Consider, the example of the square root function $\sqrt{x}$ taken from [4]. The domain of this function is the positive real numbers, and the following expression is undefined

$$((x > 0) \wedge (y = \sqrt{x})) \vee ((x \leq 0) \wedge (y = \sqrt{-x}))$$

The reason this is undefined is since the usual rules for evaluating such an expression involve evaluating each sub-expression, and then performing the Boolean operations. However, when $x < 0$ the sub-expression $y = \sqrt{x}$ is undefined, whereas when $x > 0$ the sub-expression $y = \sqrt{-x}$ is undefined. Clearly, it is desirable that such expressions be handled, and that for the example above, the expression would evaluate to true.

Classical two-valued logic does not handle this situation adequately, and there have been several proposals to deal with undefined values. Dijkstra's approach is to use the **cand** and **cor** operators in which the value of the left hand operand determines whether the right-hand operand expression is evaluated or not. Jone's logic of partial functions [5] uses a three-valued logic[2] and Parnas's[3] approach is an extension to the predicate calculus to deal with partial functions that preserves the two-valued logic.

16.5.1   Logic of Partial Functions

Jones [5] has proposed the logic of partial functions (LPFs) as an approach to deal with terms that may be undefined. This is a three-valued logic and a logical term may be true, false or undefined (denoted $\perp$). The definition of the truth functional operators used in classical two-valued logic is extended to three-valued logic. The truth tables for conjunction and disjunction are defined in Fig. 16.1.

---

[1] It is best to avoid undefinedness by taking care of the definitions of terms and expressions.

[2] The above expression would evaluate to true under Jones three-valued logic of partial functions.

[3] The above expression evaluates to true for Parnas logic (a two-valued logic).

∧ P	Q T	F	⊥
	P∧Q		
T	T	F	⊥
F	F	F	F
⊥	⊥	F	⊥

∨ P	Q T	F	⊥
	P∨Q		
T	T	T	T
F	T	F	⊥
⊥	T	⊥	⊥

Fig. 16.1 Conjunction and disjunction operators

→ P	Q T	F	⊥
	P→Q		
T	T	F	⊥
F	T	T	T
⊥	T	⊥	⊥

↔ P	Q T	F	⊥
	P↔Q		
T	T	F	⊥
F	F	T	⊥
⊥	⊥	⊥	⊥

Fig. 16.2 Implication and equivalence operators

The conjunction of P and Q is true when both P and Q are true; false if one of P or Q is false, and undefined otherwise. The operation is commutative. The disjunction of P and Q ($P \vee Q$) is true if one of P or Q is true; false if both P and Q are false; and undefined otherwise. The implication operation ($P \rightarrow Q$) is true when P is false or when Q is true; false when P is true and Q is false and undefined otherwise. The equivalence operation ($P \leftrightarrow Q$) is true when both P and Q are true or false; it is false when P is true and Q is false (and vice versa); and it is undefined otherwise (Fig. 16.2).

The not operator ($\neg$) is a unary operator such $\neg A$ is true when A is false, false when A is true and undefined when A is undefined (Fig. 16.3).

The result of an operation may be known immediately after knowing the value of one of the operands (e.g. disjunction is true if P is true irrespective of the value of Q). The law of the excluded middle: i.e. $A \vee \neg A$ does not hold in the three-valued logic, and Jones [5] argues that this is reasonable as one would not expect the following to be true

$$\left(^1/_0 = 1\right) \vee \left(^1/_0 \neq 1\right)$$

A	¬A
T	F
F	T
⊥	⊥

Fig. 16.3 Negation

Table 16.1 Examples of parnas evaluation of undefinedness

Expression	$x < 0$	$x \geq 0$
$y = \sqrt{x}$	False	True if $y = \sqrt{x}$, False otherwise
$y = {}^1/_0$	False	False
$y = x^2 + \sqrt{x}$	False	True if $y = x^2 + \sqrt{x}$, False otherwise

There are other well-known laws that fail to hold such as

(i) $E \Rightarrow E$
(ii) Deduction Theorem $E_1 \vdash E_2$ does not justify $\vdash E_1 \Rightarrow E_2$ unless it is known that E_1 is defined.

Many of the tautologies of standard logic also fail to hold.

16.5.2 Parnas Logic

Parnas's approach to logic is based on classical two-valued logic, and his philosophy is that truth values should be true or false only,[4] and that there is no third logical value. It is an extension to predicate calculus to deal with partial functions. The evaluation of a logical expression yields the value 'true' or 'false' irrespective of the assignment of values to the variables in the expression. This allows the expression: $(y = \sqrt{x}) \vee (y = \sqrt{-x})$ that is undefined in classical logic to yield the value true.

The advantages of his approach are that no new symbols are introduced into the logic, and that the logical connectives retain their traditional meaning. This makes it easier for engineers and computer scientists to understand, as it is closer to their intuitive understanding of logic.

The meaning of predicate expressions is given by first defining the meaning of the primitive expressions. These are then used as the building blocks for predicate expressions. The evaluation of a primitive expression $R_j(V)$ (where V is a comma separated set of terms with some elements of V involving the application of partial functions) is false if the value of an argument of a function used in one of the terms of V is not in the domain of that function.[5] The following examples (Tables 16.1 and 16.2) should make this clear.

These primitive expressions are used to build the predicate expressions, and the standard logical connectives are used to yield truth values for the predicate expression. Parnas logic is defined in detail in [4].

The power of Parnas logic may be seen by considering a tabular expressions example [4]. The table below specifies the behaviour of a program that searches the

---

[4]It seems strange to assign the value false to the primitive predicate calculus expression $y = {}^1/_0$.
[5]The approach avoids the undefined logical value ($\bot$) and preserves the two-valued logic.

Table 16.2 Example of Undefinedness in Array

Expression	$i \in \{1 \dots N\}$	$i \notin \{1 \dots N\}$
$B[i] = x$	*True* if $B[i] = x$	*False*
$\exists i, B[i] = x$	*True* if $B[i] = x$ for some i, *False* otherwise	*False*

		$(\exists \, i, B[i]=x)$	$\neg(\exists \, i, \ B[i]=x)$		H_2
H_1	$j'|$	$B[j']=x$	*true*		G
	present'=	*true*	*false*		

Fig. 16.4 Finding Index in Array

array B for the value x. It describes the properties of the values of j' and *present'*. There are two cases to consider (Fig. 16.4):

1. There is an element in the array with the value of x.
2. There is no such element in the array with the value of x.

Clearly, from the example above the predicate expressions $\exists i, B[i] = x$ and $\neg(\exists i, B[i] = x)$ are defined. One disadvantage of the Parnas approach is that some common relational operators (e.g., $>$, $\geq$, $\leq$, and $<$) are not primitive in the logic. However, these relational operators are then constructed from primitive operators. Further, the axiom of reflection does not hold in the logic.

16.5.3   Dijkstra and Undefinedness

The **cand** and **cor** operators were introduced by Dijkstra (Fig. 16.5) to deal with undefined values. They are non-commutative operators and allow the evaluation of predicates that contain undefined values.

Consider the following expression:

$$y = 0 \lor (x/y = 2)$$

Fig. 16.5 Edsger Dijkstra.
Courtesy of Brian Randell

Table 16.3 *a cand b*

a	*b*	*a cand b*
T	T	T
T	F	F
T	U	U
F	T	F
F	F	F
F	U	F
U	T	U
U	F	U
U	U	U

Table 16.4 *a cor b*

a	*b*	*a cor b*
T	T	T
T	F	T
T	U	T
F	T	T
F	F	F
F	U	U
U	T	U
U	F	U
U	U	U

Then this expression is undefined when $y = 0$ as x/y is undefined, since the logical disjunction operation is not defined when one of its operands is undefined. However, there is a case for giving meaning to such an expression when $y = 0$, since in that case the first operand of the logical or operation is true. Further, the logical *disjunction* operation is defined to be true if either of its operands is true. This motivates the introduction of the **cand** and **cor** operators. These operators are associative and their truth tables are defined in Tables 16.3 and 16.4:

The order of the evaluation of the operands for the **cand** operation is to *evaluate the first operand*; if the first operand is true then the result of the operation is the second operand; otherwise the result is false. The expression *a* **cand** *b* is equivalent to

$$a \, cand \, b \cong \textbf{if } a \textbf{ then } b \textbf{ else } F$$

The order of the evaluation of the operands for the **cor** operation is to evaluate the first operand. If the first operand is true then the result of the operation is true; otherwise the result of the operation is the second operand. The expression *a* **cor** *b* is equivalent to

$$a \, \textbf{cor} \, b \cong \textbf{if } a \textbf{ then } T \textbf{ else } b$$

The **cand** and **cor** operators satisfy the following laws:

- *Associativity*
  The cand and cor operators are associative.

$$(A \ \mathbf{cand} \ B) \ \mathbf{cand} \ C = A \ \mathbf{cand} \ (B \ \mathbf{cand} \ C)$$
$$(A \ \mathbf{cor} \ B) \ \mathbf{cor} \ C = A \ \mathbf{cor} \ (B \ \mathbf{cor} \ C)$$

- *Distributivity*
  The **cand** operator distributes over the **cor** operator and vice versa.

$$A \ \mathbf{cand} \ (B \ \mathbf{cor} \ C) = (A \ \mathbf{cand} \ B) \ \mathbf{cor} \ (A \ \mathbf{cand} \ C)$$
$$A \ \mathbf{cor} \ (B \wedge C) = (A \ \mathbf{cor} \ B) \ \mathbf{cand} \ (A \ \mathbf{cor} \ C)$$

De Morgan's law enables logical expressions to be simplified.

$$\neg(A \ \mathbf{cand} \ B) = \neg A \mathbf{cor} \neg B$$
$$\neg(A \ \mathbf{cor} \ B) = \neg A \ \mathbf{cand} \ \neg B$$

16.6 Logic and AI

The long-term goal of Artificial Intelligence is to create a thinking machine that is intelligent, has consciousness, has the ability to learn, has free will, and is ethical. Artificial Intelligence is a young field and John McCarthy and others coined the term in 1956. Alan Turing devised the Turing Test in the early 1950s as a way to determine whether a machine was conscious and intelligent. Turing believed that machines would eventually be developed that would stand a good chance of passing the 'Turing Test'.

There are deep philosophical problems in Artificial Intelligence, and some researchers believe that its goals are impossible or incoherent. Even if Artificial Intelligence is possible there are moral issues to consider such as the exploitation of artificial machines by humans and whether it is ethical to do this. Weizenbaum argues that AI is a threat to human dignity, and that AI should not replace humans in positions that require respect and care.

John McCarthy (Fig. 16.6) has long advocated the use of logic in AI, and mathematical logic has been used in the AI field to formalize knowledge, and in guiding the design of mechanized reasoning systems. Logic has been used as an analytic tool, as a knowledge representation formalism, and as a programming language.

Fig. 16.6 John McCarthy.
Courtesy of John McCarthy

McCarthy's long-term goal was to formalize common sense reasoning: i.e. the normal reasoning that is employed in problem solving and dealing with normal events in the real world. McCarthy [6] argues that it is reasonable for logic to play a key role in the formalization of common sense knowledgwe, and this includes the formalization of basic facts about actions and their effects; facts about beliefs and desires; and facts about knowledge and how it is obtained. His approach allows common sense problems to be solved by logical reasoning.

Its formalization requires sufficient understanding of the common sense world, and often the relevant facts to solve a particular problem are unknown. It may be that knowledge thought relevant may be irrelevant and vice versa. A computer may have millions of facts stored in its memory, and the problem is how to determine which of these should be chosen from its memory to serve as premises in logical deduction.

McCarthy's influential 1959 paper discusses various common sense problems such as getting home from the airport. Mathematical logic is the standard approach to express premises, and it includes rules of inferences that are used to deduce valid conclusions from a set of premises. Its rigorous deductive reasoning shows how new formulae may be logically deduced from a set or premises.

McCarthy's approach to programs with common sense has been criticized by Bar-Hillel and others on the grounds that common sense is fairly elusive, and the difficulty that a machine would have in determining which facts are relevant to a particular deduction from its known set of facts. However, McCarthy's approach has showed how logical techniques can contribute to the solution of specific AI problems.

Logic programming languages describe what is to be done, rather than how it should be done. These languages are concerned with the statement of the problem to be solved, rather than how the problem will be solved. These languages use mathematical logic as a tool in the statement of the problem definition. Logic is a useful tool in developing a body of knowledge (or theory), and it allows rigorous mathematical deduction to derive further truths from the existing set of truths. The theory is built up from a small set of axioms or postulates and rules of inference derive further truths logically.

The objective of logic programming is to employ mathematical logic to assist with computer programming. Many problems are naturally expressed as a theory, and the statement of a problem to be solved is often equivalent to determining if a new hypothesis is consistent with an existing theory. Logic provides a rigorous way to determine this, as it includes a rigorous process for conducting proof.

Computation in logic programming is essentially logical deduction, and logic programming languages use first-order[6] predicate calculus. They employ theorem proving to derive a desired truth from an initial set of axioms. These proofs are constructive[7] in that more an actual object that satisfies the constraints is produced rather than a pure existence theorem. Logic programming specifies the objects, the relationships between them and the constraints that must be satisfied for the problem.

- The set of objects involved in the computation
- The relationships that hold between the objects
- The constraints of the particular problem.

The language interpreter decides how to satisfy the particular constraints. Artificial Intelligence influenced the development of logic programming, and John McCarthy[8] demonstrated that mathematical logic could be used for expressing knowledge. The first logic programming language was Planner developed by Carl Hewitt at MIT in 1969. It uses a procedural approach for knowledge representation rather than McCarthy's declarative approach.

The best-known logic programming languages is Prolog, which was developed in the early 1970s by Alain Colmerauer and Robert Kowalski. It stands for *pro*gramming in *log*ic. It is a goal-oriented language that is based on predicate logic. Prolog became an ISO standard in 1995. The language attempts to solve a goal by tackling the sub-goals that the goal consists of

$$\text{goal} : -\text{subgoal}_1, \ldots, \text{subgoal}_n.$$

That is, in order to prove a particular goal it is sufficient to prove sub-goal$_1$ through sub-goal$_n$. Each line of a Prolog program consists of a rule or a fact, and the language specifies what exists rather than how. The following program fragment has one rule and two facts

---

[6]First-order logic allows quantification over objects but not functions or relations. Higher order logics allow quantification of functions and relations.

[7]For example, the statement $\exists x$ such that $x = \sqrt{4}$ states that there is an x such that x is the square root of 4, and the constructive existence yields that the answer is that $x = 2$ or $(x - (-2))$ i.e. constructive existence provides more the truth of the statement of existence, and an actual object satisfying the existence criteria is explicitly produced.

[8]John McCarthy received the Turing Award in 1971 for his contributions to Artificial Intelligence. He also developed the programming language LISP.

grandmother(G, S) : $-$parent(P, S), mother(G, P).

mother(sarah, isaac).

parent(isaac, jacob).

The first line in the program fragment is a rule that states that G is the grand-mother of S if there is a parent P of S and G is the mother of P. The next two statements are facts stating that Isaac is a parent of Jacob, and that Sarah is the mother of Isaac. A particular goal clause is true if all of its subclauses are true

$$\text{goalclause}(V_g) : - \text{clause}_1(V_1), \ldots, \text{clause}_m(V_m)$$

A Horn clause consists of a goal clause and a set of clauses that must be proven separately. Prolog finds solutions by *unification*: i.e. by binding a variable to a value. For an implication to succeed, all goal variables Vg on the left side of: must find a solution by binding variables from the clauses which are activated on the right side. When all clauses are examined and all variables in Vg are bound, the goal succeeds. But if a variable cannot be bound for a given clause, then that clause fails. Following the failure, Prolog *backtracks*, and this involves going back to the left to previous clauses to continue trying to unify with alternative bindings. Backtracking gives Prolog the ability to find multiple solutions to a given query or goal.

Logic programming languages generally use a simple searching strategy to consider alternatives

- If a goal succeeds and there are more goals to achieve, then remember any untried alternatives and go on to the next goal.
- If a goal is achieved and there are no more goals to achieve then stop with success.
- If a goal fails and there are alternative ways to solve it then try the next one.
- If a goal fails and there are no alternate ways to solve it, and there is a previous goal, then go back to the previous goal.
- If a goal fails and there are no alternate ways to solve it, and no previous goal, then stop with failure.

Constraint programming is a programming paradigm where relations between variables can be stated in the form of constraints. Constraints specify the properties of the solution, and differ from the imperative programming languages in that they do not specify the sequence of steps to execute.

16.7   Theorem Provers for Logic

The word "*proof*" is generally interpreted as facts or evidence that support a particular proposition or belief, and such proofs are conducted in natural language. The proof of a theorem in mathematics requires additional rigour, and such proofs consist of a mixture of natural language and mathematical argument. It is common to skip over the trivial steps in a mathematical proof, and independent mathematicians conduct peer reviews to provide additional confidence in the correctness of the proof, and to ensure that no unwarranted assumptions or errors in reasoning have been made. Proofs conducted in logic are extremely rigorous with every step in the proof is explicit.[9]

Herbert Simon and Alan Newell developed the first theorem prover with their work on a program called 'Logic Theorist' or 'LT' [7]. This program could independently provide proofs of various theorems in Russell's and Whitehead's Principia Mathematica[10] [8]. Russell and Whitehead had attempted to derive all mathematics from axioms and the inference rules of logic, and the LT program conducted proof from a small set of propositional axioms and deduction rules. The LT program succeeded in proving 38 of the 52 theorems in Chap. 2 of Principia Mathematica. Its approach was to start with the theorem to be proved, and to then search for relevant axioms and operators to prove the theorem.

LT was demonstrated at the Dartmouth conference in 1956 (the conference that led to the birth of the Artificial Intelligence field), and it showed that computers had the ability to encode knowledge and information, and to perform intelligent operations such as solving theorems in mathematics. The heuristic approach of the LT program tried to emulate human mathematicians, and could not guarantee that a proof could be found for every valid theorem.

The proof of theorems in formal verification of computer system often involves several million formulae and manual proof is error prone. There are several tools available to support theorem proving, and these include the Boyer–Moore theorem prover (known as NQTHM); the Isabelle theorem prover; and the HOL system.

B.S. Boyer and J.S. Moore developed the Boyer–Moore theorem prover in the early 1970s [9]. It has been improved since then and it is currently known as NQTHM (it has been superseded by ACL2 available from the University of Texas).

It has been effective in proving well-known theorems such as Goedel's Incompleteness Theorem, the insolvability of the Halting problem, a formalization of the Motorola MC 68020 Microprocessor, and many more.

---

[9]Perhaps a good analogy might be that a mathematical proof is like a program written in a high-level language such as C, whereas a formal proof in logic is like a program written in assembly language.

[10]Russell is said to have remarked that he was delighted to see that the Principia Mathematica could be done by machine, and that if he and Whitehead had known this in advance that they would not have wasted 10 years doing this work by hand in the early twentieth century.

Computational Logic Inc. was a company founded by Boyer and Moore in 1983 to share the benefits of a formal approach to software development with the wider computing community. It was based in Austin, Texas, and provided services in the mathematical modelling of hardware and software systems. This involved the use of mathematics and logic to formally specify microprocessors and other systems. The use of its theorem prover was to formally verify that the implementation meets its specification: i.e. to prove that the microprocessor or other system satisfies its specification.

Isabelle is a theorem proving environment developed at Cambridge University by Larry Paulson and Tobias Nipkow of the Technical University of Munich. It allows mathematical formulas to be expressed in a formal language and provides tools for proving those formulas. The main application is the formalization of mathematical proofs, and proving the correctness of computer hardware or software with respect to its specification, and proving properties of computer languages and protocols.

Isabelle is a generic theorem prover in the sense that it has the capacity to accept a variety of formal calculi, whereas most other theorem provers are specific to a specific formal calculus. Isabelle is available free of charge under an open source license.

The HOL system is an environment for interactive theorem proving in a higher order logic. The HOL system has been applied to the formalization of mathematics and the verification of hardware. It was originally developed at Cambridge University in the United Kingdom in the early 1980s, and HOL 4 is the latest version and is an open source project. It is used by academia and industry.

There is a steep learning curve with the theorem provers above and it generally takes a couple of months for users to become familiar with them. However, automated theorem proving has become a useful tool in the verification of integrated circuit design. Several semiconductor companies use automated theorem proving to demonstrate the correctness of division and other operators on their processors.

16.8 Review Questions

1. What is fuzzy logic?
2. What is intuitionist logic and how is it different from classical logic?
3. Discuss the problem of undefinedness and the advantages and disadvantages of three-valued logics. Describe the approaches of Parnas, Dijkstra and Jones.
4. What is temporal logic?

5. Show how the temporal operators may be expressed in classical mathematics. Discuss the merits of temporal operators.
6. Investigate the Isabelle (or another) theorem proving environment and determine the extent to which it may assist with proof.
7. Discuss the applications of logic to AI.

16.9 Summary

We discussed some advanced topics in logic in this chapter, including fuzzy logic, temporal logic, intuitionist logic, undefined values, logic and AI and theorem provers. Fuzzy logic is an extension of classical logic that acts as a mathematical model for vagueness, whereas temporal logic is concerned with the expression of properties that have time dependencies

Intuitionism was a controversial school of mathematics that aimed to provide a solid foundation for mathematics. Its adherents rejected the law of the excluded middle, and insisted that for an entity to exist that there must be a constructive proof of its existence. Martin Löf applied intuitionistic logic to type theory in the 1970s.

Partial functions arise naturally in computer science, and such functions may fail to be defined for one or more values in their domain. There are a number of approaches to deal with undefined values, including the logic of partial functions; Dijkstra's approach with his cand and cor operators; and Parnas's approach which preserves a classical two-valued logic.

We discussed temporal logic and its applications to the safety critical field, including the specification of properties with time dependencies. We discussed the application of logic to the AI field, and logic has been used to formalize knowledge in an AI systems. Finally, we discussed some of the existing theorem provers, and their applications in providing a rigorous proof of a theorem, and in avoiding errors or jumps in reasoning.

References

1. Temporal Logic. Stanford Encyclopedia of Philosophy. http://plato.stanford.edu/entries/logic-temporal/
2. Intuitionist Logic. An Introduction. A. Heyting. North-Holland Publishing. 1966.
3. Intuitionist Type Theory. Per Martin Löf. Notes by Giovanni Savin of lectures given in Padua, June, 1980. Bibliopolis. Napoli. 1984.
4. Predicate Calculus for Software Engineering. David L.Parnas. IEEE.
5. Systematic Software Development using VDM. Cliff Jones. Prentice Hall International. 1986.

6. Programs with Common Sense. John McCarthy. Proceedings of the Teddington Conference on the Mechanization of Thought Processes. 1959.
7. A. Newell and H. Simon. The Logic Theory Machine. IRE Transactions on Information Theory, 2, 61–79. 1956.
8. Principia Mathematica. B.Russell and A.N. Whitehead. Cambridge University Press. Cambridge. 1910.
9. A Computational Logic. The Boyer Moore Theorem Prover. Robert Boyer and J.S. Moore. Academic Press. 1979.

Software Engineering Mathematics

17

Key Topics

Birth of Software Engineering
Software Engineering Mathematics
Floyd
Hoare
Formal Methods
Software Inspections and Testing
Project Management
Software Process Maturity Models

17.1 Introduction

The NATO Science Committee organized two famous conferences on software engineering in the late 1960s. The first conference was held in Garmisch, Germany, in 1968, and it was followed by a second conference in Rome in 1969. The Garmisch conference was attended by over fifty people from 11 countries.

The conferences highlighted the problems that existed in the software sector in the late 1960s, and the term 'software crisis' was coined to refer to these problems. These included budget and schedule overruns of projects, and problems with the quality and reliability of the delivered software. This conference led to the birth of *software engineering* as a separate discipline, and the realization that programming is quite distinct from science and mathematics. Programmers are like engineers in the sense that they design and build products. Therefore, they need an appropriate

© Springer International Publishing Switzerland 2016   283
G. O'Regan, *Guide to Discrete Mathematics*, Texts in Computer Science,
DOI 10.1007/978-3-319-44561-8_17

software engineering education (not just on the latest technologies but on the fundamentals of engineering) in order to properly design and develop software.

The construction of bridges was problematic in the nineteenth century, and many people who presented themselves as qualified to design and construct bridges did not have the required knowledge and expertise. Consequently, many bridges collapsed, endangering the lives of the public. This led to legislation requiring an engineer to be licensed by the professional engineering prior to practicing as an engineer. These engineering associations identify a core body of knowledge that the engineer is required to possess, and the licensing body verifies that the engineer has the required qualifications and experience. The licensing of engineers by most branches of engineering ensures that only personnel competent to design and build products actually do so. This in turn leads to products that the public can safely use. In other words, the engineer has a responsibility to ensure that the products are properly built, and are safe for the public to use.

Parnas argues that traditional engineering be contrasted with the software engineering discipline where there is no licensing mechanism, and where individuals with no qualifications can participate in the design and building of software products.[1] However, best practice in modern HR places a strong emphasis on the qualification of staff.

The Standish group has conducted research since the late 1990s [1] on the extent of problems with schedule and budget overruns of IT projects. The results indicate serious problems with on-time delivery, cost overruns and quality.[2] Fred Brooks has argued that software is inherently complex, and that there is no silver bullet that will resolve all of the problems associated with software projects such as schedule overruns and software quality problems [2, 3].

Poor quality software can at best cause minor irritation to clients, and in some circumstances it may seriously disrupt the work of the client organization leading to injury or even the death of individuals (e.g. as in the case of the Therac-25[3] radiotherapy machine). The Y2K problem occurred due to poor design, as the representation of the date used two digits to record the year rather than four. Its correction required major rework, as it was necessary to examine all existing software code to determine how the date was represented, and to make appropriate

---

[1]Modern HR recruitment specifies the requirements for a particular role, and the interviews establish whether the candidate is suitably qualified, and has the appropriate experience for the role. Parnas is arguing against the content of courses that emphasize the latest technologies rather than the fundamentals of engineering.

[2]It should be noted that these are IT projects covering diverse sectors including banking, telecommunications, etc., rather than pure software companies. Mature software companies using the CMM tend to be more consistent in project delivery with high quality.

[3]Therac-25 was a radiotherapy machine produced by the Atomic Energy of Canada Limited (AECL). It was involved in at least six accidents between 1985 and 1987 in which patients were given massive overdoses of radiation. The dose given was over 100 times the intended dose and three of the patients died from radiation poisoning. These accidents highlighted the dangers of software control of safety-critical systems. The investigation subsequently highlighted the poor software design of the system and the poor software development practices employed.

corrections. Clearly, well-designed programs would have hidden the representation of the date thereby minimizing the changes required for year 2000 compliance.

Mathematics plays a key role in engineering, and it may potentially assist software engineers in delivering high-quality software products that are safe to use. Several mathematical approaches that may assist in delivering high-quality software are described in [4]. However, it is important to recognise that while the use of mathematics is suitable for some areas of software engineering (especially in the safety and security critical fields), less rigorous techniques (such as software inspections and testing) are sufficient in most other areas of software engineering.

There is a lot of industrial interest in approaches to mature software engineering practices in software organizations (e.g. the use of software process maturity models such as the CMMI). These include approaches to assess and mature the software engineering processes in software companies, and they are described in [5, 6].[4] Software process improvement focuses mainly on improving the effectiveness of the management, engineering and organization practices related to software engineering.

17.2   What Is Software Engineering?

Software engineering involves multi-person construction of multi-version programs. The IEEE 610.12 definition states that:

Definition 17.1 (*Software Engineering*) Software engineering is the application of a systematic, disciplined, quantifiable approach to the development, operation, and maintenance of software; that is, the application of engineering to software, and the study of such approaches.

Software engineering includes the following:

1. Methodologies to determine requirements, design, develop, implement and test software to meet customers' needs.
2. The philosophy of engineering: i.e. an engineering approach to developing software is adopted. That is, products are properly designed, developed, tested, with quality and safety properly addressed.
3. Mathematics[5] may be employed to assist with the design and verification of software products. The level of mathematics to be employed will depend on the

---

[4]The process maturity models focus mainly on the management, engineering and organizational practices required in software engineering. The models focus on what needs to be done rather how it should be done.

[5]There is no consensus at this time as to the appropriate role of mathematics in software engineering. The use of mathematics is invaluable in the safety critical and security critical fields as it provides an extra level of confidence in the correctness of the software.

Fig. 17.1 David Parnas

safety critical nature of the product, as systematic peer reviews and testing are often sufficient.

4. Sound project and quality management practices are employed.

Software engineering requires the engineer to state precisely the requirements that the software product is to satisfy, and then to produce designs that will meet these requirements. Engineers provide a precise description of the problem to be solved; they then proceed to producing a design and validating its correctness; finally, the design is implemented and testing is performed to verify the correctness of the implementation with respect to the requirements. The software requirements needs to be unambiguous, and should clearly state what is and what is not required.

Classical engineers produce the product design, and then analyse their design for correctness. They use mathematics in their analysis, as this is the basis of confirming that the specifications are met. The level of mathematics employed will depend on the particular application and calculations involved. The term 'engineer' is generally applied only to people who have attained the necessary education and competence to be called engineers, and who base their practice on mathematical and scientific principles. Often in computer science the term engineer is employed rather loosely to refer to anyone who builds things, rather than to an individual with a core set of knowledge, experience and competence.

Parnas[6] (Fig. 17.1) is a strong advocate of the classical engineering approach, and he argues that computer scientists should have the right education to apply scientific and mathematical principles to their work. This includes mathematics and design, to enable them to be able to build high-quality and safe products. Baber has argued [7] that "mathematics is the language of engineering". He argues that students should be shown how to turn a specification into a program using mathematics.

Parnas advocates a solid engineering approach to the teaching of mathematics with an emphasis on its application to developing and analysing product designs. He argues that software engineers need education on engineering mathematics; specification and design; converting designs into programs; software inspections, and testing. The education should enable the software engineer to produce well-designed programs that will correctly implement the requirements.

---

[6]Parnas has made important contributions to software engineering including information hiding which is used in the object-oriented world.

He argues that software engineers have individual responsibilities as professional engineers.[7] They are responsible for designing and implementing high-quality and reliable software that is safe to use. They are also accountable for their own decisions and actions,[8] and have a responsibility to object to decisions that violate professional standards. Professional engineers need to be honest about current capabilities, especially when asked to work on problems that have no appropriate technical solution. Another words, they should be honest and avoid accepting a contract for something that cannot be done.

The licensing of a professional engineer provides confidence that the engineer has the right education and experience to build safe and reliable products. Professional engineers are required to follow rules of good practice, and to object when the rules are violated.[9] The professional engineering body is responsible for enforcing standards and certification. The term 'engineer' is a title that is awarded on merit, but it also places responsibilities on its holder.

The approach used in current software engineering is to follow a well-defined software engineering process. The process includes activities such as project management, requirements gathering, requirements specification, architecture design, software design, coding, and testing. Most companies use a set of templates for the various phases. The waterfall model [8] and spiral model [9] are popular software development lifecycles.

The waterfall model (Fig. 17.2) starts with requirements, followed by specification, design, implementation, and testing. It is typically used for projects where the requirements can be identified and it is often called the 'V' life cycle model. The left-hand side of the 'V' involves requirements, specification, design, and coding and the right-hand side is concerned with unit tests, integration tests, system tests and acceptance testing. Each phase has entry and exit criteria that must be satisfied before the next phase commences. There are several variations of the waterfall model.

The spiral model (Fig. 17.3) is useful where the requirements are not fully known at project initiation. There is an evolution of the requirements during development which proceeds in a number of spirals, with each spiral typically

---

[7]The concept of accountability is not new; indeed the ancient Babylonians employed a code of laws c. 1750 B.C. known as the Hammarabi Code. This code included the law that if a house collapsed and killed the owner then the builder of the house would be executed.

[8]However, it is unlikely that an individual programmer would be subject to litigation in the case of a flaw in a program causing damage or loss of life. A comprehensive disclaimer of responsibility for problems rather than a guarantee of quality accompany most software products. Software engineering is a team-based activity involving several engineers in various parts of the project, and it could be potentially difficult for an outside party to prove that the cause of a particular problem is due to the professional negligence of a particular software engineer, as there are many others involved in the process such as reviewers of documentation and code and the various test groups. Companies are more likely to be subject to litigation, as a company is legally responsible for the actions of their employees in the workplace, and the fact that a company is a financially richer entity than one of its employees.

[9]Software companies that are following the CMMI or ISO 9000 will employ audits to verify that the rules and best practice have been followed. Auditors report their findings to management and the findings are addressed appropriately by the project team and affected individuals.

Fig. 17.2 Waterfall lifecycle
model (V-model)

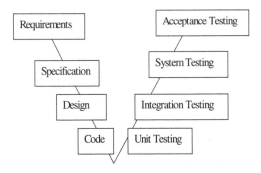

Fig. 17.3 Spiral lifecycle
model

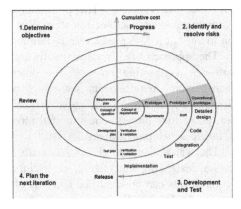

involves updates to the requirements, design, code, testing and a user review of the
particular iteration or spiral.

The spiral is, in effect, a reusable prototype and the customer examines the
current iteration and provides feedback to the development team to be included in
the next spiral. The approach is to partially implement the system. This leads to a
better understanding of the requirements of the system and it then feeds into the
next cycle in the spiral. The process repeats until the requirements and product are
fully complete.

There has been a growth of popularity among software developers in lightweight
methodologies such as *Agile*. This is a software development methodology that
claims to be more responsive to customer needs than traditional methods such as the
waterfall model. *The waterfall development model is similar to a wide and slow
moving value stream*, and halfway through the project 100 % if the requirements
are typically 50 % done. *However, for agile development 50 % of requirements are
typically 100 % done halfway through the project.*

Ongoing changes to requirements are considered normal in the Agile world, and
it is believed to be more realistic to change requirements regularly throughout the
project rather than attempting to define all of the requirements at the start of the
project. The methodology includes controls to manage changes to the requirements,

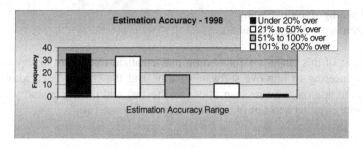

Fig. 17.4 Standish group report estimation accuracy

and good communication and early regular feedback is an essential part of the process.

A story may be a new feature or a modification to an existing feature. It is reduced to the minimum scope that can deliver business value, and a feature may give rise to several stories. Stories often build upon other stories and the entire software development lifecycle is employed for the implementation of each story. *Stories are either done or not done*, i.e. *there is such thing as a story being 80 % done.* The story is complete only when it passes its acceptance tests. For more details on Agile see [6, 10].

The challenge in software engineering is to deliver high-quality software on time to customers. The Standish Group research (Fig. 17.4) on project cost overruns in the US during 1998 showed that 33 % of projects are between 21 and 50 % over estimate, 18 % are between 51 and 100 % over estimate, and 11 % of projects are between 101 and 200 % overestimate.

The accurate estimation of project cost and effort are key challenges, and project managers need to determine how good their current estimation process actually is and to make improvements. Many companies today employ formal project management methodologies such as Prince 2 or Project Management Professional (PMP). These methodologies allow projects to be rigorously managed and include processes for initiating a project, planning a project, executing a project, monitoring and controlling a project and closing a project.

The Capability Maturity Model developed by the Software Engineering Institute (SEI) has become useful in software engineering. The SEI has collected empirical data to suggest that there is a close relationship between software process maturity and the quality and the reliability of the delivered software. The CMMI enables the organization to improve processes as follows:

- Developing and managing requirements
- Design activities
- Configuration Management
- Selection and Management of Suppliers
- Planning and Managing projects
- Building quality into the product with peer reviews

- Performing rigorous testing
- Performing independent audits

The rest of this chapter is focused on mathematical techniques to support software engineering to improve software quality, and the chapter concludes with a short discussion on software inspections and testing and process maturity models. For a more detailed account of software engineering see [6].

17.3   Early Software Engineering Mathematics

Robert Floyd was born in New York in 1936, and he did pioneering work on software engineering from the 1960s (Fig. 17.5). He made important contributions to the theory of parsing; the semantics of programming languages; program verification; and methodologies for the creation of efficient and reliable software.

Mathematics and computer science were regarded as two completely separate disciplines in the 1960s, and software development was based on the assumption that the completed code would always contain defects. It was therefore better and more productive to write the code as quickly as possible, and to then perform debugging to find the defects. Programmers then corrected the defects, made patches and re-tested and found more defects. This continued until they could no longer find defects. Of course, there was always the danger that defects remained in the code that could give rise to software failures.

Floyd believed that there was a way to construct a rigorous proof of the correctness of the programs using mathematics. He showed that mathematics could be used for program verification, and he introduced the concept of *assertions* that provided a way to verify the correctness of programs.

Flowcharts were employed in the 1960s to explain the sequence of basic steps for computer programs. Floyd's insight was to build upon flowcharts and to apply *an invariant assertion to each branch* in the flowchart. These assertions state the essential relations that exist between the variables at that point in the flowchart.

Fig. 17.5 Robert Floyd

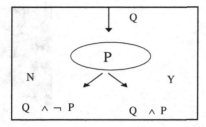

Fig. 17.6 Branch assertions in flowcharts

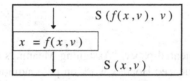

Fig. 17.7 Assignment assertions in flowcharts

An example relation is "$R = Z > 0, X = 1, Y = 0$". He devised a general flowchart language to apply his method to programming languages. The language essentially contains boxes linked by flow of control arrows [11].

Consider the assertion Q that is true on entry to a branch where the condition at the branch is P. Then, the assertion on exit from the branch is $Q \wedge \neg P$ if P is false and $Q \wedge P$ otherwise (Fig. 17.6).

The use of assertions may be employed in an assignment statement. Suppose x represents a variable and v represents a vector consisting of all the variables in the program. Suppose $f(x, v)$ represents a function or expression of x and the other program variables represented by the vector v. Suppose the assertion $S(f(x, v), v)$ is true before the assignment $x = f(x, v)$. Then the assertion $S(x, v)$ is true after the assignment (Fig. 17.7). This is given by:

Floyd used flowchart symbols to represent entry and exit to the flowchart. This included entry and exit assertions to describe the program's entry and exit conditions.

Floyd's technique showed how a computer program is a sequence of logical assertions. Each assertion is true whenever control passes to it, and statements appear between the assertions. The initial assertion states the conditions that must be true for execution of the program to take place, and the exit assertion essentially describes what must be true when the program terminates.

His key insight was the recognition that if it can be shown that the assertion immediately following each step is a consequence of the assertion immediately preceding it, then the assertion at the end of the program will be true, provided the appropriate assertion was true at the beginning of the program.

Fig. 17.8 C.A.R Hoare

He published an influential paper, "Assigning Meanings to Programs", in 1967 [11], and this paper influenced Hoare's work on preconditions and post-conditions leading to Hoare logic [12]. Floyd's paper also presented a formal grammar for flowcharts, together with rigorous methods for verifying the effects of basic actions like assignments.

Hoare logic is a formal system of logic used for programming semantics and for program verification. It was developed by C.A.R. Hoare (Fig. 17.8), and was originally published in Hoare's 1969 paper "An axiomatic basis for computer programming" [12]. Hoare and others have subsequently refined it, and it provides a logical methodology for precise reasoning about the correctness of computer programs.

Hoare was influenced by Floyd's 1967 paper that applied assertions to flowcharts, and he recognised that this provided an effective method for proving the correctness of programs. He built upon Floyd's approach to cover the familiar constructs of high-level programming languages.

This led to the axiomatic approach to defining the semantics of every statement in a programming language, and the approach consists of axioms and proof rules. He introduced what has become known as the Hoare triple, and this describes how the execution of a fragment of code changes the state. A Hoare triple is of the form:

$$P\{Q\}R$$

where P and R are assertions and Q is a program or command. The predicate P is called the *precondition*, and the predicate R is called the *postcondition*.

Definition 4.2 (*Partial Correctness*) The meaning of the Hoare triple above is that whenever the predicate P holds of the state before the execution of the command or program Q, then the predicate R will hold after the execution of Q. The brackets indicate partial correctness as if Q does not terminate then R can be any predicate. R may be chosen to be false to express that Q does not terminate.

Total correctness requires Q to terminate, and at termination R is true. Termination needs to be proved separately. Hoare logic includes axioms and rules of inference rules for the constructs of imperative programming language.

Hoare and Dijkstra were of the view that the starting point of a program should always be the specification, and that the proof of the correctness of the program should be developed along with the program itself.

That is, the starting point is the mathematical specification of what a program is to do, and mathematical transformations are applied to the specification until it is turned into a program that can be executed. The resulting program is then known to be correct by construction.

17.4   Mathematics in Software Engineering

Mathematics plays a key role in classical engineering to assist with design and verification of software products. It is therefore reasonable to apply appropriate mathematics in software engineering (especially for safety and security critical systems) to assure that the delivered systems conform to the requirements. The extent to which mathematics should be used is controversial with strong views in both camps. In many cases, peer reviews and testing will be sufficient to build quality into the software product. In other cases, and especially with safety and security critical applications, it is desirable to have the extra assurance that may be provided with mathematical techniques.

Mathematics allows a rigorous analysis to take place and avoids an over-reliance on intuition. The emphasis is on applying mathematics to solve practical problems and to develop products that are fit for use. Engineers are taught how to apply mathematics in their work, and the emphasis is always on the application of mathematics to solve practical problems.

Classical mathematics may be applied to software engineering and specialized mathematical methods and notations have also been developed. The classical mathematics employed includes sets, relations, functions, logic, graph theory, automata theory, matrix theory, probability and statistics, calculus, and matrix theory. Specialized formal specification languages such as Z and VDM have been developed, and these allow the requirements to be formally specified in precise mathematical language.

The term 'formal method' refers to various mathematical techniques used in the software field for the specification and formal development of software. Formal methods consist of formal specification languages or notations, and employ a collection of tools to support the syntax checking of the specification, as well as the proof of properties about the specification. The term 'formal method' is used to describe a formal specification language and a method for the design and implementation of computer systems.

The mathematical analysis of the formal specification allows questions to be asked about what the system does, and these questions may be answered independently of the implementation. Mathematical notation is precise, and this helps to avoid the problem of ambiguity inherent in a natural language description of a system. The formal specification may be used to promote a common understanding for all stakeholders.

Formal methods have been applied to a diverse range of applications, including the safety critical field; security critical field; the railway sector; the nuclear field; microprocessor verification; the specification of standards, and the specification and verification of programs.

There are various tools to support formal methods including syntax checkers; specialized editors; tools to support refinement; automated code generators; theorem provers; and specification animation tools. Formal methods need to mature further before they will be used in mainstream software engineering, and they are described in more detail in Chap. 18.

17.5 Software Inspections and Testing

Software inspections play an important role in building quality into software products. The Fagan Inspection Methodology was developed by Michael Fagan at IBM in the mid-1970s [13]. It is a seven-step process that identifies and removes defects in work products. The Fagan methodology mandates that requirement documents, design documents, source code, and test plans are all formally inspected.

There are several *roles* defined in the process including the *moderator* who chairs the inspection; the *reader* who reads or paraphrases the particular deliverable; the *author* who is the creator of the deliverable; and the *tester* who is concerned with the testing viewpoint.

The inspection process will consider whether a design is correct with respect to the requirements, and whether the source code is correct with respect to the design. There are several stages in the Fagan inspection process, including planning, overview, preparation, inspection, process improvement, rework, and follow-up.

Software testing plays a key role in verifying that a software product is of high quality and conforms to the customer's quality expectations. Testing is both a constructive activity in that it is verifying the correctness of functionality, and it is also a destructive activity in that the objective is to find as many defects as possible in the software. The testing verifies that the requirements are correctly implemented as well as identifying whether any defects are present in the software product.

There are various types of testing such as unit testing, integration testing, system testing, performance testing, usability testing, regression testing, and customer acceptance testing. The testing needs to be planned and test cases prepared and executed. The results of testing are reported and any issues corrected and retested. The test cases will need to be appropriate to verify the correctness of the software. Software inspection and testing are described in more detail in [6].

17.6 Process Maturity Models

The Software Engineering Institute (SEI) developed the Capability Maturity Model (CMM) in the early 1990s as a framework to help software organizations to improve their software process maturity, and to implement best practice in software and systems engineering. The SEI believes that there is a close relationship between the maturity of software processes and the quality of the delivered software product.

The CMM applied the ideas of Deming [14], Juran [15] and Crosby [16] to the software field. These quality gurus were influential in transforming manufacturing companies with quality problems to effective quality driven organizations with a reduced cost of poor quality.

Watt Humphries (Fig. 17.9) did early work on software process improvement at IBM [17]. He moved to the SEI in the late 1980s and the first version of the CMM was released in 1991. It is now called the Capability Maturity Model Integration (CMMI®) [18].

The CMMI consists of five maturity levels with each maturity level (except level one) consisting of several process areas. Each process area consists of a set of goals that are implemented by practices related to that process area leading to an effective process.

The emphasis on level two of the CMMI is on maturing management practices such as project management, requirements management, configuration management, and so on. The emphasis on level three of the CMMI is to mature engineering and organization practices. This maturity level includes peer reviews and testing, requirements development, software design and implementation practices, and so on. Level four is concerned with ensuring that key processes are performing within strict quantitative limits, and adjusting processes, where necessary, to perform within these defined limits. Level five is concerned with continuous process improvement, which is quantitatively verified.

The CMMI allows organizations to benchmark themselves against other similar organizations. This is done by appraisals conducted by an authorized SCAMPI lead appraiser. The results of an SCAMPI appraisal are generally reported back to the SEI, and there is a strict qualification process to become an authorized lead

Fig. 17.9 Watts Humphrey.
Courtesy of Watts Humphrey

appraiser. An appraisal is useful in verifying that an organization has improved, and it enables the organization to prioritize improvements for the next improvement cycle.

17.7   Review Questions

1. What is software engineering? Describe the difference between classical engineers and software engineers.
2. Describe the 'software crisis' of the late 1960s that led to the first software engineering conference in 1968.
3. Discuss the Standish Research Report and the level of success of IT projects today. In your view is there a crisis in software engineering today? Give reasons for your answer.
4. Discuss what the role of mathematics should be in current software engineering.
5. Describe the waterfall and spiral lifecycles. What are the similarities and differences between them?
6. Discuss the contributions of Floyd and Hoare.
7. Explain the difference between partial correctness and total correctness.
8. What are formal methods?
9. Discuss the process maturity models (including the CMMI). What are their advantages and disadvantages?
10. Discuss how software inspections and testing can assist in the delivery of high-quality software.

17.8   Summary

This chapter presented a short account of some important developments in software engineering. Its birth was at the Garmisch conference in 1968, and it was recognized that there was a crisis in the software field, and a need for sound methodologies to design, develop and maintain software to meet customer needs.

Classical engineering has a successful track record in building high-quality products that are safe for the public to use. It is therefore natural to consider using an engineering approach to developing software, and this involves identifying the customer requirements, carrying out a rigorous design to meet the requirements, developing and coding a solution to meet the design, and conducting appropriate inspections and testing to verify the correctness of the solution.

Mathematics plays a key role in classical engineering to assist with the design and verification of software products. It is therefore reasonable to apply appropriate mathematics in software engineering (especially for safety critical systems) to assure that the delivered systems conform to the requirements. The extent to which mathematics should be used is controversial with strong views in both camps.

There is a lot more to the successful delivery of a project than just the use of mathematics or peer reviews and testing. Sound project management and quality management practices are essential, as a project that is not properly managed will suffer from schedule, budget or cost overruns as well as problems with quality.

Maturity models such as the CMMI can assist organizations in maturing key management and engineering practices, and may help companies in their goals to deliver high-quality software systems that are consistently delivered on time and budget.

References

1. Estimating: Art or Science. Featuring Morotz Cost Expert. Standish Group Research Note. 1999.
2. The Mythical Man Month. Fred Brooks. Addison Wesley. 1975.
3. No Silver Bullet. Essence and Accidents of Software Engineering. Fred Brooks. *Information Processing*. Elsevier. Amsterdam, 1986.
4. Mathematical Approaches to Software Quality. Gerard O' Regan. Springer. 2006.
5. Introduction to Software Process Improvement. Gerard O' Regan. Springer. 2010.
6. Introduction to Software Quality. Gerard O' Regan. Springer Verlag. 2014.
7. The Language of Mathematics. Utilizing Math in Practice. Robert L. Baber. Wiley. 2011.
8. The Software Lifecycle Model (Waterfall Model). W. Royce. In Proc. WESTCON, August, 1970.
9. A Spiral Model for software development and enhancement. Barry Boehm. *Computer*. May 1988.
10. Extreme Programming Explained. Embrace Change. Kent Beck. Addison Wesley. 2000.
11. Assigning Meanings to Programs. Robert Floyd. Proceedings of Symposia in Applied Mathematics, (19):19–32. 1967.
12. An Axiomatic Basis for Computer Programming. C.A.R. Hoare. Communications of the ACM. 12(10):576–585. 1969.
13. Design and Code Inspections to Reduce Errors in Software Development. Michael Fagan. *IBM Systems Journal* 15(3). 1976.
14. Out of Crisis. W. Edwards Deming. M.I.T. Press. 1986.
15. Juran's Quality Handbook. 5th edition. Joseph Juran. McGraw Hill. 2000.
16. Quality is Free. The Art of Making Quality Certain. Philip Crosby. McGraw Hill. 1979.
17. Managing the Software Process. Watts Humphrey. Addison Wesley. 1989.
18. CMMI. Guidelines for Process Integration and Product Improvement. Third Edition. Mary Beth Chrissis, Mike Conrad and Sandy Shrum. SEI Series in Software Engineering. Addison Wesley. 2011.

Formal Methods

18

18.1 Introduction

The term 'formal methods' refers to various mathematical techniques used for the formal specification and development of software. They consist of a formal specification language, and employ a collection of tools to support the syntax checking of the specification, as well as the proof of properties of the specification. They allow questions to be asked about what the system does independently of the implementation.

The use of mathematical notation avoids speculation about the meaning of phrases in an imprecisely worded natural language description of a system. Natural language is inherently ambiguous, whereas mathematics employs a precise rigorous notation. Spivey [1] defines formal specification as:

© Springer International Publishing Switzerland 2016   299
G. O'Regan, *Guide to Discrete Mathematics*, Texts in Computer Science,
DOI 10.1007/978-3-319-44561-8_18

Definition 18.1 (*Formal Specification*) Formal specification is the use of mathematical notation to describe in a precise way the properties that an information system must have, without unduly constraining the way in which these properties are achieved.

The formal specification thus becomes the key reference point for the different parties involved in the construction of the system. It may be used as the reference point in the requirements; program implementation; testing and program documentation. It promotes a common understanding for all those concerned with the system. The term '*formal methods*' is used to describe a formal specification language and a method for the design and implementation of computer systems.

The specification is written in a mathematical language, and the implementation is derived from the specification via step-wise refinement.[1] The refinement step makes the specification more concrete and closer to the actual implementation. There is an associated proof obligation to demonstrate that the refinement is valid, and that the concrete state preserves the properties of the more abstract state. Thus, assuming that the original specification is correct, and the proofs of correctness of each refinement step are valid, then there is a very high degree of confidence in the correctness of the implemented software.

Step-wise refinement is illustrated as follows: the initial specification S is the initial model M_0; it is then refined into the more concrete model M_1, and M_1 is then refined into M_2, and so on until the eventual implementation $M_n = E$ is produced.

$$S = M_0 \subseteq M_1 \subseteq M_2 \subseteq M_3 \subseteq \cdots \subseteq M_n = E$$

Requirements are the foundation of the system to be built, and irrespective of the best design and development practices, the product will be incorrect if the requirements are incorrect. The objective of *requirements validation* is to ensure that the requirements reflect what is actually required by the customer (*in order to build the right system*). Formal methods may be employed to model the requirements, and the model exploration yields further desirable or undesirable properties. The ability to prove that certain properties are true of the specification is very valuable, especially in safety critical and security critical applications. These properties are logical consequences of the definition of the requirements, and, where appropriate, the requirements may need to be amended. Thus, formal methods may be employed in a sense to debug the requirements during requirements validation.

The use of formal methods generally leads to more robust software and to increased confidence in its correctness. The challenges involved in the deployment

---

[1]It is debatable whether step-wise refinement is cost effective in mainstream software engineering, as it involves re-writing a specification ad nauseam. It is time-consuming, as significant time is required to prove that each refinement step is valid.

Table 18.1 Criticisms of formal methods

No.	Criticism
1	Often the formal specification is as difficult to read as the program[a]
2	Many formal specifications are wrong[b]
3	Formal methods are strong on syntax but provide little assistance in deciding on what technical information should be recorded using the syntax[c]
4	Formal specifications provide a model of the proposed system. However, a precise unambiguous mathematical statement of the requirements is what is needed[d]
5	Step-wise refinement is unrealistic.[e] It is like, for example, deriving a bridge from the description of a river and the expected traffic on the bridge. There is always a need for a creative step in design
6	Much unnecessary mathematical formalisms have been developed rather than using the available classical mathematics[f]

[a]Of course, others might reply by saying that some of Parnas's tables are not exactly intuitive, and that the notation he employs in some of his tables is quite unfriendly. The usability of all of the mathematical approaches needs to be enhanced if they are to be taken seriously by industrialists
[b]Obviously, the formal specification must be analysed using mathematical reasoning and tools to provide confidence in its correctness. The validation may be carried out using mathematical proof of key properties of the specification; software inspections; or specification animation
[c]VDM includes a method for software development as well as the specification language
[d]Models are extremely valuable as they allow simplification of the reality. A mathematical study of the model demonstrates whether it is a suitable representation of the system. Models allow properties of the proposed requirements to be studied prior to implementation
[e]Step-wise refinement involves rewriting a specification with each refinement step producing a more concrete specification (that includes code and formal specification) until eventually the detailed code is produced. However, tool support may make refinement easier
[f]Approaches such as VDM or Z are useful in that they add greater rigour to the software development process. Classical mathematics is familiar to students and therefore it is desirable that new formalisms are introduced only where absolutely necessary

of formal methods in an organization include the education of staff in formal specification, as the use of these mathematical techniques may be a culture shock to many staff.

Formal methods have been applied to a diverse range of applications, including the security critical field; the safety critical field; the railway sector; microprocessor verification; the specification of standards, and the specification and verification of programs.

Parnas and others have criticized formal methods on the following grounds (Table 18.1).

However, formal methods are potentially quite useful and reasonably easy to use. The use of a formal method such as Z or VDM forces the software engineer to be precise and helps to avoid ambiguities present in natural language. Clearly, a formal specification should be subject to a peer review to provide confidence in its correctness. New formalisms need to be intuitive to be usable by practitioners. The advantage of classical mathematics is that it is familiar to students.

18.2 Why Should We Use Formal Methods?

There is a strong motivation to use best practice in software engineering in order to produce software adhering to high-quality standards. Quality problems with software may cause minor irritations or major damage to a customer's business including loss of life.[2] Formal methods are a leading-edge technology that may help companies to reduce the occurrence of defects in software products. Brown [2] argues that for the safety critical field that:

Comment 18.1 (Missile Safety) *Missile systems must be presumed dangerous until shown to be safe, and that the absence of evidence for the existence of dangerous errors does not amount to evidence for the absence of danger.*

This suggests that companies in the safety critical field need to demonstrate that every reasonable practice was taken to prevent the occurrence of defects. One such practice is the use of formal methods, and its exclusion may need to be justified in some domains. It is quite possible that a software company may be sued for software which injures a third party,[3] and this suggests that companies will need a rigorous quality assurance system to prevent the occurrence of defects.

There is some evidence to suggest that the use of formal methods provides savings in the cost of the project. For example, a 9 % cost saving is attributed to the use of formal methods during the CICS project; the T800 project attributes a 12-month reduction in testing time to the use of formal methods. These are discussed in more detail in Chap. 1 of [3].

The use of formal methods is mandatory in certain circumstances. The Ministry of Defence in the United Kingdom issued two safety-critical standards[4] in the early 1990s related to the use of formal methods in the software development lifecycle.

The first is Defence Standard 00-55, "The Procurement of safety critical software in defense equipment" [4] which makes it mandatory to employ formal methods in safety-critical software development in the UK; and mandates the use of formal proof that the most crucial programs correctly implement their specifications.

The second is Def Stan 00-56 "Hazard analysis and safety classification of the computer and programmable electronic system elements of defense equipment" [5]. The objective of this standard is to provide guidance to identify which systems or parts of systems being developed are safety-critical and thereby require the use of formal methods. This proposed system is subject to an initial hazard analysis to determine whether there are safety-critical parts.

---

[2]We mentioned the serious problems with the Therac-25 radiotherapy machine in Chap. 17.

[3]A comprehensive disclaimer of responsibility for problems (rather than a guarantee of quality) accompany most software products, and so the legal aspects of licensing software may protect software companies from litigation. However, greater legal protection for the customer can be built into the contract between the supplier and the customer for bespoke-software development.

[4]The U.K. Defence Standards 0055 and 0056 have been revised in recent years to be less prescriptive on the use of formal methods.

The reaction to these defence standards 00-55 and 00-56 was quite hostile initially, as most suppliers were unlikely to meet the technical and organization requirements of the standard [6]. The standards were subsequently revised to be less prescriptive on the use of formal methods.

18.3   Applications of Formal Methods

Formal methods have been employed to verify correctness in the nuclear power industry, the aerospace industry, the security technology area, and the railroad domain. These sectors are subject to stringent regulatory controls to ensure safety and security. Several organizations have piloted formal methods with varying degrees of success. These include IBM, who developed VDM at its laboratory in Vienna; IBM (Hursley) piloted the Z formal specification language on the CICS (Customer Information Control System) project.

The mathematical techniques developed by Parnas (i.e., tabular expressions) have been employed to specify the requirements of the A-7 aircraft as part of a research project for the US Navy.[5] Tabular expressions have also been employed for the software inspection of the automated shutdown software of the Darlington Nuclear power plant in Canada.[6] These are two successful uses of mathematical techniques in software engineering.

There are examples of the use of formal methods in the railway domain, and examples dealing with the modeling and verification of a railroad gate controller and railway signaling are described in [3]. Clearly, it is essential to verify safety critical properties such as "when the train goes through the level crossing then the gate is closed".

18.4   Tools for Formal Methods

A key criticism of formal methods is the limited availability of tools to support the software engineer in writing a formal specification or in conducting proof. Many of the early tools were criticized as not being of industrial strength. However, in recent years more advanced tools to support the software engineer's work in formal specification and formal proof have become available, and this should continue in the coming years.

---

[5]However, the resulting software was never actually deployed on the A-7 aircraft.

[6]This was an impressive use of mathematical techniques and it has been acknowledged that formal methods must play an important role in future developments at Darlington. However, given the time and cost involved in the software inspection of the shutdown software some managers have less enthusiasm in shifting from hardware to software controllers [7].

The tools include syntax checkers that determine whether the specification is syntactically correct; specialized editors which ensure that the written specification is syntactically correct; tools to support refinement; automated code generators that generate a high-level language corresponding to the specification; theorem provers to demonstrate the presence or absence of key properties and to prove the correctness of refinement steps, and to identify and resolve proof obligations; and specification animation tools where the execution of the specification can be simulated.

The *B*-Toolkit from *B*-Core is an integrated set of tools that supports the *B*-Method. These include syntax and type checking, specification animation, proof obligation generator, an auto-prover, a proof assistant, and code generation. This allows, in theory, a complete formal development from initial specification to final implementation to be achieved, with every proof obligation justified, leading to a provably correct program.

The IFAD Toolbox[7] is a support tool for the VDM-SL specification language, and it includes support for syntax and type checking, an interpreter and debugger to execute and debug the specification, and a code generator to convert from VDM-SL to C++. It also includes support for graphical notations such as the OMT/UML design notations.

18.5 Approaches to Formal Methods

There are two key approaches to formal methods: namely the model-oriented approach of VDM or Z, and the algebraic or axiomatic approach of the process calculi such as the calculus communicating systems (CCS) or communicating sequential processes (CSP).

18.5.1 Model-Oriented Approach

The *model-oriented* approach to specification is based on mathematical models, and a model is a mathematical representation or abstraction of a physical entity or system. The model aims to provide a mathematical explanation of the behaviour of the physical world, and it is considered suitable if its properties closely match those of the system being modeled. A model will allow predictions of future behaviour to be made, and many models are employed in the physical world (e.g., weather forecasting system).

It is fundamental to explore the model to determine its *adequacy*, and to determine the extent to which it explains the underlying physical behaviour, and allows predictions of future behaviour to be made. This will determine its

---

[7]The IFAD Toolbox has been renamed to VDMTools as IFAD sold the VDM Tools to CSK in Japan.

acceptability as a representation of the physical world. Models that are ineffective will be replaced with models that offer a better explanation of the manifested physical behaviour. There are many examples in science of the replacement of one theory by a newer one. For example, the Copernican model of the universe replaced the older Ptolemaic model, and Newtonian physics was replaced by Einstein's theories on relativity [8].

The model-oriented approach to software development involves defining an abstract model of the proposed software system. The model acts as a representation of the proposed system, and the model is then explored to assess its suitability. The exploration of the model takes the form of model interrogation, i.e., asking questions and determining the effectiveness of the model in answering the questions. The modeling in formal methods is typically performed via elementary discrete mathematics, including set theory, sequences, functions and relations.

VDM and Z are model-oriented approaches to formal methods. VDM arose from work done in the IBM laboratory in Vienna in formalizing the semantics for the PL/1 compiler, and it was later applied to the specification of software systems. The origin of the Z specification language is in work done at Oxford University in the early 1980s.

18.5.2 Axiomatic Approach

The *axiomatic approach* focuses on the properties that the proposed system is to satisfy, and there is no intention to produce an abstract model of the system. The required properties and behaviour of the system are stated in mathematical notation. The difference between the axiomatic specification and a model-based approach is may be seen in the example of a stack.

The stack includes operators for pushing an element onto the stack and popping an element from the stack. The properties of *pop* and *push* are explicitly defined in the axiomatic approach. The model-oriented approach constructs an explicit model of the stack and the operations are defined in terms of the effect that they have on the model. The specification of the *pop* operation on a stack is given by axiomatic properties, for example, $pop(push(s, x)) = s$.

Comment 18.2 (Axiomatic Approach) *The property-oriented approach has the advantage that the implementer is not constrained to a particular choice of implementation, and the only constraint is that the implementation must satisfy the stipulated properties.*

The emphasis is on the required properties of the system, and implementation issues are avoided. The focus is on the specification of the underlying behaviour, and properties are typically stated using mathematical logic or higher-order logics. Mechanized theorem-proving techniques may be employed to prove results.

One potential problem with the axiomatic approach is that the properties specified may not be satisfiable in any implementation. Thus, whenever a 'formal axiomatic theory' is developed a corresponding 'model' of the theory must be identified, in order to ensure that the properties may be realized in practice. *That is, when proposing a system that is to satisfy some set of properties, there is a need to prove that there is at least one system that will satisfy the set of properties.*

18.6 Proof and Formal Methods

A mathematical proof typically includes natural language and mathematical symbols, and often many of the tedious details of the proof are omitted. The proof of a conjecture may be by a 'divide and conquer' technique; i.e., breaking the conjecture down into subgoals and then attempting to prove the subgoals. Many proofs in formal methods are concerned with crosschecking the details of the specification, or checking the validity of refinement steps, or checking that certain properties are satisfied by the specification. There are often many tedious lemmas to be proved, and theorem provers[8] are essential in assisting with this. Machine proof needs to be explicit, and reliance on some brilliant insight is avoided. Proofs by hand are notorious for containing errors or jumps in reasoning, while machine proofs are explicit but are often extremely lengthy and unreadable (e.g., the actual machine proof of correctness of the VIPER microprocessor[9] [6] consisted of several million formulae).

A formal mathematical proof consists of a sequence of formulae, where each element is either an axiom or derived from a previous element in the series by applying a fixed set of mechanical rules.

Theorem provers are invaluable in resolving many of the thousands of proof obligations that arise from a formal specification, and it is not feasible to apply formal methods in an industrial environment without the use of machine-assisted proof. Automated theorem proving is difficult, as often mathematicians prove a theorem with an initial intuitive feeling that the theorem is true. Human intervention to provide guidance or intuition improves the effectiveness of the theorem prover.

The proof of various properties about a program increases confidence in its correctness. However, an absolute proof of correctness[10] is unlikely except for the most trivial of programs. A program may consist of legacy software that is assumed to work; a compiler that is assumed to work correctly creates it. Theorem provers

---

[8]Many existing theorem provers are difficult to use and are for specialist use only. There is a need to improve their usability.

[9]This verification was controversial with RSRE and Charter overselling VIPER as a chip design that conforms to its formal specification.

[10]This position is controversial with others arguing that if correctness is defined mathematically then the mathematical definition (i.e. formal specification) is a theorem, and the task is to prove that the program satisfies the theorem. They argue that the proofs for non-trivial programs exist, and that the reason why there are not many examples of such proofs is due to a lack of mathematical specifications.

are programs that are assumed to function correctly. The best that formal methods can claim is increased confidence in correctness of the software, rather than an absolute proof of correctness.

18.7   The Future of Formal Methods

The debate concerning the level of use of mathematics in software engineering is still ongoing. Most practitioners are against the use of mathematics and avoid its use. They tend to employ methodologies such as software inspections and testing to improve confidence in the correctness of the software. Industrialists often need to balance conflicting needs such as quality; cost; and aggressive time pressures. They argue that commercial realities dictate that appropriate methodologies and techniques are required that allow them to achieve their business goals in a timely manner.

The other camp argues that the use of mathematics is essential in the delivery of high-quality and reliable software, and that if a company does not place sufficient emphasis on quality it will pay the price in terms of a poor reputation in the market place.

It is generally accepted that mathematics and formal methods must play a role in the safety critical and security critical fields. Apart from that the extent of the use of mathematics is a hotly disputed topic. The pace of change in the world is extraordinary, and companies face major competitive pressures in a global market place. It is unrealistic to expect companies to deploy formal methods unless they have clear evidence that it will support them in delivering commercial products to the market place ahead of their competition, at the right price and with the right quality. Formal methods need to prove that it can do this if it wishes to be taken seriously in mainstream software engineering. The issue of technology transfer of formal methods to industry is discussed in [9].

18.8   The Vienna Development Method

VDM dates from work done by the IBM research laboratory in Vienna. This group was specifying the semantics of the PL/1 programming language using an operational semantic approach (discussed in Chap. 12). That is, the semantics of the language were defined in terms of a hypothetical machine, which interprets the programs of that language [10, 11]. Later work led to the Vienna Development Method (VDM) with its specification language, Meta IV. This was used to give the denotational semantics of programming languages; i.e., a mathematical object (set, function, etc.) is associated with each phrase of the language [11]. The mathematical object is termed the *denotation* of the phrase.

VDM is a *model-oriented approach* and this means that an explicit model of the state of an abstract machine is given, and operations are defined in terms of this state. Operations may act on the system state, taking inputs, and producing outputs as well as a new system state. Operations are defined in a precondition and post-condition style. Each operation has an associated proof obligation to ensure that if the precondition is true, then the operation preserves the system invariant. The initial state itself is, of course, required to satisfy the system invariant.

VDM uses keywords to distinguish different parts of the specification, e.g., preconditions, postconditions, as introduced by the keywords *pre* and *post*, respectively. In keeping with the philosophy that formal methods specifies *what* a system does as distinct from *how*, VDM employs post-conditions to stipulate the effect of the operation on the state. The previous state is then distinguished by employing *hooked variables*, e.g., v^-, and the postcondition specifies the new state which is defined by a logical predicate relating the pre-state to the post-state.

VDM is more than its specification language VDM-SL, and is, in fact, a software development method, with rules to verify the steps of development. The rules enable the executable specification, i.e., the detailed code, to be obtained from the initial specification via refinement steps. Thus, we have a sequence $S = S_0, S_1, ...,$ $S_n = E$ of specifications, where S is the initial specification, and E is the final (executable) specification.

Retrieval functions enable a return from a more concrete specification to the more abstract specification. The initial specification consists of an initial state, a system state, and a set of operations. The system state is a particular domain, where a domain is built out of primitive domains such as the set of natural numbers, etc., or constructed from primitive domains using domain constructors such as Cartesian product, disjoint union, etc. A domain-invariant predicate may further constrain the domain, and a *type* in VDM reflects a domain obtained in this way. Thus, a type in VDM is more specific than the signature of the type, and thus represents values in the domain defined by the signature, which satisfy the domain invariant. In view of this approach to types, it is clear that VDM types may not be 'statically type checked'.

VDM specifications are structured into modules, with a module containing the module name, parameters, types, operations, etc. Partial functions occur frequently in computer science as many functions, may be undefined, or fail to terminate for some arguments in their domain. VDM addresses partial functions by employing nonstandard logical operators, namely the logic of partial functions (LPFs), which was discussed in Chap. 16.

VDM has been used in industrial projects, and its tool support includes the IFAD Toolbox.[11] There are several variants of VDM, including VDM++, the object-oriented extension of VDM, and the Irish school of the VDM, which is discussed in the next section.

---

[11]The VDM Tools are now available from the CSK Group in Japan.

18.9   VDM$^+$, the Irish School of VDM

The Irish School of VDM is a variant of standard VDM, and is characterized by
[12] its constructive approach, classical mathematical style, and its terse notation.
This method aims to combine the *what* and *how* of formal methods in that its terse
specification style stipulates in concise form *what* the system should do; further-
more, the fact that its specifications are constructive (or functional) means that the
how is included with the *what*. However, it is important to qualify this by stating
that the how as presented by VDM$^+$ is not directly executable, as several of its
mathematical data types have no corresponding structure in high-level programming languages or functional languages. Thus, a conversion or reification of the
specification into a functional or higher level language must take place to ensure a
successful execution. Further, the fact that a specification is constructive is no
guarantee that it is a good implementation strategy, if the construction itself is
naive.

The Irish school follows a similar development methodology as in standard
VDM, and is a model-oriented approach. The initial specification is presented, with
initial state and operations defined. The operations are presented with preconditions; however, no postcondition is necessary as the operation is 'functionally' (i.e.,
explicitly) constructed.

There are proof obligations to demonstrate that the operations preserve the
invariant. That is, if the precondition for the operation is true, and the operation is
performed, then the system invariant remains true after the operation. The philosophy is to exhibit existence *constructively* rather than a theoretical proof of existence that demonstrates the existence of a solution without presenting an algorithm
to construct the solution.

The school avoids the existential quantifier of predicate calculus and reliance on
logic in proof is kept to a minimum, and emphasis instead is placed on equational
reasoning. Structures with nice algebraic properties are sought, and one nice
algebraic structure employed is the monoid, which has closure, associativity, and a
unit element. The concept of isomorphism is powerful, reflecting that two structures
are essentially identical, and thus we may choose to work with either, depending on
which is more convenient for the task in hand.

The school has been influenced by the work of Polya and Lakatos. The former
[13] advocated a style of problem solving characterized by first considering an
easier sub-problem, and considering several examples. This generally leads to a
clearer insight into solving the main problem. Lakatos's approach to mathematical
discovery [14] is characterized by heuristic methods. A primitive conjecture is
proposed and if global counter-examples to the statement of the conjecture are
discovered, then the corresponding *hidden lemma* for which this global coun-
terexample is a local counter example is identified and added to the statement of the
primitive conjecture. The process repeats, until no more global counterexamples are
found. A skeptical view of absolute truth or certainty is inherent in this.

Partial functions are the norm in VDM$^{\clubsuit}$, and as in standard VDM, the problem is that functions may be undefined, or fail to terminate for several of the arguments in their domain. The logic of partial functions (LPFs) is avoided, and instead care is taken with recursive definitions to ensure termination is achieved for each argument. Academic and industrial projects have been conducted using the method of the Irish school, but at this stage tool support is limited.

18.10   The *Z* Specification Language

Z is a formal specification language founded on Zermelo set theory, and it was developed by Abrial at Oxford University in the early 1980s. It is a model-oriented approach where an explicit model of the state of an abstract machine is given, and the operations are defined in terms of the effect on the state. It includes a mathematical notation that is similar to VDM, and it employs the visually striking schema calculus, which consists essentially of boxes, with these boxes or schemas used to describe operations and states. The schema calculus enables schemas to be used as building blocks and combined with other schemas. The *Z* specification language was published as an ISO standard (ISO/IEC 13568:2002) in 2002.

The schema calculus is a powerful means of decomposing a specification into smaller pieces or schemas. This helps to make *Z* specification highly readable, as each individual schema is small in size and self-contained. The exception handling is done by defining schemas for the exception cases, and these are then combined with the original operation schema. Mathematical data types are used to model the data in a system and these data types obey mathematical laws. These laws enable simplification of expressions and are useful with proofs.

Operations are defined in a precondition/postcondition style. However, the precondition is implicitly defined within the operation; i.e., it is not separated out as in standard VDM. Each operation has an associated proof obligation to ensure that if the precondition is true, then the operation preserves the system invariant. The initial state itself is, of course, required to satisfy the system invariant. Postconditions employ a logical predicate which relates the pre-state to the post-state, and the post-state of a variable v is given by priming, e.g., v'. Various conventions are employed, e.g., $v?$ indicates that v is an input variable and $v!$ indicates that v is an output variable. The symbol Ξ Op operation indicates that this operation does not affect the state, whereas Δ Op indicates that this operation that affects the state.

Many data types employed in *Z* have no counterpart in standard programming languages. It is therefore important to identify and describe the concrete data structures that will ultimately represent the abstract mathematical structures. The operations on the abstract data structures may need to be refined to yield operations on the concrete data structure that yield equivalent results. For simple systems, direct refinement (i.e., one step from abstract specification to implementation) may be possible; in more complex systems, deferred refinement is employed, where a sequence of increasingly concrete specifications are produced to yield the executable specification eventually. Z has been successfully applied in industry, and one of its well-known successes is the CICS project at IBM Hursley in England. Z is described in more detail in Chap. 19.

18.11   The *B* Method

The *B-Technologies* [15] consist of three components: a method for software development, namely the *B*-Method; a supporting set of tools, namely, the *B*-Toolkit; and a generic program for symbol manipulation, namely, the *B*-Tool (from which the *B*-Toolkit is derived). The *B*-Method is a model-oriented approach and is closely related to the *Z* specification language. Abrial developed the *B* specification language, and every construct in the language has a set theoretic counterpart, and the method is founded on Zermelo set theory. Each operation has an explicit precondition.

One key purpose [15] of the *abstract machine* in the *B*-Method is to provide encapsulation of variables representing the state of the machine and operations that manipulate the state. Machines may refer to other machines, and a machine may be introduced as a refinement of another machine. The abstract machines are specification machines, refinement machines, or implementable machines. The *B*-Method adopts a layered approach to design where the design is gradually made more concrete by a sequence of design layers. Each design layer is a refinement that involves a more detailed implementation in terms of abstract machines of the previous layer. The design refinement ends when the final layer is implemented purely in terms of library machines. Any refinement of a machine by another has associated proof obligations, and proof is required to verify the validity of the refinement step.

Specification animation of the Abstract Machine Notation (AMN) specification is possible with the *B*-Toolkit, and this enables typical usage scenarios of the AMN specification to be explored for requirements validation. This is, in effect, an early form of testing, and it may be used to demonstrate the presence or absence of desirable or undesirable behavior. Verification takes the form of a proof to demonstrate that the invariant is preserved when the operation is executed within its precondition, and this is performed on the AMN specification with the *B*-Toolkit.

The *B*-Toolkit provides several tools that support the *B*-Method, and these include syntax and type checking; specification animation, proof obligation generator, auto prover, proof assistor, and code generation. Thus, in theory, a complete formal development from initial specification to final implementation may be achieved, with every proof obligation justified, leading to a provably correct program.

The *B*-Method and toolkit have been successfully applied in industrial applications, including the CICS project at IBM Hursley in the United Kingdom. The automated support provided has been cited as a major benefit of the application of the *B*-Method and the *B*-Toolkit.

18.12   Predicate Transformers and Weakest Preconditions

The precondition of a program S is a predicate, i.e., a statement that may be true or false, and it is usually required to prove that if the precondition Q is true,; i.e., $\{Q\}$ $S\,\{R\}$, then execution of S is guaranteed to terminate in a finite amount of time in a state satisfying R.

The weakest precondition of a command S with respect to a postcondition R represents the set of all states such that if execution begins in any one of these states, then execution will terminate in a finite amount of time in a state with R true [16]. These set of states may be represented by a predicate Q', so that $wp(S, R) = wp_S(R) = Q'$, and so wp_S is a predicate transformer, i.e., it may be regarded as a function on predicates. The weakest precondition is the precondition that places the fewest constraints on the state than all of the other preconditions of (S,R). That is, all of the other preconditions are stronger than the weakest precondition.

The notation $Q\{S\}R$ is used to denote partial correctness and indicates that if execution of S commences in any state satisfying Q, and if execution terminates, then the final state will satisfy R. Often, a predicate Q which is stronger than the weakest precondition $wp(S,R)$ is employed, especially where the calculation of the weakest precondition is nontrivial. Thus, a stronger predicate Q such that $Q \Rightarrow wp(S, R)$ is sometimes employed.

There are many properties associated with the weakest preconditions, and these may be used to simplify expressions involving weakest preconditions, and in determining the weakest preconditions of various program commands such as assignments, iterations, etc. Weakest preconditions may be used in developing a proof of correctness of a program in parallel with its development [17].

An imperative program may be regarded as a predicate transformer. This is since a predicate P characterizes the set of states in which the predicate P is true, and an imperative program may be regarded as a binary relation on states, which may be extended to a function F, leading to the Hoare triple $P\{F\}Q$. That is, the program F acts as a predicate transformer with the predicate P regarded as an input assertion, i.e., a Boolean expression that must be true before the program F is executed, and the predicate Q is the output assertion, which is true if the program F terminates (where F commenced in a state satisfying P).

18.13   The Process Calculi

The objectives of the process calculi [18] are to provide mathematical models that provide insight into the diverse issues involved in the specification, design, and implementation of computer systems which continuously act and interact with their environment. These systems may be decomposed into sub-systems that interact with each other and their environment.

The basic building block is the *process*, which is a mathematical abstraction of the interactions between a system and its environment. A process that lasts indefinitely may be specified recursively. Processes may be assembled into systems; they may execute concurrently; or communicate with each other. Process communication may be synchronized, and this takes the form of a process outputting a message simultaneously to another process inputting a message. Resources may be shared among several processes. Process calculi such as CSP [18] and CCS [19] have been developed to enrich the understanding of communication and concurrency, and these calculi obey a rich collection of mathematical laws.

The expression (a ? P) in CSP describes a process which first engages in event a, and then behaves as process P. A recursive definition is written as (μX) · F(X), and the example of a simple chocolate vending machine is given recursively as:

$$\text{VMS} = \mu X : \{coin, choc\} \cdot (coin ? (choc ? X))$$

The simple vending machine has an alphabet of two symbols, namely, *coin* and *choc*. The behaviour of the machine is that a coin is entered into the machine, and then a chocolate selected and provided.

CSP processes use channels to communicate values with their environment, and input on channel c is denoted by ($c?.x\, P_x$). This describes a process that accepts any value x on channel c, and then behaves as process P_x. In contrast, ($c!e\, P$) defines a process which outputs the expression e on channel c and then behaves as process P.

The π-calculus is a process calculus based on names. Communication between processes takes place between known channels, and the name of a channel may be passed over a channel. There is no distinction between channel names and data values in the π-calculus. The output of a value v on channel a is given by $\bar{a}v$; i.e., output is a negative prefix. Input on a channel a is given by $a(x)$, and is a positive prefix. Private links or restrictions are given by (x)P in the π-calculus.

18.14 The Parnas Way

Parnas has been influential in the computing field, and his ideas on the specification, design, implementation, maintenance, and documentation of computer software remain important. He advocates a solid classical engineering approach to developing software, and he argues that the role of an engineer is to apply scientific principles and mathematics in designing and developing software products. His main contributions to software engineering are summarized in Table 18.2.

Table 18.2 Parnas's contributions to software engineering

Area	Description
Tabular expressions	These are mathematical tables for specifying requirements, and enable complex predicate logic expressions to be represented in a simpler form
Mathematical documentation	He advocates the use of precise mathematical documentation
Requirements specification	He advocates the use of mathematical relations to specify the requirements precisely
Software design	He developed information hiding which is used in object-oriented design[a], and allows software to be designed for change [21]. Every information-hiding module has an interface that provides the only means to access the services provided by the modules. The interface hides the module's implementation
Software inspections	His approach requires the reviewers to take an active part in the inspection. They are provided with a list of questions by the author and their analysis involves the production of mathematical table to justify the answers
Predicate logic	He developed an extension of the predicate calculus to deal with partial functions. This approach preserves the classical two-valued logic and deals with undefined values that may occur in predicate logic expressions

[a]It is surprising that many in the object-oriented world seem unaware that information hiding goes back to the early 1970s and many have never heard of Parnas

18.15 Usability of Formal Methods

There are practical difficulties associated with the use of formal methods. It seems to be assumed that programmers and customers are willing to become familiar with the mathematics used in formal methods. There is little evidence to suggest that customers in mainstream organizations would be prepared to use formal methods.[12] Customers are concerned with their own domain and speak the technical language of that domain.[13] Often, the use of mathematics is an alien activity that bears little resemblance to their normal work. Programmers are interested in programming rather than in mathematics, and generally have no interest in becoming mathematicians.[14]

---

[12]The domain in which the software is being used will influence the willingness or otherwise of the customers to become familiar with the mathematics required. There is very little interest from customers in mainstream software engineering, and the perception is that formal methods are difficult to use. However, in some domains such as the regulated sector there is a greater willingness of customers to become familiar with the mathematical notation.

[13]The author's experience is that most customers have a very limited interest in using mathematics.

[14]Mathematics that is potentially useful to software engineers was discussed in Chap. 17.

Table 18.3 Techniques for validation of formal specification

Technique	Description
Proof	This involves demonstrating that the formal specification adheres to key properties of the requirements. The implementation will need to preserve these properties also
Software inspections	This involves a Fagan like inspection to perform the validation. It may involve comparing an informal set of requirements (unless the customer has learned the formal method) with the formal specification
Specification animation	This involves program (or specification) execution as a way to validate the formal specification. It is similar to testing

However, the mathematics involved in most formal methods is reasonably elementary, and, in theory, if both customers and programmers are willing to learn the formal mathematical notation, then a rigorous validation of the formal specification can take place to verify its correctness. Both parties can review the formal specification to verify its correctness, and the code can be verified to be correct with respect to the formal specification. It is usually possible to get a developer to learn a formal method, as a programmer has some experience of mathematics and logic; however, in practice, it is more difficult to get a customer to learn a formal method.

This means that often a formal specification of the requirements and an informal definition of the requirements using a natural language are maintained. It is essential that both of these documents are consistent and that there is a rigorous validation of the formal specification. Otherwise, if the programmer proves the correctness of the code with respect to the formal specification, and the formal specification is incorrect, then the formal development of the software is incorrect. There are several techniques to validate a formal specification (Table 18.3) and these are described in [20]:

Why are Formal Methods difficult?

Formal methods are perceived as being difficult to use and of offering limited value in mainstream software engineering. Programmers receive some training in mathematics as part of their education. However, in practice, most programmers who learn formal methods at university never use formal methods again once they take an industrial position.

It may well be that the very nature of formal methods is such that it is suited only for specialists with a strong background in mathematics. Some of the reasons why formal methods are perceived as being difficult are (Table 18.4)

Characteristics of a Usable Formal Method

It is important to investigate ways by which formal methods can be made more usable to software engineers. This may involve designing more usable notations and better tools to support the process. Practical training and coaching to employees can help also. Some of the characteristics of a usable formal method are (Table 18.5).

Table 18.4 Factors in difficulty of formal methods

Factor	Description
Notation/intuition	The notation employed differs from that used in mathematics. Many programmers find the notation in formal methods to be unintuitive
Formal specification	It is easier to read a formal specification than to write one
Validation of formal specification	The validation of a formal specification using proof techniques or a Fagan like inspection is difficult
Refinement[a]	The refinement of a formal specification into successive more concrete specifications with proof of validity of each refinement step is difficult and time consuming
Proof	Proof can be difficult and time consuming
Tool support	Many of the existing tools are difficult to use

[a]It is highly unlikely that refinement is cost effective for mainstream software engineering. However, it may be useful in the regulated environment

Table 18.5 Characteristics of a usable formal method

Characteristic	Description
Intuitive	A formal method should be intuitive
Teachable	A formal method needs to be teachable to the average software engineer. The training should include (at least) writing practical formal specifications
Tool support	Good tools to support formal specification, validation, refinement and proof are required
Adaptable to change	Change is common in a software engineering environment. A usable formal method should be adaptable to change
Technology transfer path	The process for software development needs to be defined to include formal methods. The migration to formal methods needs to be managed
Cost[a]	The use of formal methods should be cost effective with a return on investment. There should be benefits in time, quality and productivity

[a]A commercial company will expect a return on investment from the use of a new technology. This may be reduced software development costs, improved quality, improved timeliness of projects or improvements in productivity

18.16 Review Questions

1. What are formal methods and describe their potential benefits? How essential is tool support?
2. What is stepwise refinement and is it realistic in mainstream software engineering?
3. Discuss Parnas's criticisms of formal methods and discuss whether his views are justified.

4. Discuss the applications of formal methods and which areas have benefited most from their use? What problems have arisen?
5. Describe a technology transfer path for the potential deployment of formal methods in an organization.
6. Explain the difference between the model-oriented approach and the axiomatic approach.
7. Discuss the nature of proof in formal methods and tools to support proof.
8. Discuss the Vienna Development Method and explain the difference between standard VDM and VDM♣.
9. Discuss Z and B. Describe the tools in the B-Toolkit.
10. Discuss process calculi such as CSP, CCS or π–calculus.

18.17 Summary

This chapter discussed formal methods, which are a rigorous approach to the development of high-quality software. Formal methods employ mathematical techniques for the specification and formal development of software, and are very useful in the safety critical field. They consist of formal specification languages or notations; a methodology for formal software development; and a set of tools to support the syntax checking of the specification, as well as the proof of properties of the specification.

Formal methods allow questions to be asked and answered about what the system does independently of the implementation. The use of formal methods generally leads to more robust software and to increased confidence in its correctness. There are challenges involved in the deployment of formal methods, as the use of these mathematical techniques may be a culture shock to many staff.

Formal methods may be model oriented or axiomatic oriented. The model-oriented approach includes formal methods such as VDM, Z and B. The axiomatic approach includes the process calculi such as CSP, CCS and the π calculus.

The usability of formal methods was considered as well as an examination of why formal methods are difficult and what the characteristics of a usable formal method would be.

References

1. The Z Notation. A Reference Manual. J.M. Spivey. Prentice Hall. International Series in Computer Science. 1992.
2. Rational for the development of the U.K. Defence Standards for Safety Critical software. M.J. D Brown. Compass Conference. 1990.

3. Applications of Formal Methods. Edited by Michael Hinchey and Jonathan Bowen. Prentice Hall International Series in Computer Science. 1995.

4. 00-55 (Part 1)/ Issue 1. The Procurement of Safety Critical Software in Defence Equipment. Part 1: Requirements. Ministry of Defence. Interim Defence Standard. UK. 1991.

5. 00-55 (Part 2)/ Issue 1. The Procurement of Safety Critical Software in Defence Equipment. Part 2: Guidance. Ministry of Defence. Interim Defence Standard. UK. 1991.

6. The Evolution of Def Stan 00-55 and 00-56. An intensification of the formal methods debate in the UK. Margaret Tierney. Research Centre for Social Sciences. University of Edinburgh. 1991.

7. Experience with Formal Methods in Critical Systems. Susan Gerhart, Dan Craighen and Ted Ralston. IEEE Software. January 1994.

8. The Structure of Scientific Revolutions. Thomas Kuhn. University of Chicago Press. 1970.

9. Mathematical Approaches to Software Quality. Gerard O' Regan.Springer. 2006.

10. The Vienna Development Method. The Meta language. Dines Bjørner and Cliff Jones. *Lecture Notes in Computer Science* (61). Springer Verlag. 1978.

11. Formal Specification and Software Development. Dines Bjørner and Cliff Jones. Prentice Hall International Series in Computer Science. 1982.

12. Computation Models and Computing. PhD Thesis. Mícheál Mac An Airchinnigh. Dept.·of Computer Science. Trinity College Dublin.

13. How to Solve It. A New Aspect of Mathematical Method. Georges Polya. Princeton University Press. 1957.

14. Proof and Refutations. The Logic of Mathematical Discovery. Imre Lakatos. Cambridge University Press. 1976.

15. MSc. Thesis. Eoin McDonnell. Dept. of Computer Science. Trinity College Dublin. 1994.

16. The Science of Programming. David Gries. Springer Verlag. Berlin. 1981.

17. A Disciple of Programming. E.W. Dijkstra. Prentice Hall. 1976.

18. Communicating Sequential Processes. C.A.R. Hoare. Prentice Hall International Series in Computer Science. 1985.

19. A Calculus of Mobile Processes. Part 1. Robin Milner et al. LFCS Report Series. ECS-LFCS-89-85. Department of Computer Science. University of Edinburgh.

20. A Personal View of Formal Methods. B.A. Wichmann. National Physical Laboratory. March 2000.

21. On the Criteria to be used in Decomposing Systems into Modules. David Parnas. Communications of the ACM, 15(12). 1972.

Z Formal Specification Language

19

Keywords

Sets, relations and functions
Bags and sequences
Data reification
Refinement
Schema calculus
Proof in Z

19.1 Introduction

Z is a formal specification language based on Zermelo set theory. It was developed at the Programming Research Group at Oxford University in the early 1980s [1], and became an ISO standard in 2002. Z specifications are mathematical and employ a classical two-valued logic. The use of mathematics ensures precision, and allows inconsistencies and gaps in the specification to be identified. Theorem provers may be employed to demonstrate that the software implementation meets its specification.

Z is a '*model oriented*' approach with an explicit model of the state of an abstract machine given, and operations are defined in terms of this state. Its mathematical notation is used for formal specification, and the schema calculus is used to structure the specifications. The schema calculus is visually striking, and consists essentially of boxes, with these boxes or schemas used to describe operations and states. The schemas may be used as building blocks and combined with other

© Springer International Publishing Switzerland 2016   319
G. O'Regan, *Guide to Discrete Mathematics*, Texts in Computer Science,
DOI 10.1007/978-3-319-44561-8_19

schemas. The simple schema below (Fig. 19.1) is the specification of the positive square root of a real number.

The schema calculus is a powerful means of decomposing a specification into smaller pieces or schemas. This helps to make Z specifications highly readable, as each individual schema is small in size and self-contained. Exception handling is addressed by defining schemas for the exception cases. These are then combined with the original operation schema. Mathematical data types are used to model the data in a system, these data types obey mathematical laws. These laws enable simplification of expressions, and are useful with proofs.

Operations are defined in a precondition/postcondition style. A precondition must be true before the operation is executed, and the postcondition must be true after the operation has executed. The precondition is implicitly defined within the operation. Each operation has an associated proof obligation to ensure that if the precondition is true, then the operation preserves the system invariant. The system invariant is a property of the system that must be true at all times. The initial state itself is, of course, required to satisfy the system invariant.

The precondition for the specification of the square root function above is that $num? \geq 0$; i.e., the function $SqRoot$ may be applied to positive real numbers only. The postcondition for the square root function is $root!^2 = num?$ and $root! \geq 0$. That is, the square root of a number is positive and its square gives the number. Postconditions employ a logical predicate which relates the pre-state to the post-state, with the post-state of a variable being distinguished by priming the variable, e.g., v'.

Z is a typed language and whenever a variable is introduced its type must be given. A type is simply a collection of objects, and there are several standard types in Z. These include the natural numbers $\mathbb{N}$, the integers $\mathbb{Z}$ and the real numbers $\mathbb{R}$. The declaration of a variable x of type X is written $x: X$. It is also possible to create your own types in Z.

Various conventions are employed within Z specification, for example $v?$ indicates that v is an input variable; $v!$ indicates that v is an output variable. The variable $num?$ is an input variable and $root!$ is an output variable for the square root example above. The notation Ξ in a schema indicates that the operation Op does not affect the state; whereas the notation Δ in the schema indicates that Op is an operation that affects the state.

$$
\begin{array}{|l}
\hline
\text{--} SqRoot \text{----------} \\
num?, root! : \mathbb{R} \\
\hline
num? \geq 0 \\
root!^2 = num? \\
root! \geq 0 \\
\hline
\end{array}
$$

Fig. 19.1 Specification of positive square root

$$
\begin{array}{|l}
\hline
\text{--Library---} \\
\text{on-shelf, missing, borrowed : } \mathbb{P} \ \textit{Bkd-Id} \\
\hline
\textit{on-shelf} \cap \textit{missing} = \varnothing \\
\textit{on-shelf} \cap \textit{borrowed} = \varnothing \\
\textit{borrowed} \cap \textit{missing} = \varnothing \\
\hline
\end{array}
$$

Fig. 19.2 Specification of a library system

$$
\begin{array}{|l}
\hline
\text{--Borrow---} \\
\Delta \ \textit{Library} \\
b? : \textit{Bkd-Id} \\
\hline
b? \in \textit{on-shelf} \\
\textit{on-shelf}' = \textit{on-shelf} \setminus \{b?\} \\
\textit{borrowed}' = \textit{borrowed} \cup \{b?\} \\
\hline
\end{array}
$$

Fig. 19.3 Specification of borrow operation

Many of the data types employed in Z have no counterpart in standard programming languages. It is therefore important to identify and describe the concrete
data structures that ultimately will represent the abstract mathematical structures. As
the concrete structures may differ from the abstract, the operations on the abstract
data structures may need to be refined to yield operations on the concrete data that
yield equivalent results. For simple systems, direct refinement (i.e., one step from
abstract specification to implementation) may be possible; in more complex systems, deferred refinement[1] is employed, where a sequence of increasingly concrete
specifications are produced to yield the executable specification. There is a calculus
for combining schemas to make larger specifications, and this is discussed later in
the chapter.

Example 6.1 The following is a Z specification to borrow a book from a library
system. The library is consists of books that are on the shelf; books that are
borrowed; and books that are missing (Fig. 19.2). The specification models a
library with sets representing books on the shelf, on loan or missing. These are three
mutually disjoint subsets of the set of books *Bkd-Id*.

The system state is defined in the *Library* schema below, and operations such as
Borrow and *Return* affect the state. The *Borrow* operation is specified in (Fig. 19.3).

The notation $\mathbb{P}$*Bkd-Id* is used to represent the power set of *Bkd-Id* (i.e., the set of
all subsets of *Bkd-Id*). The disjointness condition for the library is expressed by the

---

[1]Step-wise refinement involves producing a sequence of increasingly more concrete specifications
until eventually the executable code is produced. Each refinement step has associated proof
obligations to prove that the refinement step is valid.

requirement that the pair wise intersection of the subsets *on-shelf, borrowed, missing* is the empty set.

The precondition for the *Borrow* operation is that the book must be available on the shelf to borrow. The postcondition is that the borrowed book is added to the set of borrowed books and is removed from the books on the shelf.

Z has been successfully applied in industry including the CICS project at IBM Hursley in the UK.[2] Next, we describe key parts of Z including sets, relations, functions, sequences and bags.

19.2   Sets

Sets were discussed in Chap. 2 and this section focuses on their use in Z. Sets may be enumerated by listing all of their elements. Thus, the set of all even natural numbers less than or equal to 10 is

$$\{2, 4, 6, 8, 10\}$$

Sets may be created from other sets using set comprehension: i.e., stating the properties that its members must satisfy. For example, the set of even natural numbers less than 10 is given by set comprehension as

$$\{n : \mathbb{N} | n \neq 0 \wedge n < 10 \wedge n \bmod 2 = 0 \ \cdot \ n\}$$

There are three main parts to the set comprehension above. The first part is the signature of the set and this is given by $n: \mathbb{N}$ above. The first part is separated from the second part by a vertical line. The second part is given by a predicate, and for this example the predicate is $n \neq 0 \wedge n < 10 \wedge n \bmod 2 = 0$. The second part is separated from the third part by a bullet. The third part is a term, and for this example it is simply n. The term is often a more complex expression: e.g., $\log(n^2)$.

In mathematics, there is just one empty set. However, since Z is a typed set theory, there is an empty set for each type of set. Hence, there are an infinite number of empty sets in Z. The empty set is written as $\varnothing$ [X] where X is the type of the empty set. In practice, X is omitted when the type is clear.

Various operations on sets such as union, intersection, set difference and symmetric difference are employed in Z. The power set of a set X is the set of all subsets of X, and is denoted by $\mathbb{P}$ X. The set of non-empty subsets of X is denoted by $\mathbb{P}_1$X where

$$\mathbb{P}_1 X = \{U : \mathbb{P} X | U \neq \varnothing[X]\}$$

---

[2]This project claimed a 9 % increase in productivity attributed to the use of formal methods.

A finite set of elements of type X (denoted by F X) is a subset of X that cannot be put into a one to one correspondence with a proper subset of itself. This is defined formally as

$$F\,X = \{U : \mathbb{P}X | \neg \exists V : \mathbb{P}U \cdot V \neq U \wedge (\exists f : V \rightarrowtail U)\}$$

The expression $f : V \rightarrowtail U$ denotes that f is a bijection from U to V and injective, surjective and bijective functions were discussed in Chap. 2.

The fact that Z is a typed language means that whenever a variable is introduced (e.g., in quantification with $\forall$ and $\exists$) it is first declared. For example, $\forall j: J \cdot P \Rightarrow Q$. There is also the unique existential quantifier $\exists_1 j: J \mid P$ which states that there is exactly one j of type J that has property P.

19.3 Relations

Relations are used extensively in Z and were discussed in Chap. 2. A relation R between X and Y is any subset of the Cartesian product of X and Y; i.e., $R \subseteq (X \times Y)$, and a relation in Z is denoted by R: $X \leftrightarrow Y$. The notation $x \mapsto y$ indicates that the pair $(x, y) \in R$.

Consider, the relation *home_owner: Person* $\leftrightarrow$ *Home* that exists between people and their homes. An entry *daphne* $\mapsto$ *mandalay* $\in$ *home_owner* if *daphne* is the owner of *mandalay*. It is possible for a person to own more than one home:

$$rebecca \mapsto nirvana \in home\_owner$$
$$rebecca \mapsto tivoli \in home\_owner$$

It is possible for two people to share ownership of a home:

$$rebecca \mapsto nirvana \in home\_owner$$
$$lawrence \mapsto nirvana \in home\_owner$$

There may be some people who do not own a home, and there is no entry for these people in the relation *home_owner*. The type *Person* includes every possible person, and the type *Home* includes every possible home. The domain of the relation *home_owner* is given by

$$x \in \text{dom}\, home\_owner \Leftrightarrow \exists h : Home \cdot x \mapsto h \in home\_owner.$$

The range of the relation *home_owner* is given by

$$h \in \text{ran}\, home\_owner \Leftrightarrow \exists x : Person \cdot x \mapsto h \in home\_owner.$$

The composition of two relations $home\_owner$: $Person \leftrightarrow Home$ and $home\_value$: $Home \leftrightarrow Value$ yields the relation $owner\_wealth$: $Person \leftrightarrow Value$ and is given by the relational composition $home\_owner; home\_value$ where

$$p \mapsto v \in home\_owner; home\_value \Leftrightarrow$$
$$(\exists h : Home \cdot \mapsto h \in home\_owner \wedge h \mapsto v \in home\_value)$$

The relational composition may also be expressed as

$$owner\_wealth = home\_value \circ home\_owner$$

The union of two relations often arises in practice. Suppose a new entry $aisling \mapsto muckross$ is to be added. Then this is given by

$$home\_owner' = home\_owner \cup \{aisling \mapsto muckross\}$$

Suppose that we are interested in knowing all females who are house owners. Then we restrict the relation $home\_owner$ so that the first element of all ordered pairs has to be female. Consider $female$: $\mathbb{P} \; Person$ with $\{aisling, rebecca\} \subseteq female$.

$$home\_owner = \{aisling \mapsto muckross, rebecca \mapsto nirvana,$$
$$lawrence \mapsto nirvana\}$$

$$female \triangleleft home\_owner = \{aisling \mapsto muckross, rebecca \mapsto nirvana\}$$

That is, $female \triangleleft home\_owner$ is a relation that is a subset of $home\_owner$, and the first element of each ordered pair in the relation is female. The operation $\triangleleft$ is termed domain restriction and its fundamental property is

$$x \mapsto y \in U \triangleleft R \Leftrightarrow (x \in U \wedge x \mapsto y \in R\}$$

where R: $X \leftrightarrow Y$ and U: $\mathbb{P} \; X$.

There is also a domain anti-restriction (subtraction) operation and its fundamental property is

$$x \mapsto y \in U \vartriangleleft\!\!\!- R \Leftrightarrow (x \notin U \wedge x \mapsto y \in R\}$$

where R: $X \leftrightarrow Y$ and U: $\mathbb{P}X$.

There are also range restriction (the $\triangleright$ operator) and the range anti-restriction operator (the $\vartriangleright\!\!\!-$ operator). These are discussed in [1].

19.4 Functions

A function [1] is an association between objects of some type X and objects of another type Y such that given an object of type X, there exists only one object in Y associated with that object. A function is a set of ordered pairs where the first element of the ordered pair has at most one element associated with it. A function is therefore a special type of relation, and a function may be *total* or *partial*.

A total function has exactly one element in Y associated with each element of X, whereas a partial function has at most one element of Y associated with each element of X (there may be elements of X that have no element of Y associated with them).

A partial function from X to Y ($f : X \nrightarrow Y$) is a relation $f : X \leftrightarrow Y$ such that:

$$\forall x : X; y, z : Y \cdot (x \mapsto y \in f \wedge x \mapsto z \in f \Rightarrow y = z)$$

The association between x and y is denoted by $f(x) = y$, and this indicates that the value of the partial function f at x is y. A total function from X to Y (denoted $f : X \rightarrow Y$) is a partial function such that every element in X is associated with some value of Y.

$$f : X \rightarrow Y \Leftrightarrow f : X \nrightarrow Y \wedge \operatorname{dom} f = X$$

Clearly, every total function is a partial function but not vice versa.

One operation that arises quite frequently in specifications is the function override operation. Consider the following specification of a temperature map:

$$
\begin{array}{|l}
\hline
\textit{--TempMap}\text{---------} \\
\textit{CityList} : \mathbb{P}\textit{City} \\
\textit{temp} : \textit{City} \nrightarrow \mathbb{Z} \\
\hline
\operatorname{dom} \textit{temp} = \textit{CityList} \\
\hline
\end{array}
$$

Suppose the temperature map is given by $temp = \{Cork \mapsto 17, Dublin \mapsto 19, London \mapsto 15\}$. Then consider the problem of updating the temperature map if a new temperature reading is made in Cork: e.g., $\{Cork \mapsto 18\}$. Then the new temperature chart is obtained from the old temperature chart by function override to yield $\{Cork \mapsto 18, Dublin \mapsto 19, London \mapsto 15\}$. This is written as:

$$temp' = temp \oplus \{Cork \mapsto 18\}$$

The function override operation combines two functions of the same type to give a new function of the same type. The effect of the override operation is that the entry $\{Cork \mapsto 17\}$ is removed from the temperature chart and replaced with the entry $\{Cork \mapsto 18\}$.

Suppose $f, g: X \nrightarrow Y$ are partial functions then $f \oplus g$ is defined and indicates that f is overridden by g. It is defined as follows:

$$(f \oplus g)(x) = g(x) \text{ where } x \in \text{dom } g$$
$$(f \oplus g)(x) = f(x) \text{ where } x \notin \text{dom } g \wedge x \in \text{dom} f$$

This may also be expressed (using domain anti-restriction) as

There is notation in Z for injective, surjective and bijective functions. An injective function is one to one: i.e.,

$$f(x) = f(y) \Rightarrow x = y.$$

A surjective function is onto: i.e.,

$$\text{Given } y \in Y, \exists x \in X \text{ such that } f(x) = y$$

A bijective function is one to one and onto, and it indicates that the sets X and Y can be put into one to one correspondence with one another. Z includes lambda calculus notation (λ-calculus was discussed in Chap. 12) to define functions. For example, the function cube = $\lambda x: \mathbf{N} \cdot x * x * x$. Function composition $f ; g$ is similar to relational composition.

19.5 Sequences

The type of all sequences of elements drawn from a set X is denoted by seq X. Sequences are written as $\langle x_1, x_2, \ldots x_n \rangle$ and the empty sequence is denoted by $\langle \rangle$. Sequences may be used to specify the changing state of a variable over time, with each element of the sequence representing the value of the variable at a discrete time instance.

Sequences are functions and a sequence of elements drawn from a set X is a finite function from the set of natural numbers to X. A partial finite function f from X to Y is denoted by $f: X \nrightarrow Y$. A finite sequence of elements of X is given by $f: \mathbf{N} \nrightarrow X$, and the domain of the function consists of all numbers between 1 and # f (where #f is the cardinality of f). It is defined formally as

$$\text{seq } X == \{f : \mathbf{N} \nrightarrow X \mid \text{dom } f = 1 .. \#f \cdot f\}$$

The sequence $\langle x_1, x_2, \ldots x_n \rangle$ above is given by:

$$\{1 \mapsto x_1, 2 \mapsto x_2, \ldots n \mapsto x_n\}$$

There are various functions to manipulate sequences. These include the sequence concatenation operation. Suppose $\sigma = \langle x_1, x_2, \ldots x_n \rangle$ and $\tau = \langle y_1, y_2, \ldots y_m \rangle$ then:

$$\sigma \,^\frown \tau = \langle x_1, x_2, \ldots x_n, y_1, y_2, \ldots y_m \rangle$$

The head of a non-empty sequence gives the first element of the sequence.

$$\text{head } \sigma = \text{head } \langle x_1, x_2, \ldots x_n \rangle = x_1$$

The tail of a non-empty sequence is the same sequence except that the first element of the sequence is removed.

$$\text{tail}\sigma = \text{tail}\langle x_1, x_2, \ldots x_n \rangle = \langle x_2, \ldots x_n \rangle$$

Suppose $f\colon X \to Y$ and a sequence σ: seq X then the function map applies f to each element of σ:

$$\text{map} f \; \sigma = \text{map} f \langle x_1, x_2, \ldots x_n \rangle = \langle f(x_1), f(x_2), \ldots f(x_n) \rangle$$

The map function may also be expressed via function composition as

$$\text{map} f \; \sigma = \sigma; f$$

The reverse order of a sequence is given by the rev function:

$$\text{rev } \sigma = \text{rev}\langle x_1, x_2, \ldots x_n \rangle = \langle x_n, \ldots x_2, x_1 \rangle$$

19.6 Bags

A bag is similar to a set except that there may be multiple occurrences of each element in the bag. A bag of elements of type X is defined as a partial function from the type of the elements of the bag to positive whole numbers. The definition of a bag of type X is

$$\text{bag} X = X \nrightarrow \mathbb{N}_1.$$

For example, a bag of marbles may contain 3 blue marbles, 2 red marbles, and 1 green marble. This is denoted by B = [b, b, b, g, r, r]. The bag of marbles is thus denoted by

$$\text{bag} \, Marble = Marble \nrightarrow \mathbb{N}_1.$$

$$
\begin{array}{|l}
\hline
-\Delta Vending\ Machine \text{------} \\
stock : \text{bag}\ Good \\
price : Good \rightarrow \mathbb{N}_1 \\
\hline
\text{dom}\ stock \subseteq \text{dom}\ price \\
\hline
\end{array}
$$

Fig. 19.4 Specification of vending machine using bags

The function count determines the number of occurrences of an element in a bag. For the example above, count *Marble b* = 3, and count *Marble y* = 0 since there are no yellow marbles in the bag. This is defined formally as

$$
\begin{aligned}
\text{count bag}\, X\, y &= 0 & y \notin \text{bag}\, X \\
\text{count bag}\, X\, y &= (\text{bag}\, X)(y) & y \in \text{bag}\, X
\end{aligned}
$$

An element *y* is in bag X if and only if *y* is in the domain of bag X.

$$
y\ \text{in bag}\, X \Leftrightarrow y \in \text{dom}(\text{bag}\, X)
$$

The union of two bags of marbles $B_1 = [b, b, b, g, r, r]$ and $B_2 = [b, g, r, y]$ is given by $B_1 \uplus B_2 = [b, b, b, b, g, g, r, r, y]$. It is defined formally as

$$
\begin{aligned}
(B_1 \uplus B_2)\,(y) &= B_2(y) & y \notin \text{dom}\, B_1 \wedge y \in \text{dom}\, B_2 \\
(B_1 \uplus B_2)\,(y) &= B_1(y) & y \in \text{dom}\, B_1 \wedge y \notin \text{dom}\, B_2 \\
(B_1 \uplus B_2)\,(y) &= B_1(y) + B_2(y) & y \in \text{dom}\, B_1 \wedge y \in \text{dom}\, B_2
\end{aligned}
$$

A bag may be used to record the number of occurrences of each product in a warehouse as part of an inventory system. It may model the number of items remaining for each product in a vending machine (Fig. 19.4).

The operation of a vending machine would require other operations such as identifying the set of acceptable coins, checking that the customer has entered sufficient coins to cover the cost of the good, returning change to the customer, and updating the quantity on hand of each good after a purchase. A more detailed examination is in [1].

19.7 Schemas and Schema Composition

The schemas in Z are visually striking and the specification is presented in two-dimensional graphic boxes. Schemas are used for specifying states and state transitions, and they employ notation to represent the before and after state (e.g., *s* and *s'* where *s'* represents the after state of *s*). They group all relevant information that belongs to a state description.

There are a number of useful schema operations such as schema inclusion, schema composition and the use of propositional connectives to link schemas together. The Δ convention indicates that the operation affects the state whereas the Ξ convention indicates that the state is not affected. These operations and conventions allow complex operations to be specified concisely, and assist with the readability of the specification. Schema composition is analogous to relational composition, and allows new schemas to be derived from existing schemas.

A schema name S_1 may be included in the declaration part of another schema S_2. The effect of the inclusion is that the declarations in S_1 are now part of S_2 and the predicates of S_1 are S_2 are joined together by conjunction. If the same variable is defined in both S_1 and S_2, then it must be of the same type in both schemas.

$$
\begin{array}{|l}
\text{-}S_1\text{------} \\
x,y : \mathbb{N} \\
\hline
x + y > 2 \\
\hline
\end{array}
\qquad
\begin{array}{|l}
\text{-}S_2\text{------} \\
S_1 ; z : \mathbb{N} \\
\hline
z = x + y \\
\hline
\end{array}
$$

The result is that S_2 includes the declarations and predicates of S_1 (Fig. 19.5):

Two schemas may be linked by propositional connectives such as $S_1 \wedge S_2$, $S_1 \vee S_2$, $S_1 \Rightarrow S_2$, and $S_1 \Leftrightarrow S_2$. The schema $S_1 \vee S_2$ is formed by merging the declaration parts of S_1 and S_2, and then combining their predicates by the logical $\vee$ operator. For example, $S = S_1 \vee S_2$ yields (Fig. 19.6):

Schema inclusion and the linking of schemas use normalization to convert sub-types to maximal types, and predicates are employed to restrict the maximal type to the sub-type. This involves replacing declarations of variables (e.g., u: 1.35 with u: Z, and adding the predicate $u > 0$ and $u < 36$ to the predicate part of the schema).

$$
\begin{array}{|l}
\text{-}S_2\text{------} \\
x,y : \mathbb{N} \\
z : \mathbb{N} \\
\hline
x + y > 2 \\
z = x + y \\
\hline
\end{array}
$$

Fig. 19.5 Schema inclusion

$$
\begin{array}{|l}
\text{-}S\text{------} \\
x,y : \mathbb{N} \\
z : \mathbb{N} \\
\hline
x + y > 2 \vee z = x + y \\
\hline
\end{array}
$$

Fig. 19.6 Merging schemas ($S_1 \vee S_2$)

The Δ and Ξ conventions are used extensively, and the notation Δ *TempMap* is used in the specification of schemas that involve a change of state. The notation Δ *TempMap* represents:

$$\Delta\,TempMap = TempMap \land TempMap'$$

The longer form of Δ *TempMap* is written as

$$
\begin{array}{|l}
-\Delta TempMap\text{———} \\
CityList,\ CityList' : \mathbb{P}\ City \\
temp,\ temp' : City \nrightarrow Z \\
\hline
\text{dom } temp = CityList \\
\text{dom } temp' = CityList' \\
\end{array}
$$

The notation Ξ *TempMap* is used in the specification of operations that do not involve a change to the state.

$$
\begin{array}{|l}
-\Xi\ TempMap\text{————} \\
\Delta TempMap \\
\hline
CityList = CityList' \\
temp = temp' \\
\end{array}
$$

Schema composition is analogous to relational composition and it allows new specifications to be built from existing ones. It allows the after state variables of one schema to be related with the before variables of another schema. The composition of two schemas S and T (S; T) is described in detail in [1] and involves 4 steps (Table 19.1):

The example below should make schema composition clearer. Consider the composition of S and T where S and T are defined as follows

$$
\begin{array}{|l}
-S\text{———} \\
x,x',y? : \mathbb{N} \\
\hline
x' = y? - 2 \\
\end{array}
\qquad
\begin{array}{|l}
-T\text{———} \\
x,x' : \mathbb{N} \\
\hline
x' = x + 1 \\
\end{array}
$$

$$
\begin{array}{|l}
-S_1\text{———} \\
x,x^+,y? : \mathbb{N} \\
\hline
x^+ = y? - 2 \\
\end{array}
\qquad
\begin{array}{|l}
-T_1\text{———} \\
x^+,x' : \mathbb{N} \\
\hline
x' = x^+ + 1 \\
\end{array}
$$

S_1 and T_1 represent the results of Step 1 and Step 2, with x' renamed to x^+ in S, and x renamed to x^+ in T. Step 3 and Step 4 yield (Fig. 19.7).:

Schema composition is useful as it allows new specifications to be created from existing ones.

Table 19.1 Schema composition

Step	Procedure
1.	Rename all *after* state variables in S to something new: S $[s^+/s']$
2.	Rename all *before* state variables in T to the same new thing: i.e., T $[s^+/s]$
3.	Form the conjunction of the two new schemas: S $[s^+/s'] \wedge T [s^+/s]$
4.	Hide the variable introduced in Steps 1 and 2. S; T = (S $[s^+/s'] \wedge T [s^+/s])\backslash(s^+)$

$$\frac{-S_1 \wedge T_1}{x,x^+,x',y? : \mathbb{N}}$$

$$\frac{-S ; T}{x, x', y? : \mathbb{N}}$$

$$\begin{vmatrix} x^+ = y? - 2 \\ x' = x^+ + 1 \end{vmatrix}$$

$$\begin{vmatrix} \exists x^+ : \mathbb{N} \bullet \\ (x^+ = y? - 2 \\ x' = x^+ + 1) \end{vmatrix}$$

Fig. 19.7 Schema composition

19.8   Reification and Decomposition

A Z specification involves defining the state of the system and then specifying the required operations. The Z specification language employs many constructs that are not part of conventional programming languages, and a Z specification is therefore not directly executable on a computer. A programmer implements the formal specification, and mathematical proof may be employed to prove that a program meets its specification.

Often, there is a need to write an intermediate specification that is between the original Z specification and the eventual program code. This intermediate specification is more algorithmic and uses less abstract data types than the Z specification. The intermediate specification is termed the design and the design needs to be correct with respect to the specification, and the program needs to be correct with respect to the design. The design is a refinement (reification) of the state of the specification, and the operations of the specification have been decomposed into those of the design.

The representation of an abstract data type such as a set by a sequence is termed data reification, and data reification is concerned with the process of transforming an abstract data type into a concrete data type. The abstract and concrete data types are related by the retrieve function, and the retrieve function maps the concrete data type to the abstract data type. There are typically several possible concrete data types for a particular abstract data type (i.e., refinement is a relation), whereas there is one abstract data type for a concrete data type (i.e., retrieval is a function). For example, sets are often reified to unique sequences; however, more than one unique sequence can represent a set whereas a unique sequence represents exactly one set.

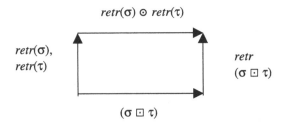

Fig. 19.8 Refinement commuting diagram

The operations defined on the concrete data type are related to the operations defined on the abstract data type. That is, the commuting diagram property is required to hold (Fig. 19.8). That is, for an operation $\boxdot$ on the concrete data type to correctly model the operation $\odot$ on the abstract data type the following diagram must commute, and the commuting diagram property requires proof. That is, it is required to prove that:

$$ret(\sigma \boxdot \tau) = (ret\ \sigma) \odot (ret\ \tau)$$

In Z, the refinement and decomposition is done with schemas. It is required to prove that the concrete schema is a valid refinement of the abstract schema, and this gives rise to a number of proof obligations. It needs to be proved that the initial states correspond to one another, and that each operation in the concrete schema is correct with respect to the operation in the abstract schema, and also that it is applicable (i.e., whenever the abstract operation may be performed the concrete operation may also be performed).

19.9 Proof in Z

Mathematicians perform rigorous proof of theorems using technical and natural language. Logicians employ formal proofs to prove theorems using propositional and predicate calculus. Formal proofs generally involve a long chain of reasoning with every step of the proof justified. Rigorous proofs involve precise reasoning using a mixture of natural and mathematical language. Rigorous proofs [1] have been described as being analogous to high level programming languages, whereas formal proofs are analogous to machine language.

A mathematical proof includes natural language and mathematical symbols, and often many of the tedious details of the proof are omitted. Many proofs in formal methods such as Z are concerned with crosschecking on the details of the specification, or on the validity of the refinement step, or proofs that certain properties are satisfied by the specification. There are often many tedious lemmas to be proved, and tool support is essential as proof by hand often contain errors or jumps in reasoning. Machine proofs are lengthy and largely unreadable; however, they provide extra confidence as every step in the proof is justified.

The proof of various properties about the programs increases confidence in its correctness.

19.10 Review Questions

1. Describe the main features of the Z specification language.

2. Explain the difference between $\mathbb{P}_1 X$, $\mathbb{P} X$ and FX.
3. Give an example of a set derived from another set using set comprehension. Explain the three main parts of set comprehension in Z.
4. Discuss the applications of Z and which areas have benefited most from their use? What problems have arisen?
5. Give examples to illustrate the use of domain and range restriction operators and domain and range anti-restriction operators with relations in Z.
6. Give examples to illustrate relational composition.
7. Explain the difference between a partial and total function, and give examples to illustrate function override.
8. Give examples to illustrate the various operations on sequences including concatenation, head, tail, map and reverse operations.
9. Give examples to illustrate the various operations on bags.
10. Discuss the nature of proof in Z and tools to support proof.
11. Explain the process of refining an abstract schema to a more concrete representation, the proof obligations that are generated, and the commuting diagram property.

19.11 Summary

Z is a formal specification language that was developed in the early 1980s at Oxford University in England. It has been employed in both industry and academia, and it was use successfully on the IBM's CICS project. Its specifications are mathematical, and this leads to more rigorous software development. Its mathematical approach allows properties to be proved about the specification, and any gaps or inconsistencies in the specification may be identified.

Z is a 'model oriented' approach and an explicit model of the state of an abstract machine is given, and the operations are defined in terms of their effect on the state. Its main features include a mathematical notation that is similar to VDM, and the schema calculus. The latter consists essentially of boxes and are used to describe operations and states.

The schema calculus enables schemas to be used as building blocks to form larger specifications. It is a powerful means of decomposing a specification into smaller pieces, and helps with the readability of Z specifications, as each individual schema is small in size and self-contained.

Z is a highly expressive specification language, and it includes notation for sets, functions, relations, bags, sequences, predicate calculus and schema calculus. Z specifications are not directly executable as many of its data types and constructs are not part of modern programming languages. Therefore, there is a need to refine the Z specification into a more concrete representation, and prove that the refinement is valid.

Reference

1. Z. An Introduction to Formal Methods. Antoni Diller. John Wiley and Sons. England. 1990.

Probability, Statistics and Applications

20

Key Topics

Sample Spaces
Random Variables
Mean, Mode and Median
Variance
Normal Distributions
Histograms
Hypothesis Testing
Software Reliability Models
Queueing Theory

20.1 Introduction

Statistics is an empirical science that is concerned with the collection, organization, analysis, interpretation and presentation of data. The data collection needs to be planned and this may include surveys and experiments. Statistics are widely used by government and industrial organizations, and they may be employed for forecasting as well as for presenting trends. They allow the behaviour of a population to be studied and inferences to be made about the population. These inferences may be tested (*hypothesis testing*) to ensure their validity.

The analysis of statistical data allows an organization to understand its performance in key areas, and to identify problematic areas. Organizations will often examine performance trends over time, and will devise appropriate plans and

© Springer International Publishing Switzerland 2016
G. O'Regan, *Guide to Discrete Mathematics*, Texts in Computer Science,
DOI 10.1007/978-3-319-44561-8_20

actions to address problematic areas. The effectiveness of the actions taken will be judged by improvements in performance trends over time.

It is often not possible to study the entire population, and instead a representative subset or sample of the population is chosen. This *random sample* is used to make inferences regarding the entire population, and it is essential that the sample chosen is indeed random and representative of the entire population. Otherwise, the inferences made regarding the entire population will be invalid.

A statistical experiment is a causality study that aims to draw a conclusion regarding values of a *predictor variable*(s) on a *response variable*(s). For example, a statistical experiment in the medical field may be conducted to determine if there is a causal relationship between the use of a particular drug and the treatment of a medical condition such as lowering of cholesterol in the population. A statistical experiment involves the following:

- Planning the research
- Designing the experiment
- Performing the experiment
- Analyzing the results
- Presenting the results

Probability is a way of expressing the likelihood of a particular event occurring. It is normal to distinguish between the frequency interpretation and the subjective interpretation of probability [1]. For example, if a geologist states that 'there is a 70 % chance of finding gas in a certain region' then this statement is usually interpreted in two ways:

- The geologist is of the view that over the long run 70 % of the regions whose environment conditions are very similar to the region under consideration have gas (*Frequency Interpretation*).
- The geologist is of the view that it is likely that the region contains gas, and that 0.7 is a measure of the geologist's belief in this hypothesis (*Personal Interpretation*).

However, the mathematics of probability is the same for both the frequency and personal interpretation.

20.2　Probability Theory

Probability theory provides a mathematical indication of the likelihood of an event occurring, and the probability of an event is a numerical value between 0 and 1. A probability of 0 indicates that the event cannot occur whereas a probability of 1 indicates that the event is guaranteed to occur. If the probability of an event is greater than 0.5 then this indicates that the event is more likely to occur than not to occur.

A *sample space* is the set of all possible outcomes of an experiment, and an *event* E is a subset of the sample space. For example, the sample space for the experiment of tossing a coin is the set of all possible outcomes of this experiment: i.e., head or tails. The event that the toss results a tail is a subset of the sample space.

$$S = \{h, t\} \qquad E = \{t\}$$

Similarly, the sample space for the gender of a newborn baby is the set of outcomes: i.e., the newborn baby is a boy or a girl. The event that the baby is a girl is a subset of the sample space.

$$S = \{b, g\} \qquad E = \{g\}$$

For any two events E and F of a sample space S, we can also consider the union and intersection of these events. That is,

- E $\cup$ F consists of all outcomes that are in E or F or both.
- E $\cap$ F (normally written as EF) consists of all outcomes that are in both E and F.
- E^c denotes the complement of E with respect to S and represents the outcomes of S that are not in E.

If EF = $\emptyset$ then there are no outcomes in both E and F, and so the two events E and F are mutually exclusive. The union and intersection of two events can be extended to the union and intersection of a family of events E_1, E_2, ..., E_n (i.e., $\cup_{i=1}^{n} E_i$ and $\cap_{i=1}^{n} E_i$).

20.2.1 Laws of Probability

The laws of probability essentially state that the probability of an event is between 0 and 1, and that the probability of the union of a mutually disjoint set of events is the sum of their individual probabilities.

i. P(S) = 1
ii. P($\emptyset$) = 0
iii. $0 \leq P(E) \leq 1$
iv. For any sequence of mutually exclusive events E_1, E_2, ..., E_n. (i.e., $E_i E_j = \emptyset$ where $i \neq j$) then the probability of the union of these events is the sum of their individual probabilities: i.e.,

$$P\left(\bigcup_{i=1}^{n} E_i\right) = \sum_{i=1}^{n} P(E_i).$$

The probability of the union of two events (not necessarily disjoint) is given by:

$$P(E \cup F) = P(E) + P(F) - P(EF)$$

The probability of an event E not occurring is denoted by E^c and is given by $1 - P(E)$. The probability of an event E occurring given that an event F has occurred is termed the *conditional probability* (denoted by $P(E|F)$) and is given by

$$P(E|F) = \frac{P(EF)}{P(F)} \qquad \text{where } P(F) > 0$$

This formula allows us to deduce that

$$P(EF) = P(E|F)P(F)$$

Bayes formula enables the probability of an event E to be determined by a weighted average of the conditional probability of E given that the event F occurred and the conditional probability of E given that F has not occurred:

$$E = E \cap S = E \cap (F \cup F^c)$$
$$= EF \cup EF^c$$

$$\begin{aligned} P(E) &= P(EF) + P(EF^c) \qquad (\text{since } EF \cap EF^c = \varnothing) \\ &= P(E|F)P(F) + P(E|F^c)P(F^c) \\ &= P(E|F)P(F) + P(E|F^c)(1 - P(F)) \end{aligned}$$

Two events E, F are *independent* if the knowledge that F has occurred does not change the probability that E has occurred. That is, $P(E|F) = P(E)$ and since $P(E|F) = P(EF)/P(F)$ we have that two events E, F are independent if:

$$P(EF) = P(E)P(F)$$

Two events E and F that are not independent are said to be *dependent*.

20.2.2   Random Variables

Often, some numerical quantity determined by the result of the experiment is of interest rather than the result of the experiment itself. These numerical quantities are termed *random variables*. A random variable is termed *discrete* if it can take on a finite or countable number of values; otherwise it is termed *continuous*.

The *distribution function* of a random variable is the probability that the random variable X takes on a value less than or equal to x. It is given by

$$F(x) = P\{X \le x\}$$

All probability questions about X can be answered in terms of its distribution function F. For example, the computation of P $\{a < X < b\}$ is given by

$$P\{a < X < b\} = P\{X \le b\} - P\{X \le a\}$$
$$= F(b) - F(a)$$

The probability mass function for a discrete random variable X (denoted by $p(a)$) is the probability that it is a certain value. It is given by

$$p(a) = P\{X = a\}$$

Further, $F(a)$ can also be expressed in terms of the probability mass function

$$F(a) = \sum_{\forall x \le a} p(x)$$

We may also define a probability density function and a probability distribution function X for a continuous random variable X [2], and all probability statements about X can be answered in terms of its density function $f(x)$, and the derivative of the probability distribution function yields the probability density function.

The expected value (i.e., the *mean*) of a discrete random variable X (denoted E[X]) is given by the weighted average of the possible values of X, and the expected value of a function of a random variable is given by E[g(X)]. These are given by

$$E[X] = \sum_{iX_i} P\{X = x_i\}$$
$$E[g(X)] = \sum_{i} g(x_i) P\{X = x_i\}$$

The *variance* of a random variable is a measure of the spread of values from the mean, and is defined by

$$Var(X) = E[X^2] - (E[X])^2$$

The standard deviation σ is given by the square root of the variance. That is,

$$\sigma = \sqrt{Var(X)}$$

The *covariance* of two random variables is a measure of the relationship between two random variables X and Y, and indicates the extent to which they both change (in either similar or opposite ways) together. It is defined by

$$Cov(X, Y) = E[XY] - E[X]E[Y].$$

It follows that the covariance of two independent random variables is zero. Variance is a special case of covariance (when the two random variables are identical). This follows since $Cov(X, X) = E[X \cdot X] - (E[X])(E[X]) = E[X^2] - (E[X])^2 = Var(X)$.

A positive covariance $(Cov(X, Y) \geq 0)$ indicates that Y tends to increase as X does, whereas a negative covariance indicates that Y tends to decrease as X increases.

The *correlation* of two random variables is an indication of the relationship between two variables X and Y. If the correlation is negative then Y tends to decrease as X increases, and if it is positive number then Y tends to increase as X increases. The correlation coefficient is a value that is between ± 1 and it is defined by

$$Corr(X, Y) = \frac{Cov(X,Y)}{\sqrt{Var(X)Var(Y)}}$$

Once the correlation between two variables has been calculated the probability that the observed correlation was due to chance can be computed. This is to ensure that the observed correlation is a real one and not due to a chance occurrence.

There are a number of special random variables, and these include the Bernoulli trial, where there are just two possible outcomes of an experiment: i.e., success or failure. The probability of success and failure is given by

$$P\{X = 0\} = 1 - p$$
$$P\{X = 1\} = p$$

The mean of the Bernoulli distribution is given by p and the variance by $p(1 - p)$. The *Binomial distribution* involves n Bernoulli trials, each of which results in success or failure. The probability of i successes from n trials is then given by

$$P\{X = i\} = \binom{n}{i}p^i(1 - p)^{n-i}$$

with the mean of the Binomial distribution given by np, and the variance is given by $np(1 - p)$.

The *Poisson distribution* may be used as an approximation to the Binomial Distribution when n is large and p is small. The probability of i successes is given by

$$P\{X = i\} = e^{-\lambda}\lambda^i/i!$$

and the mean and variance of the Poisson distribution is given by λ.

There are many other well-known distributions such as the *hypergeometric distribution* that describes the probability of i successes in n draws from a finite population without replacement; the *uniform distribution*; the *exponential distribution*, the *normal distribution* and the *gamma* distribution. The mean and variance of important probability distributions are summarized in Table 20.1.

The reader is referred to [1] for a more detailed account of probability theory.

Table 20.1 Probability distributions

Distribution name	Density function	Mean/variance
Binomial	$P\{X = i\} = \binom{n}{i}p^i(1-p)^{n-i}$	$np, np(1-p)$
Poisson	$P\{X = i\} = e^{-\lambda}\lambda^i/i!$	λ, λ
Hypergeometric	$P\{X = i\} = \binom{N}{i}\binom{M}{n-i}/\binom{N+M}{n}$	$nN/N+M, np(1-p)[1-(n-1)/N+M-1]$
Uniform	$f(x) = 1/(\beta - \alpha)\alpha \leq x \leq \beta,\ 0$	$(\alpha + \beta)/2, (\beta - \alpha)^2/12$
Exponential	$f(x) = \lambda e^{-\lambda x}$	$1/\lambda, 1/\lambda^2$
Normal	$f(x) = \frac{1}{\sqrt{2\pi}\sigma}e^{-(x-\mu)^2/2\sigma^2}$	μ, σ^2
Gamma	$f(x) = \lambda e^{-\lambda x}(\lambda x)^{\alpha-1}/\Gamma(\alpha)\ (x \geq 0).$	$\alpha/\lambda, \alpha/\lambda^2$

20.3   Statistics

The field of statistics is concerned with summarizing, digesting and extracting information from large quantities of data. Statistics provide a collection of methods for planning an experiment, and analyzing data to draw accurate conclusions from the experiment. We distinguish between descriptive statistics and inferential statistics:

Descriptive Statistics
This is concerned with describing the information in a set of data elements in graphical format, or by describing its distribution.

Inferential Statistics
This is concerned with making inferences with respect to the population by using information gathered in the sample.

20.3.1   Abuse of Statistics

Statistics are extremely useful in drawing conclusions about a population. However, it is essential that the random sample is valid and that the experiment is properly conducted to enable valid conclusions to be inferred. Some examples of the abuse of statistics include

- The sample size may be too small to draw conclusions.
- It may not be a genuine random sample of the population.
- Graphs may be drawn to exaggerate small differences.
- Area may be misused in representing proportions.
- Misleading percentages may be used.

The quantitative data used in statistics may be discrete or continuous. *Discrete data* is numerical data that has a finite number of possible values, and *continuous data* is numerical data that has an infinite number of possible values.

20.3.2   Statistical Sampling

Statistical sampling is concerned with the methodology of choosing a random sample of a population, and the study of the sample with the goal of drawing valid conclusions about the entire population. The assumption is that if a genuine representative sample of the population is chosen, then a detailed study of the sample will provide insight into the whole population. This helps to avoid a lengthy expensive (and potentially infeasible) study of the entire population.

The sample chosen must be random and the sample size must be sufficiently large to enable valid conclusions to be made for the entire population.

Random Sample

A *random sample* is a sample of the population such that each member of the population has an equal chance of being chosen.

There are various ways of generating a random sample from the population including (Table 20.2).

Once the random sample group has been chosen the next step is to obtain the required information from the sample. This may be done by interviewing each member in the sample; calling each member; conducting a mail survey and so on (Table 20.3).

Table 20.2 Sampling techniques

Sampling technique	Description
Systematic sampling	Every kth member of the population is sampled
Stratified sampling	The population is divided into two or more strata and each subpopulation (stratum) is then sampled. Each element in the subpopulation shares the same characteristics (e.g., age groups, gender)
Cluster sampling	A population is divided into clusters and a few of these clusters are exhaustively sampled (i.e., every element in the cluster is considered)
Convenience sampling	Sampling is done as convenient and often allows the element to choose whether or not it is sampled

Table 20.3 Types of survey

Survey type	Description
Direct measurement	This may involve a direct measurement of all in the sample (e.g., the height of students in a class)
Mail survey	This involves sending a mail survey to the sample. This may have a lower response rate and may thereby invalidate the findings
Phone survey	This is a reasonably efficient and cost effective way to gather data. However, refusals or hang-ups may affect the outcome
Personal interview	This tends to be expensive and time consuming, but it allows detailed information to be collected
Observational study	An observational study allows individuals to be studied, and the variables of interest to be measured
Experiment	An experiment imposes some treatment on individuals in order to study the response

20.3.3 Averages in a Sample

The term 'average' generally refers to the arithmetic *mean* of a sample, but it may also refer to the statistical *mode* or *median* of the sample. These terms are defined below:

Mean
The *arithmetic mean* of a set of n numbers is defined to be the sum of the numbers divided by n. That is, the arithmetic mean for a sample of size n is given by

$$\bar{x} = \frac{\sum_{i=1}^{n} x_i}{n}$$

The actual mean of the population is denoted by μ, and it may differ from the sample mean.

Mode
The mode is the data element that occurs most frequently in the sample. It is possible that two elements occur with the same frequency, and if this is the case then we are dealing with a bi-modal or possibly a multi-modal sample.

Median
The median is the middle element when the data set is arranged in increasing order of magnitude.
   If there are an odd number of elements in the sample the median is the middle element. Otherwise, the median is the arithmetic mean of the two middle elements.

Mid Range
The midrange is the arithmetic mean of the highest and lowest data elements in the sample. That is, $(x_{max} + x_{min})/2$.
   The arithmetic mean is the most widely used average in statistics.

20.3.4 Variance and Standard Deviation

An important characteristic of a sample is its distribution, and the spread of each element from some measure of central tendency (e.g., the mean). One elementary measure of dispersion is that of the sample *range*, and it is defined to be the difference between the maximum and minimum value in the sample. That is, the sample range is defined to be

$$\text{range} = x_{max} - x_{min}.$$

The sample range is not a reliable measure of dispersion as only two elements in the sample are used, and extreme values in the sample can distort the range to be very large even if most of the elements are quite close to one another.

The standard deviation is the most common way to measure dispersion, and it gives the average distance of each element in the sample from the mean. The sample standard deviation is denoted by s and is defined by

$$s = \sqrt{\frac{\sum (x_i - \bar{x})^2}{n - 1}}$$

The population standard deviation is denoted by σ and is defined by

$$\sigma = \sqrt{\frac{\sum (x_i - \mu)^2}{N}}$$

Variance is another measure of dispersion and it is defined as the square of the standard deviation. The sample variance is given by

$$s^2 = \frac{\sum (x_i - \bar{x})^2}{n - 1}$$

The population variance is given by

$$\sigma^2 = \frac{\sum (x_i - \mu)^2}{N}$$

20.3.5 Bell-Shaped (Normal) Distribution

The German mathematician Gauss (Fig. 20.1) originally studied the normal distribution, and it is also known as the *Gaussian distribution* (Fig. 20.2). It is shaped like a bell and so is popularly known as the *bell-shaped* distribution. The empirical frequencies of many natural populations exhibit a bell-shaped (*normal*) curve.

The *normal distribution* N has mean μ, and standard deviation σ. Its density function $f(x)$ where (where $-\infty < x < \infty$) is given by

Fig. 20.1 Carl Friedrich Gauss

Fig. 20.2 Standard normal
bell curve (Gaussian
distribution)

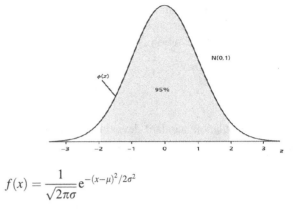

$$f(x) = \frac{1}{\sqrt{2\pi}\sigma} e^{-(x-\mu)^2/2\sigma^2}$$

The *unit* (or *standard*) normal distribution $Z(0, 1)$ has mean 0 and standard deviation of 1. Every normal distribution may be converted to the unit normal distribution by $Z = (X - \mu)/\sigma$, and every probability statement about X

$$f(y) = \frac{1}{\sqrt{2\pi}} e^{-y^2/2}$$

has an equivalent probability statement about Z. The unit normal density function is given by

For a normal distribution 68.2 % of the data elements lie within one standard deviation of the mean; 95.4 % of the population lies within two standard deviations of the mean; and 99.7 % of the data lies within three standard deviations of the mean. For example, the shaded area under the curve within two standard deviations of the mean represents 95 % of the population.

A fundamental result in probability theory is the *Central Limit Theorem*, and this theorem essentially states that the sum of a large number of independent and identically distributed random variables has a distribution that is approximately normal. That is, suppose X_1, X_2, ..., X_n is a sequence of independent random variables each with mean μ and variance σ^2. Then for large n the distribution of

$$\frac{x_1 + x_2 + \cdots + x_n - n\mu}{\sigma\sqrt{n}}$$

is approximately that of a unit normal variable Z. One application of the central limit theorem is in relation to the binomial random variables, where a binomial random variable with parameters (n, p) represents the number of successes of n independent trials, where each trial has a probability of p of success. This may be expressed as

$$X = X_1 + X_2 + \cdots + X_n$$

where $X_i = 1$ if the ith trial is a success and is 0 otherwise. $E(X_i) = p$ and $Var(X_i) = p(1 - p)$, and then by applying the central limit theorem it follows that for large n

$$\frac{X - np}{\sqrt{np(1 - p)}}$$

will be approximately a unit normal variable (which becomes more normal as n becomes larger).

The sum of independent normal random variables is normally distributed, and it can be shown that the sample average of X_1, X_2, ..., X_n is normal, with a mean equal to the population mean but with a variance reduced by a factor of $1/n$.

$$E(\bar{X}) = \sum_{i=1}^{n} \frac{E(X_i)}{n} = \mu$$

$$Var(\bar{X}) = \frac{1}{n^2} \sum_{i=1}^{n} Var(X_i) = \frac{\sigma^2}{n}$$

It follows that from this that the following is a unit normal random variable.

$$\sqrt{n} \frac{(X - \mu)}{\sigma}$$

The term *six-sigma* (6σ) is a methodology concerned with continuous process improvement and aims for very high quality (close to perfection). A 6σ process is one in which 99.9996 % of the products are expected to be free from defects (3.4 defects per million).

20.3.6 Frequency Tables, Histograms and Pie Charts

A frequency table is used to present or summarize data (Tables 20.4 and 20.5). It lists the data classes (or categories) in one column and the frequency of the category in another column.

A histogram is a way to represent data in bar chart format (Fig. 20.3). The data is divided into intervals where an interval is a certain range of values. The horizontal axis of the histogram contains the intervals (also known as buckets) and the vertical axis shows the frequency (or relative frequency) of each interval. The bars represent the frequency and there is no space between the bars.

Table 20.4 Frequency table —salary

Profession	Salary	Frequency
Project manager	65,000	3
Architect	65,000	1
Programmer	50,000	8
Tester	45,000	2
Director	90,000	1

Table 20.5 Frequency table
—test results

Mark	Frequency
0–24	3
25–49	10
50–74	15
75–100	2

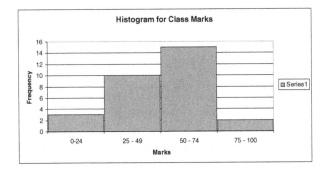

Fig. 20.3   Histogram test results

Fig. 20.4   Pie chart test
results

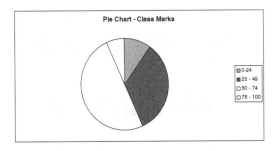

A histogram has an associated shape. For example, it may resemble a normal distribution, a bi-modal or multi-modal distribution. It may be positively or negatively skewed. The construction of a histogram first involves the construction of a frequency table where the data is divided into disjoint classes and the frequency of each class is determined.

A pie chart (Fig. 20.4) offers an alternate way to histograms in the presentation of data. A frequency table is first constructed, and the pie chart presents a visual representation of the percentage in each data class.

20.3.7   Hypothesis Testing

The basic concept of inferential statistics is *hypothesis testing*, where a hypothesis is a statement about a particular population whose truth or falsity is unknown.

Hypothesis testing is concerned with determining whether the values of the random sample from the population are consistent with the hypothesis. There are two mutually exclusive hypotheses: one of these is the null hypothesis H_0 and the other is the alternate research hypothesis H_1. The null hypothesis H_0 is what the researcher is hoping to reject, and the research hypothesis H_1 is what the researcher is hoping to accept.

Statistical testing is then employed to test the hypothesis, and the result of the test is that we either reject the null hypothesis (and therefore accept the alternative hypothesis), or that we fail to reject (i.e., we accept) the null hypothesis. The rejection of the null hypothesis means that the null hypothesis is highly unlikely to be true, and that the research hypothesis should be accepted.

Statistical testing is conducted at a certain level of significance, with the probability of the null hypothesis H_0 being rejected when it is true never greater than α. The value α is called the level of significance of the test, with α usually being 0.1, 0.05, 0.005. A significance level β may also be applied to with respect to accepting the null hypothesis H_0 when H_0 is false, and usually $\alpha = \beta$.

The objective of a statistical test is not to determine whether or not H_0 is actually true, but rather to determine whether its validity is consistent with the observed data. That is, H_0 should only be rejected if the resultant data is very unlikely if H_0 is true.

The errors that can occur with hypothesis testing include type 1 and type 2 errors. Type 1 errors occur when we reject the null hypothesis when the null hypothesis is actually true. Type 2 errors occur when we accept the null hypothesis when the null hypothesis is false (Table 20.6).

For example, an example of a false positive is where the results of a blood test comes back positive to indicate that a person has a particular disease when in fact the person does not have the disease. Similarly, an example of a false negative is where a blood test is negative indicating that a person does not have a particular disease when in fact the person does. Both errors can potentially be very serious.

The terms α and β represent the level of significance that will be accepted, and normally $\alpha = \beta$. In other words, α is the probability that we will reject the null hypothesis when the null hypothesis is true, and β is the probability that we will accept the null hypothesis when the null hypothesis is false.

Testing a hypothesis at the $\alpha = 0.05$ level is equivalent to establishing a 95 % confidence interval. For 99 % confidence α will be 0.01, and for 99.999 % confidence then α will be 0.00001.

Table 20.6 Hypothesis testing

Action	H_0 true, H_1 false	H_0 false, H_1 true
Reject H_1	Correct	False positive—type 2 error $P(\text{accept } H_0 | H_0 \text{ false}) = \beta$
Reject H_0	False negative—type 1 error $P(\text{reject } H_0 | H_0 \text{ true}) = \alpha$	Correct

The hypothesis may be concerned with testing a specific statement about the value of an unknown parameter θ of the population. This test is to be done at a certain level of significance, and the unknown parameter may, for example, be the mean or variance of the population. An estimator for the unknown parameter is determined, and the hypothesis that this is an accurate estimate is rejected if the random sample is not consistent with it. Otherwise, it is accepted.

The steps involved in hypothesis testing include the following:

1. Establish the null and alternative hypothesis,
2. Establish error levels (significance),
3. Compute the test statistics (often a t-test),
4. Decide on whether to accept or reject the null hypothesis.

The difference between the observed and expected test statistic, and whether the difference could be accounted for by normal sampling fluctuations is the key to the acceptance or rejection of the null hypothesis.

20.4 Software Reliability

The design and development of high-quality software has become increasingly important for society. Many software companies desire a sound mechanism to predict the reliability of their software prior to its deployment at the customer site, and this has led to a growing interest in software reliability models.

Definition 12.1 (*Software Reliability*) *Software reliability* is defined as the probability that the program works without failure for a specified length of time, and is a statement of the future behaviour of the software. It is generally expressed in terms of the *mean time to failure* (MTTF) or the *mean time between failure* (MTBF).

Statistical sampling techniques are often employed to predict the reliability of hardware, as it is not feasible to test all items in a production environment. The quality of the sample is then used to make inferences on the quality of the entire population, and this approach is effective in manufacturing environments where variations in the manufacturing process often lead to defects in the physical products.

There are similarities and differences between hardware and software reliability. A hardware failure may arise due to a component wearing out due to its age, and often a replacement is required. Most hardware components are expected to last for a certain period of time, and the variation in the failure rate of a hardware component are often due to the manufacturing process and to the operating environment of the component. Good hardware reliability predictors have been developed, and each hardware component has an expected mean time to failure. The reliability of a product may be determined from the reliability of the individual components of the hardware.

Software is an intellectual undertaking involving a team of designers and programmers. It does not physically wear out and software failures manifest themselves from particular user inputs. Each copy of the software code is identical and the software is either correct or incorrect. That is, software failures are due to design and implementation errors rather than to physically wearing out. The software community has not yet developed a sound software reliability predictor model.

The software population to be sampled consists of all possible execution paths of the software, and since this is potentially infinite it is generally not possible to perform exhaustive testing.

The way in which the software is used (i.e., the inputs entered by the users) will impact upon its perceived reliability. Let I_f represent the fault set of inputs (i.e., $i_f \in I_f$ if and only if the input of i_f by the user leads to failure). The randomness of the time to software failure is due to the unpredictability in the selection of an input $i_f \in I_f$... It may be that the elements in I_f are inputs that are rarely used, and that therefore the software will be perceived as reliable.

Statistical testing may be used to make inferences on the future performance of the software. This requires an understanding of the expected usage profile of the system, as well as the population of all possible usages of the software. The sampling is done in accordance with the expected usage profile.

20.4.1 Software Reliability and Defects

The release of an unreliable software product may result in damage to property or injury (including loss of life) to a third party. Consequently, companies need to be confident that their software products are fit for use prior to their release. The project team needs to conduct extensive inspections and testing of the software prior to its release.

Objective product quality criteria may be set (e.g., 100 % of tests performed and passed) to be satisfied prior to release. This provides a degree of confidence that the software has the desired quality, and is safe and fit for purpose. However, these results are historical in the sense that they are a statement of past and present quality. The question is whether the past behaviour provides a sound indication of future behaviour.

Software reliability models are an attempt to predict the future reliability of the software, and to assist in deciding on whether the software is ready for release.

A defect does not always result in a failure, as it may be benign and may occur on a rarely used execution path. Many observed failures arise from a small proportion of the existing defects. Adam's 1984 case study [3] indicated that over 33 % of the defects led to an observed failure with mean time to failure greater than 5000 years; whereas less than 2 % of defects led to an observed failure with a mean time to failure of less than 50 years. This suggests that a small proportion of defects led to almost all of the observed failures (Table 20.7).

Table 20.7 Adam's 1984 study of software failures of IBM products

	Rare				Frequent			
	1	2	3	4	5	6	7	8
MTTF (years)	5,000	1,580	500	158	50	15.8	5	1.58
Avg. % fixes	33.4	28.2	18.7	10.6	5.2	2.5	1.0	0.4
Prob failure	0.008	0.021	0.044	0.079	0.123	0.187	0.237	0.300

The analysis shows that 61.6 % of all fixes (Group 1. and 2.) were made for failures that will be observed less than once in 1580 years of expected use, and that these constitute only 2.9 % of the failures observed by typical users. On the other hand, groups 7 and 8 constitute 53.7 % of the failures observed by typical users and only 1.4 % of fixes.

This showed that *coverage testing* is not cost effective in increasing MTTF. *Usage testing*, in contrast, would allocate 53.7 % of the test effort to fixes that will occur 53.7 % of the time for a typical user. Harlan Mills has argued [4] that the data in the table shows that usage testing is 21 times more effective than coverage testing.

There is a need to be careful with *reliability growth models*, as there is no tangible growth in reliability unless the corrected defects are likely to manifest themselves as a failure.[1] Many existing software reliability growth models assume that all remaining defects in the software have an equal probability of failure, and that the correction of a defect leads to an increase in software reliability. These assumptions are questionable.

The defect count and defect density may be poor predictors of operational reliability, and an emphasis on removing a large number of defects from the software may not be sufficient in itself to achieve high reliability.

The correction of defects in the software leads to newer versions of the software, and reliability models assume reliability growth: i.e., the new version is more reliable than the older version as several identified defects have been corrected. However, in some sectors such as the safety critical field the view is that the new version of a program is a new entity, and that no inferences may be drawn until further investigation has been done. The relationship between the new version and the previous version of the software needs to be considered (Table 20.8).

The safety critical industry (e.g., the nuclear power industry) takes the conservative viewpoint that any change to a program creates a new program. The new program is therefore required to demonstrate its reliability.

---

[1]We are assuming that the defect has been corrected perfectly with no new defects introduced by the changes made.

Table 20.8 New and old version of software	Similarities and differences between new/old version
	• The new version of the software is identical to the previous version except that the identified defects have been corrected
	• The new version of the software is identical to the previous version, except that the identified defects have been corrected, but the developers have introduced some new defects
	• No assumptions can be made about the behaviour of the new version of the software until further data is obtained

20.4.2 Cleanroom Methodology

Harlan Mills and others at IBM developed the Cleanroom methodology to assist in the development of high-quality software. The software is released only when the probability of zero-defects is very high.

The way in which the software is used will impact upon its perceived quality and reliability. Failures will manifest themselves on certain input sequences only, and as users will generally employ different input sequences, each user will have a different perception of the reliability of the software. Knowledge of the way that the software will be used allows the software testing to be focused on verifying the correctness of the common everyday tasks carried out by users.

This means that it is important to determine the operational profile of users to allow effective testing of the software to take place. The operational environment may not be stable as users may potentially change their behaviour over time. The collection of operational data involves identifying the operations to be performed and the probability of that operation being performed.

The Cleanroom approach [4] applies statistical techniques to enable a software reliability measure to be calculated, and it is based on the expected usage of the software. It employs *statistical usage testing* rather than coverage testing, and applies statistical quality control to certify the mean time to failure of the software. The statistical usage testing involves executing tests chosen from the population of all possible uses of the software in accordance with the probability of expected use.

Coverage testing involves designing tests that cover every path through the program, and this type of testing is as likely to find a rare execution failure as well as a frequent execution failure. It is highly desirable to find failures that occur on frequently used parts of the system.

The advantage of usage testing (that matches the actual execution profile of the software) is that it has a better chance of finding execution failures on frequently used parts of the system. This helps to maximize the expected mean time to failure.

20.4.3 Software Reliability Models

Models are simplifications of the reality and a good model allows accurate predictions of future behaviour to be made. The adequacy of the model is judged by

Table 20.9 Characteristics
of good software reliability
model

Characteristics of good software reliability model
Good theoretical foundation
Realistic assumptions
Good empirical support
As simple as possible (Ockham's razor)
Trustworthy and accurate

model exploration, and determining if its predictions are close to the actual manifested behaviour. More accurate models are sought to replace inadequate models.

A model is judged effective if there is good empirical evidence to support it. Models are often modified (or replaced) over time, as further facts and observations lead to aberrations that cannot be explained by the current model. A good software reliability model will have the following characteristics (Table 20.9):

There are several software reliability predictor models employed (with varying degrees of success). Some of them just compute defect counts rather than estimating software reliability in terms of mean time to failure. They include (Table 20.10):

- *Size and Complexity Metrics*

These are used to predict the number of defects that a system will reveal in operation or testing.

- *Operational Usage Profile*

These predict failure rates based on the expected operational usage profile of the system. The number of failures encountered is determined and the software reliability predicted.

- *Quality of the Development Process*

These predict failure rates based on the process maturity of the software development process in the organization.

The extent to which the software reliability model can be trusted depends on the accuracy of its predictions. Empirical data will need to be gathered to determine the accuracy of the predictions. It may be acceptable to have a little inaccuracy during the early stages of prediction, provided the predictions of operational reliability are close to the observations. A model that gives overly optimistic results is termed 'optimistic,' whereas a model that gives overly pessimistic results is termed 'pessimistic.'

Table 20.10 Software reliability models

Model	Description	Comments
Jelinski/moranda model	The failure rate is a Poisson process and is proportional to the current defect content of program. The initial defect count is N; the initial failure rate is $N\varphi$; it decreases to $(N - 1)\varphi$ after the first fault is detected and eliminated, and so on. The constant φ is termed the proportionality constant	Assumes defects corrected perfectly and no new defects are introduced Assumes each fault contributes the same amount to failure rate
Littlewood/verrall model	Successive execution time between failures independent exponentially distributed random variables. Software failures are the result of the particular inputs and faults introduced from the correction of defects	Does not assume perfect correction of defects
Seeding and Tagging	This is analogous to estimating the fish population of a lake (Mills). A known number of defects is inserted into a software program and the proportion of these identified during testing determined Another approach (Hyman) is to regard the defects found by one tester as tagged and then to determine the proportion of tagged defects found by a second independent tester	Estimate of the total number of defects in the software but not a not s/w reliability predictor Assumes all faults equally likely to be found and introduced faults representative of existing
Generalized Poisson Model	The number of failures observed in ith time interval τ_i has a Poisson distribution with mean $\phi(N - M_{i-1})$ τ_i^{α} where N is the initial number of faults; M_{i-1} is the total number of faults removed up to the end of the $(i - 1)$th time interval; and ϕ is the proportionality constant	Assumes faults removed perfectly at end of time interval

The assumptions in the reliability model need to be examined to determine whether they are realistic. Several software reliability models have questionable assumptions such as

- All defects are corrected perfectly
- Defects are independent of one another
- Failure rate decreases as defects are corrected.
- Each fault contributes the same amount to the failure rate

20.5 Queuing Theory

The term '*queue*' refers to waiting in line for a service, such as waiting in line at a bakery or a bank, and *queuing theory* is the mathematical study of waiting lines or queues. The origins of queuing theory are in work done by Erlang at the Copenhagen Telephone Exchange in the early twentieth century where he modelled the number of telephone calls arriving as a Poisson process.

Queuing theory has been applied to many fields including telecommunications and traffic management. This section aims to give a flavour and a very short introduction to queuing theory, and it has been adapted from [5]. The interested reader may consult the many other texts available for more detailed information [e.g., 6].

A supermarket may be used to illustrate the ideas of queuing theory, as it has a large population of customers some of whom may enter the supermarket and queuing system (the checkout queues). Customers will generally wait for a period of time in a queue before receiving service at the checkout, and they wait for a further period of time for the actual service to be carried out. Each service facility (the checkouts) contains identical servers, and each server is capable of providing the desired service to the customer (Fig. 20.5).

Clearly, if there are no waiting lines then immediate service is obtained. However, in general, there are significant costs associated with the provision of an immediate service, and so there is a need to balance cost with a certain amount of waiting.

Some queues are *bounded* (i.e., they can hold only a fixed number of customers), whereas others are *unbounded* and can grow as large as is required to hold all waiting customers. The customer source may be finite or infinite, and where the customer source is finite but very large it is often considered to be infinite.

Random variables (described by probability distribution functions) arise in queuing problems, and these include the random variable q, which represents the time that a customer spends in the queue waiting for service; the random variable s, which represents the amount of time that a customer spends in service; and the

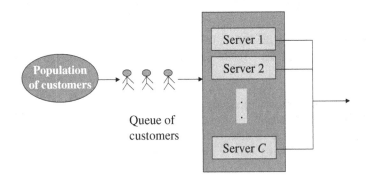

Fig. 20.5 Basic queuing system

random variable w, which represents the total time that a customer spends in the queuing system. Clearly,

$$w = q + s$$

It is assumed that the customers arrive at a queuing system one at a time at random times $(t_0 < t_1 < \cdots < t_n)$ with the random variable $\tau_k = t_k - t_{k-1}$ representing the *interarrival times* (i.e., it measures the times between successive arrivals). It is assumed that these random variables are independent and identically distributed, and it is usually assumed that arrivals form a Poisson arrival process (Fig. 20.6).

A Poisson arrival process is characterized by the fact that the interarrival times are distributed exponentially. That is,

$$P(\tau \le t) = 1 - e^{-\lambda t}$$

Further, the probability that exactly n customers will arrive in any time interval of length t is given by

$$\frac{e^{-\lambda t}(\lambda t)^n}{n!} \quad (\text{where } n = 0, 1, 2, \ldots)$$

where λ is a constant average arrival rate of customers per unit time, and the number of arrivals per unit time is Poisson distributed with mean λ.

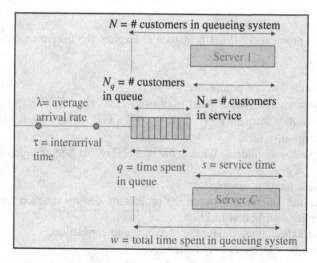

Fig. 20.6 Sample random variables in queuing theory

Similarly, it is usual to assume in queuing theory that the service times are random with μ denoting the average service rate, and let s_k denote the service time that the kth customer requires from the system. The distribution of service times is given by

$$W_s(t) = P(s \le t) = 1 - e^{-\mu t}$$

The capacity of the queues may be *infinite* (where every arriving customer is allowed to enter the queuing system no matter how many waiting customers are present), or finite (where arriving customers may wait only if there is still room in the queue).

Queuing systems may be *single server* (one server serving one customer at a time) systems or *multiple servers* (several identical servers that can service c customers at a time). The method by which the next customer is chosen from the queue to be serviced is termed the *queue discipline*, and the most common method is *first-come-first-served* (FCFS). Other methods include the last-in-first-out (LIFO); the shortest job first; or the highest priority job next.

Customers may exhibit various behaviours in a queuing system such as deciding not to join a queue if it is too long; switching between queues to try to obtain faster service; or leaving the queuing system if they have waited too long. There are many texts on queuing theory and for a more detailed account on queuing theory see [6].

20.6   Review Questions

1. What is probability? What is statistics? Explain the difference between them.

2. Explain the laws of probability.
3. What is a sample space? What is an event?
4. Prove Boole's inequality $P\left(\cup_{i=1}^{n} E_i\right) \le \sum_{i=1}^{n} P(E_i)$ where the E_i are not necessarily disjoint.
5. A couple has 2 children. What is the probability that both are girls if the eldest is a girl?
6. What is a random variable?
7. Explain the difference between the probability density function and the probability distribution function
8. Explain expectation, variance, covariance and correlation.
9. Describe how statistics may be abused.

10. What is a random sample? Describe methods available to generate a random sample from a population. How may information be gained from a sample?
11. Explain how the average of a sample may be determined, and discuss the mean, mode and median of a sample.
12. Explain sample variance and sample standard deviation.
13. Describe the normal distribution and the central limit theorem.
14. Explain hypothesis testing and acceptance or rejection of the null hypothesis.
15. What is software reliability? Describe various software reliability models.
16. Explain queuing theory and describe its applications to the computing field.

20.7 Summary

Statistics is an empirical science that is concerned with the collection, organization, analysis and interpretation and presentation of data. The data collection needs to be planned and this may include surveys and experiments. Statistics are widely used by government and industrial organizations, and they may be used for forecasting as well as for presenting trends. Statistical sampling allows the behaviour of a random sample to be studied, and inferences to be made about the population.

Probability theory provides a mathematical indication of the likelihood of an event occurring, and the probability is a numerical value between 0 and 1. A probability of 0 indicates that the event cannot occur, whereas a probability of 1 indicates that the event is guaranteed to occur. If the probability of an event is greater than 0.5, then this indicates that the event is more likely to occur than not to occur.

Software has become increasingly important for society and professional software companies aspire to develop high-quality and reliable software. Software reliability is the probability that the program works without failure for a specified length of time, and is a statement on the future behaviour of the software. It is generally expressed in terms of the mean time to failure (MTTF) or the mean time between failure (MTBF), and the software reliability measurements are an attempt to provide an objective judgment of the fitness for use of the software.

There are many reliability models in the literature and the question as to which is the best model or how to evaluate the effectiveness of the model arises. A good model will have good theoretical foundations and will give useful predictions of the reliability of the software.

Queuing theory is the mathematical study of waiting lines or queues, and its origins are in work done Erlang in the early twentieth century. Customers will generally wait for a period of time in a queue before receiving service at, and they wait for a further period of time for the actual service to be carried out. Each service facility (the checkouts) contains identical servers, and each server is capable of providing the desired service to the customer. Queuing theory has been applied to many fields including telecommunications and traffic management.

References

1. Introduction to Probability and Statistics for Engineers and Scientists. Sheldon M. Ross. Wiley Publications. New York. 1987.
2. Mathematics in Computing. Second Edition, Gerard O'Regan. Springer. 2012.
3. Optimizing preventive service of software products. E. Adams. IBM Research Journal, 28(1), pp. 2–14, 1984.
4. Engineering Software under Statistical Quality Control. Richard H. Cobb and Harlan D. Mills. IEEE Software. 1990.
5. Operating Systems. H.M. Deitel. 2nd Edition. Addison Wesley.1990.
6. Fundamentals of Queueing Theory. 4th Edition. Donald Gross and John Shortle. Wiley Interpress. 2008.

Glossary

AECL	Atomic Energy Canada Ltd.
AES	Advanced Encryption Standard
AI	Artificial Intelligence
AMN	Abstract Machine Notation
BCH	BoseChauduri and Hocquenghem
BNF	Backus Naur Form
CCS	Calculus Communicating Systems
CICS	Customer Information Control System
CMM	Capability Maturity Model
CMMI®	Capability Maturity Model Integration
CPO	Complete Partial Order
CSP	Communicating Sequential Processes
CTL	Computational Tree Logic
DAG	Directed Acyclic Graph
DES	Data Encryption Standard
DOD	Department of Defence
DPDA	Deterministic Pushdown automata
DSA	Digital Signature Algorithm
DSS	Digital Signature Standard
FCFS	First Come, First Served
FSM	Finite State Machine
GCD	Greatest Common Divisor
GCHQ	General Communications Headquarters

© Springer International Publishing Switzerland 2016
G. O'Regan, *Guide to Discrete Mathematics*, Texts in Computer Science,
DOI 10.1007/978-3-319-44561-8

GSM	Global System Mobile
HOL	Higher Order Logic
IBM	International Business Machines
IEC	International Electrotechnical Commission
IEEE	Institute of Electrical and Electronics Engineers
ISO	International Standards Organization
LCM	Least Common Multiple
LD	Limited Domain
LEM	Law Excluded Middle
LIFO	Last In, First Out
LPF	Logic of Partial Partial Functions
LT	Logic Theorist
LTL	Linear Temporal Logic
MIT	Massachusetts Institute of Technology
MTBF	Mean time between failure
MTTF	Mean time to failure
MOD	Ministry of Defence
NATO	North Atlantic Treaty Organization
NBS	National Bureau of Standards
NFA	Non Deterministic Finite State Automaton
NIST	National Institute of Standards &Technology
NP	Non-deterministic polynomial
OM	Object Modelling Technique
PDA	Pushdown Automata
PMP	Project Management Professional
RDBM	Relational Database Management System
RSA	RivestShamir and Adleman
SCAMPI	Standard CMM Appraisal Method for Process Improvement
SECD	StackEnvironmentCode, Dump
SEI	Software Engineering Institute

SQL	Structured Query Language
TM	Turing Machine
UML	Unified Modelling Language
UMTS	Universal Mobile Telecommunications System
VDM	Vienna Development Method
VDM♣	Irish School of VDM
VDM-SL	VDM specification language
WFF	Well-formed formula
YACC	Yet Another Compiler Compiler

Index

A

Abstract algebra, 109
Abuse of statistics, 342
Abu Simbel, 2
Agile development, 288
Alexander the Great, 9
Algebra, 99
Algorithm, 213
Al-Khwarizmi, 20
Alphabets and words, 186
Annuity, 91
Antikythera, 17
Application of functions, 46
Applications of relations, 40
Aquinas, 16
Archimedes, 14
Aristotle, 16
Arithmetic sequence, 87
Arithmetic series, 88
Artificial intelligence, 274
Athenian democracy, 9
Augustus, 19
Automata theory, 117
Axiomatic approach, 305
Axiomatic semantics, 193

B

Babylonians, 4, 5
Backus Naur Form, 189
Bags, 327
Bijective, 44
Binary relation, 25, 26, 34, 40, 50
Binary system, 71
Binary trees, 149
Binomial distribution, 340
Bletchey park, 158
Block codes, 174
B method, 311
Brouwer, L. E. J., 267
Bombe, 159, 161, 165

Boole's symbolic logic, 225
Boole, 225
Bush, Vannevar, 229

C

Caesar, Julius, 156
Caesar cipher, 156
Capability Maturity Model Integration
        (CMMI), 285, 287, 289, 295–297
Cayley–Hamilton theorem, 135
CCS, 313
Central limit theorem, 346
Chinese remainder theorem, 23
Chomsky hierarchy, 189
Church, Alonzo, 46, 212
Church–Turing thesis, 213
CICS, 303
Classical engineers, 286
Classical mathematics, 293
Cleanroom methodology, 353
Codd, Edgar, 40
Coding theory, 171
Combination, 95
Commuting diagram property, 332
Competence set, 40
Completeness, 212
Complete partial orders, 202
Compound interest, 85
Computability, 212
Computability and decidability, 207
Computable function, 47, 197
Computer representation of numbers, 71
Computer representation of sets, 33
Conditional probability, 338
Correlation, 340
Covariance, 339
Cramer's rule, 135
Cryptographic systems, 160
Cryptography, 155
CSP, 313

Printed in the United States
By Bookmasters